Emerging Diseases of Animals

Emerging Diseases of Animals

EDITED BY
Corrie Brown
Department of Pathology
College of Veterinary Medicine
University of Georgia
Athens, Georgia

AND
Carole Bolin
Animal Health Diagnostic Laboratory
Michigan State University
East Lansing, Michigan

ASM
PRESS WASHINGTON, D.C.

Address editorial correspondence to ASM Press, 1752 N. St. NW, Washington, DC 20036-2804, USA

Send orders to ASM Press, P.O. Box 605, Herndon, VA 20172, USA
Phone: 800-546-2416 or 703-661-1593
Fax: 703-661-1501
E-mail: books@asmusa.org
Online: www. asmpress.org

Library of Congress Cataloging-in-Publication Data

Emerging diseases of animals / edited by Corrie Brown and Carole Bolin.
p. cm.
Includes bibliographical references (p.).
ISBN 1-55581-201-5 (hardcover)
1. Communicable diseases in animals. I. Brown, Corrie, II. Bolin, Carole

SF781.E53 2000
636.089'69—dc21 00-040167

Cover illustration by Kip Carter, College of Veterinary Medicine, University of Georgia.

Contents

Contributors

Thomas Besser • Department of Veterinary Microbiology and Pathology, College of Veterinary Medicine, Washington State University, Pullman, WA 99164-6610

Carole Bolin • Animal Health Diagnostic Laboratory, Department of Veterinary Pathology, Michigan State University, East Lansing, MI 48824

Cathy A. Brown • Athens Diagnostic Laboratory, College of Veterinary Medicine, University of Georgia, Athens, GA 30602

Corrie Brown • Department of Veterinary Pathology, College of Veterinary Medicine, University of Georgia, Athens, GA 30602-7388

Margaret Davis • Field Disease Investigation Unit, College of Veterinary Medicine, Washington State University, Pullman, WA 99164-6610

Linda A. Detwiler • Veterinary Services, Animal and Plant Health Inspection Service, U.S. Department of Agriculture, 320 Corporate Blvd., Robbinsville, NJ 08691

Michael P. Doyle • Center for Food Safety and Quality Enhancement, College of Agriculture and Environmental Sciences, University of Georgia, Griffin, GA 30223

Clive Gay • Field Disease Investigation Unit, College of Veterinary Medicine, Washington State University, Pullman, WA 99164-6610

John Gay • Field Disease Investigation Unit, College of Veterinary Medicine, Washington State University, Pullman, WA 99164-6610

Craig E. Greene • Departments of Small Animal Medicine and Medical Microbiology, College of Veterinary Medicine, University of Georgia, Athens, GA 30602

Dale Hancock • Field Disease Investigation Unit, College of Veterinary Medicine, Washington State University, Pullman, WA 99164-6610

Barry G. Harmon • Department of Pathology, College of Veterinary Medicine, University of Georgia, Athens, GA 30602

Peter T. Hooper • CSIRO Australian Animal Health Laboratory, P.O. Bag 24, Geelong, Victoria 3220, Australia

Barry Kellman • International Criminal Justice and Weapons Control Center, DePaul University College of Law, Chicago, IL 60604

Duncan C. Krause • Department of Microbiology, University of Georgia, Athens, GA 30602

Linda Logan-Henfrey • Agricultural Research Service, U.S. Department of Agriculture, 5601 Sunnyside Ave., Beltsville, MD 20705-5138

Laurie G. O'Rourke • Preclinical Safety—USA, Department of Toxicology and Pathology, Novartis Pharmaceuticals Corporation, 59 Route 10, 406/185, East Hanover, NJ 07936-1080

Mitchell V. Palmer • Bovine Tuberculosis Research Project, Bacterial Diseases of Livestock Research Unit, National Animal Disease Center, Agricultural Research Service, U.S. Department of Agriculture, Ames, IA 50010

Jack C. Rhyan • National Animal Health Programs, Veterinary Services, National Wildlife Research Center, Animal and Plant Health Inspection Service, U.S. Department of Agriculture, Fort Collins, CO 80521

Daniel Rice • Field Disease Investigation Unit, College of Veterinary Medicine, Washington State University, Pullman, WA 99164-6610

Richard Rubenstein • Department of Virology, Institute for Basic Research in Developmental Disabilities, New York State Office of Mental Retardation and Developmental Disabilities, 1050 Forest Hill Rd., Staten Island, NY 10314

J. Scott Rusk • National Animal Disease Center, Agricultural Research Service, U.S. Department of Agriculture, Ames, IA 50010

David E. Swayne • Southeast Poultry Research Laboratory, Agricultural Research Service, U.S. Department of Agriculture, Athens, GA 30605

James R. Swearengen • Veterinary Medicine Division, United States Army Medical Research Institute of Infectious Diseases, Fort Detrick, MD 21702-5011

Richard Weller • Molecular Biosciences Department, Pacific Northwest National Laboratory, Richland, WA 99352

Diana L. Whipple • Bovine Tuberculosis Research Project, Bacterial Diseases of Livestock Research Unit, National Animal Disease Center, Agricultural Research Service, U.S. Department of Agriculture, Ames, IA 50010

Elizabeth S. Williams • Department of Veterinary Sciences, University of Wyoming, Laramie, 1174 Snowy Range Rd., Laramie, WY 82070

Terrance M. Wilson • Emergency Programs, Veterinary Services, Animal and Plant Health Inspection Service, U.S. Department of Agriculture, 4700 River Rd., Riverdale, MD 20737-1231

Patricia L. Worsham • Bacteriology Division, United States Army Medical Research Institute of Infectious Diseases, Fort Detrick, MD 21702-5011

Tong Zhao • Center for Food Safety and Quality Enhancement, College of Agriculture and Environmental Sciences, University of Georgia, Griffin, GA 30223

Preface

Over the last 10 years, there has been a great deal of focus on emerging infectious diseases, with numerous conferences, symposia, and publications. Most of the resulting literature has focused on emerging diseases of humans. However, a parallel but lesser-known phenomenon is that of emerging diseases of animals. Animal diseases are emerging at unprecedented rates, for many of the same reasons underlying the emergence of human diseases. These animal diseases have pleiotropic effects—on the animals themselves, on the environment, and on the health of humans, both directly and indirectly. The factors responsible for animal disease emergence are varied but also synergistic, and all ultimately relate back to the ever-enlarging human population. With a greater mass of humanity, there is increasing traffic of people and animals, bringing bodily ecosystems, with all their microfloras and potential pathogens, to new areas and animals. Along with this increasing traffic, there is considerable habitat destruction, causing animal populations to cluster in evolutionarily incompatible areas and/or where novel disease possibilities abound. New species are being brought into contact with one another for a variety of reasons—because of ecological disruption, for show, for trade, or for efficiency of production. In order to maximize efficiency of resources and produce adequate protein for human consumption, husbandry situations and novel technologies are constantly being refined and redrawn. Some of these new technologies are creating unpredictable "keystone" interruptions that serve to modify bionetworks, presaging the onset of new disease.

ASM Press has been a leader in publishing information on emerging infectious diseases, including *Emerging Infections 1, 2,* and *3, Pathology of Emerging Infections,* and *Pathology of Emerging Infections 2*. Each of these volumes has minor portions devoted to zoonotic aspects of emerging human diseases or to emerging diseases of animals per se. On the basis of the readers' responses to this disseminated information, it was deemed timely to produce an entire volume dedicated to emerging diseases in animals—hence this book. The authors were selected as experts in their fields and were considered the optimal individuals to bring the information forward to the greater infectious disease community. Most of the authors were present at the conference on which this book was based, the International Conference on Emerging Diseases: Veterinary Medicine, Agriculture and Human Health, held at the University of Georgia in August 1999. Each chapter was written with the aim of taking information that has primarily been investigated in the veterinary community and presenting it for greater understanding by infectious disease specialists in human health. Certainly, both communities will benefit through greater understanding of disease ecology, risk, and transfer possibilities. It was over a century ago that Rudolf Virchow, a physician considered to be the father of modern pathology, first articulated a theme of "one medicine." He formulated his views on the importance of comparative medicine on the basis of his work with trichinosis, a zoonotic disease. He concluded that disease in one species cannot be studied in isolation. This one-medicine theme becomes ever more relevant as pathogens from the animal and human worlds not only continue to proliferate and mutate but also collide in a multitude of unpredictable ways.

Acknowledgments

We owe a large debt of gratitude to Greg Payne, Senior Editor at ASM Press, who provided initial encouragement for this undertaking and then ample support throughout the process. His unflagging enthusiasm for our endeavors was most appreciated. In addition, Ken April of ASM Press (assisted by Dwight Christenbury and Pamela Lacey) proved to be a very capable coordinator of production and kept us moving through the final laps. At the University of Georgia, Karen Holbrook, Senior Vice President and Provost, was instrumental in providing funding for an international conference on emerging diseases of animals, which was held in August 1999. This conference presented a unique opportunity to bring together many experts in the field who shared ideas and provided the nidus on which this book was built. These experts deserve our heartfelt thanks for producing manuscripts that were thorough, insightful, illuminating, and timely. We also acknowledge our respective employers, the University of Georgia College of Veterinary Medicine and the USDA Agricultural Research Service National Animal Disease Center, for allowing us time for organizing and editing. Last but definitely not least, we thank our families for their understanding and support during this process.

Emerging Diseases of Animals
Edited by C. Brown and C. Bolin

Chapter 1

Emerging Infectious Diseases of Animals: an Overview

Corrie Brown

Emerging diseases have been moving to center stage in human medicine for more than a decade, with numerous conferences, papers, volumes, and even an entire journal devoted to their study. While it is often overlooked, there is in fact a parallel phenomenon of new diseases occurring in animal populations. These new emerging animal diseases are having pleiotropic effects—on animals, on the environment, and on the health of humans, both directly, through transfer of zoonotic agents, and also indirectly, through potential compromise of the food supply. Many of these new animal diseases are covered in this volume dedicated to the subject. This chapter will serve to construct a framework for exploring the underlying reasons for the emergence of new animal diseases.

The overall increase in the global human population and all the attendant implications of that increase are the reasons for the emergence of many new animal diseases. First, with a greater mass of humanity, there is increasing traffic of people and animals, bringing bodily ecosystems with all their microflora and potential pathogens to new areas and animals. Second, there is considerable habitat destruction, causing animal populations to cluster in evolutionarily incompatible areas or where novel disease possibilities abound. Third, new species are being brought into contact with one another for a variety of reasons—ecological disruption, show, trade, or efficiency of production. Fourth, husbandry situations and novel technologies are constantly being refined and redrawn to help maximize efficiency, minimize loss, and feed the sum total of people. Each of these factors, and examples of animal disease emergence resulting from

Corrie Brown • Department of Veterinary Pathology, College of Veterinary Medicine, University of Georgia, Athens, GA 30602-7388.

them, will be considered below. However, it should be noted that many diseases are due to more than one factor, and there is a definite and unfortunate synergy occurring as the factors overlap and collide.

MOVEMENT OF ANIMALS AND PEOPLE AS A CAUSATIVE FACTOR

Moving a disease agent to a susceptible population has been the cause of animal outbreaks throughout history. Genghis Khan, Attila the Hun, and Napoleon all took contagious bovine pleuropneumonia and rinderpest (cattle plague) into conquered territories, where animals had minimal specific immunity, and outbreaks ensued. Such instances are not confined to the history books but continue to occur in the modern world. For example, subsequent to the Gulf War, Kurds fleeing from Iraq to Turkey took rinderpest with them, harbored in their goats. Turkey suffered an outbreak of rinderpest such as Europe had not experienced in decades.

There are many diseases of animals that cause significantly decreased production, and countries with developed systems of agriculture strive to remain free of these diseases in order to ensure the health of their national herds and to maintain global trade in animals and animal products. The Office International des Epizooties, an international animal disease reporting agency, maintains lists of such diseases and tracks their movement around the globe. The diseases that cause the most serious economic devastation are considered "list A" diseases. There have been numerous examples in recent years of these list A diseases making incursions into new areas. Below are some examples.

Foot-and-mouth disease (FMD), the most contagious disease known to exist for either animals or humans, causes a transient but severe drop in production. Susceptible animals, which include all cloven-hoofed species, develop big painful blisters in the oral cavity and on the feet. Affected animals refuse to eat, forage, or nurse for 1 to 3 weeks, a period of lost production that is enough to wipe out any potential profits for intensively reared animals. Once FMD emerges in an area, controlling and eradicating the disease is extremely problematic because of its incredibly contagious nature. In 1993, FMD broke out in Italy, engendering the destruction of 8,000 head of cattle to control the disease, with economic losses estimated at $11 million (27). The source was a herd of cattle purchased from Croatia, from a producer wishing to liquidate his assets to purchase arms for the struggle there. One or more of the cattle were subclinically infected, and an outbreak ensued.

In 1997, FMD broke out in Taiwan, perhaps due to an infected piglet smuggled in from the mainland. Taiwan, which was a major exporter of pigs, suffered very severe losses. The disease spread rapidly throughout the country, necessitating the slaughter of more than 8 million hogs and losses of greater than $8 billion. It is estimated that Taiwan's pork industry will not recover for at least a decade (29).

Another list A disease, classical swine fever (CSF), caused by a pestivirus, erupted in the Netherlands, also in 1997, as a result of some frozen pig manure carried in a German truck that was inadequately decontaminated at a border inspection station. The Netherlands, like Taiwan, has an intensive pork industry. The disease spread rapidly throughout the country, necessitating the slaughter of more than 8 million hogs, with economic damages assessed at $2.3 billion (23). As was the case in Taiwan, infection spread rapidly among densely clustered swine-raising premises. In the lag period between introduction and diagnosis of the disease, it was later discovered that pigs, possibly infected, had been shipped from Holland to Spain and Italy (14), engendering retroactive control efforts in those recipient countries. In addition, the ability of an animal disease to massively disseminate in a manner not normally considered in human medicine was well illustrated in this outbreak in the Netherlands. Specifically, the discovery of infected boars at two artificial-insemination stations led to a suspicion of CSF in 1,680 pig herds which had received the semen and actual infection in 146 of those herds.

Highly pathogenic avian influenza (HPAI) made headlines recently through its mutation to a form that caused fatal disease in humans in Hong Kong (see chapter 6). However, even without this genetic modification enabling human infection, HPAI is a list A disease that causes grave concern for the global veterinary regulatory community simply because of its dramatic effect on poultry. In 1983, HPAI emerged in poultry flocks in the eastern part of the United States and caused approximately $400 million in damages and increased poultry prices (19). HPAI continues to move to new poultry populations around the world, with considerable economic impact and large-scale eradication programs.

Velogenic Newcastle disease is the only other list A disease of poultry. Outbreaks of this disease occur sporadically throughout the world, usually due to importation of subclinically infected birds, and the response is eradication through depopulation, often costly but eminently justifiable in terms of long-term economic health. An introduction of velogenic Newcastle disease occurred in California in 1997 and was traced to infected fighting game cocks (10). Rapid action by regulatory authorities prevented the spread to large poultry operations, but nevertheless, even this limited focus of disease resulted in significant international embargoes against

U.S. poultry. In Australia in 1998, velogenic Newcastle disease broke out not as a result of an introduction, but due to a mutation from an existing low-pathogenicity strain. From this single mutation, eradication efforts including $9.85 million and the destruction of 4 million birds resulted (26). These examples underscore the dramatic effect that emerging diseases can have on economics, animal populations, and the potential for sudden decreases in available animal protein for human consumption.

Translocation of animal products or disease vectors may also result in disease spread or emergence in a new location. Infected sausages or hides could bring etiologic agents to new areas. African swine fever is an example of a disease that was taken to many parts of the world in airline garbage. Now, most countries have restrictions on feeding garbage from airliners to pigs to prevent such an occurrence. Movement of pathogen-carrying vectors can result in disease emergence. Lyme disease has become problematic in novel areas as the incriminated tick vector is transported closer to susceptible populations. Similarly, West Nile encephalitis virus was carried to susceptible equine populations on Long Island and in Connecticut through translocation of mosquito vectors.

Until recently, we were only worried about accidental introductions, and cadres of federally employed regulatory veterinarians were in place to monitor for the introduction of a disease that could cause massive economic damage. However, as a result of numerous threats of criminal dissemination of human disease, an increased awareness has developed concerning the possibility of nefarious introduction of animal disease, or "agroterrorism" (see chapter 3). Because so many animal disease agents spread quite readily on their own, sophisticated microbiological manipulation of "weaponization" is not a requirement. Consequently, the low-tech nature of such an assault combined with the considerable economic damage that would be engendered make this a cause for concern in our increasingly terrorist-conscious society.

One issue that needs to be mentioned in the context of disease introduction and eradication is the problem of carcass disposal. In the event of a large mortality or depopulation event, suitable means of disposal will have to be found. As an illustration, disposal of pigs depopulated in the CSF outbreak in the Netherlands was quite problematic due to environmental regulations against burning and the presence of a high water table, which precluded burial. However, in the Dutch situation, most of the 8 million pigs destroyed were piglets, whereas in the FMD outbreak in Taiwan, most of the 8 million swine destroyed were fully grown. It is estimated that there were 500,000 metric tons of dead pigs to be disposed of in Taiwan. There is no doubt that hygienic disposal within existing environmental regulations will be problematic in many countries.

ENVIRONMENTAL DISRUPTION

Habitat alterations can lead to new diseases by a variety of means. First, destruction of existing habitat will cause a change in an animal's or a population's behavior. For instance, North American waterfowl populations are congregating in larger volumes because of the decreased availability of wetlands. In the last decade, tens of thousands of birds have been dying of diseases that never before occurred in outbreak form like this—velogenic Newcastle disease in cormorants in northern Canada and the Great Lakes region; fowl cholera (*Pasteurella multocida*) in Chesapeake Bay; and duck plague in Texas and California (3, 4, 5). Increasing urbanization of fruit bats secondary to deforestation has been cited as a factor in the emergence of various viral diseases, including Hendra virus, Menangle virus, lyssavirus, and Nipah virus diseases, in Australia and Malaysia (see chapter 5). Similarly, urban expansion in the western United States probably underlies the increase in the number of cases of feline and human plague in that region seen over the last decade (see chapter 13).

On a related note, overfishing on the North American side of the Atlantic Ocean may have been the seminal event in the forced migration of North American seals across the Arctic to the North Sea, carrying with them a previously unknown morbillivirus that, when allowed access to immunologically naive populations, engendered a massive epizootic leading to the death of thousands of European seals (12, 13). Since the recognition of this new disease, now known as phocine distemper, several other morbilliviruses of marine mammals have been documented, with altered migration and habitat destruction playing a large role in each (18).

Second, environmental disruption can cause a change in vector populations and patterns. Perhaps the best example of this phenomenon is Rift Valley fever (RVF). Caused by a bunyavirus, RVF is spread by mosquitoes and will cause disease in a number of species, including humans. Ruminants are considered the amplifier host, and often the first sign of an RVF outbreak is an increase in ruminant abortions and deaths of neonatal lambs and calves. The most recent RVF epizootic/epidemic was in East Africa in late 1997 and early 1998; it claimed the lives of thousands of livestock and wildlife, with significant human morbidity and mortality as well (9). A major factor in this outbreak was determined to be El Niño-Southern Oscillation phenomenon creating increased precipitation in East Africa and amplification of mosquito vector populations (20). Human leptospirosis is another example of a disease on the increase due to environmental changes. In this case, leptospirosis is emerging as animal reservoirs and high rainfall or flooding coincide (see chapter 9).

Third, a novel environment can result in the emergence of a toxic microbe. *Pfiesteria piscicida* is a newly described disease agent causing morbidity in both fish and humans (7). First described as a cause of fish die-offs in estuaries of the eastern seaboard of the United States in 1992, concern about this agent escalated rapidly when it was found in 1997 that commercial fishermen and recreational boaters exposed to *Pfiesteria* toxins developed skin rashes and temporary cognitive deficits (16). *Pfiesteria* is an unusual dinoflagellate with a variety of life forms. One of these life forms, the toxic zoospore, emits toxins in response to the presence of fish secretions. An increase in the fish populations was thought to be the reason for the increase in this dinoflagellate (7). The increase in fish, in turn, was attributed to anthropogenic loading of phosphorus and nitrogen—presumably due to agricultural runoff, specifically from swine and poultry operations in that region.

Novel microenvironments can impact disease emergence through evolutionary change. The theory of evolution holds that genetic changes continue to occur at an unchanged rate and have since the beginning of life. Whether or not those changes are incorporated into the population is entirely dependent on the local microenvironment (11). So, as the microenvironment changes, the process of evolution is accelerating or, at the least, being modified. It is worth noting that evolutionary changes can be logarithmically faster for viruses than for their vertebrate hosts, rendering a distinct advantage on the former. This issue of novel microenvironments and how that may facilitate disease emergence is of paramount importance in many discussions involving proactive attention to animal and public health and should certainly be considered in any decisions involving habitat change.

CROSSING SPECIES BOUNDARIES

In order for an agent to cross species boundaries and cause disease, the initial event must be to move into that species. Because of the nature of animal populations and the sheer number of species, the opportunities for this transfer to take place are abundant in the animal world and continue to increase as we move species around and crowd them into ever-dwindling natural available spaces.

Canine distemper struck African lions in the Serengeti ecosystem in the summer of 1994. Approximately one-third of the area's lions succumbed to the infection, which localized in the brain, causing an encephalitis. Canine distemper was not previously known for its ability to infect and kill members of the cat family, but in fact the virus that was circulating in the lion population was very virulent for cats (8). It is theorized that the virus moved into the lion population on multiple occasions, and on one of these occasions, the virus "took," and a variant arose which was lethal for cats. Why was this not seen before? There is increasing human population pressure on the edges of these parks. Even so, the dogs living with the humans outside the park do not venture into the interior of the park where the lions are. Presumably, the dogs gave the virus to the hyenas which travel from the edges of the park into the center. This virus probably crossed not one but two species boundaries.

Conjunctivitis and blindness due to *Mycoplasma gallisepticum* in house finches are the result of a pathogen transferred from domestic poultry, where the agent usually causes a subclinical to moderate upper respiratory disease (15). The disease in house finches was first noted in 1994, and by 1996, this disease had been reported in all states in the eastern half of the country, demonstrating the speed with which a new disease of wild animals can effectively disseminate.

The plethora of diseases that have emerged from fruit bats underscores not only the ease with which diseases can move from one area to another but also the tremendous repository of viruses contained within sylvatic species of animals. The emergence of one virus, Hendra virus, from bats, and the associated infection of humans and horses, led to a rather serendipitous discovery of other significant viral pathogens that moved from bats to other species, including Menangle virus, which created illness in pigs and humans; lyssavirus, which caused problems for bats and humans; and probably Nipah virus, which has virulence for a variety of species, most notably humans (see chapter 5).

Zoological collections are prime areas for engendering the transfer of an agent from one species to another. Examples include callitrichid hepatitis virus, given to captive golden lion tamarins through the practice of feeding "pinky" mice, and the acquisition of malignant catarrhal fever by many species of Asiatic deer, who contracted it from the African wildebeest reservoir. The recent emergence of West Nile encephalitis virus emphasizes the importance of thorough pathologic investigations of wild animals that died in captivity and the necessity of an integrated network for reporting animal and human diseases. To ignore these ecologically global microcosms is to turn our backs on prime learning opportunities and to close the door on a vital early warning system (21).

Influenza viruses should never be ignored when dealing with emerging disease problems or the possibility of agents moving from one species to another. The discovery of a severe avian-origin influenza in humans in Hong Kong and the sporadic transmission of swine influenza to humans is a constant concern (see chapter 6). An underlying tenet of influenza biology and epidemiology has always been interspecies mixing of strains and the subsequent emergence of a pandemic.

Xenotransplantation and its attendant possibilities for animal-origin viruses to make intimate contact with human cells and create new disease is an area that requires close monitoring. The likelihood of as-yet-undiscovered swine viruses emerging in transplant recipients has received careful scrutiny (see chapter 4). Of particular concern is that, just since the beginning of studies to create genetically modified swine with organs suitable for transplantation, several new viruses of swine have been described. These include porcine circovirus, swine hepatitis E virus, Menangle virus, and Nipah virus (1, 22; see chapter 5). The recent demonstration that porcine endogenous retroviruses can productively infect human cells in culture has sounded alarm bells throughout the xenotransplantation community (24). Safeguards against inadvertent etiologic transfer from graft to recipient are dependent upon the development of detection technologies that keep apace of emerging virus possibilities. However, the ability to detect viruses which have not yet been discovered is a daunting, if not impossible, task.

Ehrlichial agents are very difficult to culture, and so diagnosis often rests on sophisticated nucleic acid methodologies. Recent technologic advances allow more accurate diagnosis and also indicate that some ehrlichial diseases of humans may have their origins in deer and dogs (2, 6). As our understanding of these agents and the diseases they cause becomes more defined, it seems likely that there will be abundant evidence of species-crossing capabilities. Similarly, the ability to define and diagnose bartonellosis has advanced our understanding of cat scratch disease and its cycle among cats and fleas, with subsequent and often very damaging transfer to humans (see chapter 12).

The emergence of transmissible spongiform encephalopathies in various animals is a striking example of disease that can cross species boundaries with very alarming results (see chapter 7). The presence and amplification of bovine spongiform encephalopathy led to a similar disease, variant Creutzfeldt-Jakob disease, in humans. The recent awareness of the extent of chronic wasting disease, a spongiform encephalopathy of elk and deer, is creating great concern about possible spread to other species.

As a final note in this section, ecotourism has plenty of possibilities for creating new diseases. As the practice of traveling to remote areas to view endangered species becomes more popular, it is a certainty that we will see new diseases, either in the endangered populations or in the humans observing them. Endangered primate populations are particularly vulnerable because they live in very isolated and immunologically remote areas and are susceptible to so many pathogens carried by humans, who are now making significant incursions into these remote areas. Alternatively, many primate diseases can easily be transferred to humans, and so people need to act accordingly. For example, while on safari in eastern Africa, I observed a vervet monkey urinating into the bowl of pineapple on the breakfast buffet, visual proof of the dramatic possibilities for transfer of disease agents through bodily fluids.

HUSBANDRY CHANGES

The increasing human population needs to be adequately nourished. It is imperative to refine animal protein production to maximize yields and minimize waste. However, this must be done carefully, and sometimes changes can have devastating results. Many advocate eliminating the use of animal protein altogether and surviving only on plant products, using existing grains fed to animals to feed the global population. However, this premise is not grounded in reality. Much of livestock production in developed countries and virtually all livestock rearing in the developing world involves the utilization of plant material that has no nutritional value for human consumption. Globally, grazing lands tend to be areas that are nonarable, and so there is a great efficiency of protein production. Consequently, there is little doubt that animal husbandry will continue to provide a large proportion of the protein for human consumption. However, care needs to be taken to prevent novel husbandry technologies from leading to situations that predispose to new diseases.

The efficiency of the feeding practices for livestock have long been studied, with continual refinement aimed at increasing the amount of protein available for human consumption. Sometimes efficiency modifications have untoward and explosive results. Specifically, a change in the rendering process of animal feed in Great Britain may have contributed to the survival of a small amount of transmissible spongiform encephalopathy agent, which was then subsequently disseminated and amplified within the cattle population, resulting in the whole disease dilemma known as mad cow disease (see chapter 7). The practice of feeding subtherapeutic levels of antibiotics to livestock to maximize productivity has been in effect for decades, with equal admixtures of proponents and opponents internationally. The latter have been quick to point to this as a reason for the emergence of the pentaresistant strain of *Salmonella* known as *Salmonella* DT104, a clone currently responsible for disease in humans and animals in many parts of the world. However, although the emergence of the pentaresistant strain can be easily linked with this practice, careful scientific scrutiny is unable to establish a causal connection (see chapter 11).

Artificially feeding wild animals, while seeming to be an innocuous and perhaps altruistic practice, can have dramatic effects on animal and even human health interests. Bovine tuberculosis in groups of deer, either farmed or winter fed, has become an extremely inflammatory issue in many areas. The practice of clustering groups of deer has allowed *Mycobacterium bovis* to spread and amplify, with transmission to previously unaffected nearby cattle populations (see chapter 14). In a parallel situation, gathering of elk on winter feeding grounds has allowed very efficient transfer of *Brucella abortus,* with subsequent infection of surrounding cattle populations and volatile clashes between the two very productive industries of ranching and hunting (see chapter 8).

Aquaculture and the stocking of streams for anglers are not exempt from disease emergence. The spread of pathogens within an aquaculture system is facilitated by an environment in which food, feces, and water are intricately intertwined. *Streptococcus iniae,* a newly described bacterial organism, has been associated with epizootics of meningitis in farmed fish over the last decade (25, 30). This disease has been documented in both fresh and marine fish culture. Transfer to wild fish has also been seen with increased pathologic severity in wild species. Recently, septicemia due to *S. iniae* has been seen in humans handling whole fish infected with this agent (28). Whirling disease, caused by *Myxobolus cerebralis,* has become a major threat to the survival of wild rainbow trout in many streams in the western United States (17). This myxozoan parasite tunnels through cartilage, causing deformities and impacting the organ of equilibrium. Transmission from one area to another is believed to be due to movement of infected hatchery fish.

CONCLUSIONS

Paralleling the situation in human medicine, new animal diseases are emerging at an alarming rate, with the underlying factors all being issues that ultimately trace back to the increases in the human population. Unfortunately, it is not possible to turn the clock back and go back to a simpler time, a time when environmental and resource demands were lower, a time when "global village" was still an oxymoron, a time when "emerging disease" meant a curiosity and not a whole new discipline of study. Furthermore, in the absence of a cataclysmic event, it is unlikely that there will be a large decrease in the human population, and so there will continue to be pressures on animal populations, heralding new diseases.

Any decisions made concerning unusual animal movements, environmental disruption, or altered husbandry should be thoroughly reviewed prior to implementation to determine potential impacts on animal and human health. The two health systems—animal and human—are often inextricably intertwined, as illustrated throughout this volume. More than a hundred years ago, Rudolf Virchow first proposed the idea of "one medicine," basing his conclusions on work with zoonotic diseases and his observations of the ease with which etiologic agents could move from animals to humans and back again. Today, in our crowded world of future shock, this theme of one medicine becomes ever more relevant. Ensuring security from emerging diseases for one species requires considering security for all.

REFERENCES

1. **Allan, G. M., F. McNeilly, S. Kennedy, B. Daft, E. G. Clarke, J. A. Ellis, D. M. Haines, B. M. Meehan, and B. M. Adair.** 1998. Isolation of porcine circovirus-like viruses from pigs with a wasting disease in the USA and Europe. *J. Vet. Diagn. Invest.* **10:**3–10.
2. **Bakken, J. S., J. S. Cumler, S. M. Chen, M. R. Eckman, L. L. Van Etta, and D. H. Walker.** 1994. Human granulocytic ehrlichiosis in the upper Midwest United States: a new species emerging? *J. Am. Med. Assoc.* **272:**212–218.
3. **Ban, B. D.** 1994. Avian cholera hits Chesapeake Bay. *J. Am. Vet. Med. Assoc.* **204:**1121.
4. **Bannerjee, M., W. M. Reed, S. D. Fitzgerald, and B. Panigrahy.** 1994. Neurotropic velogenic Newcastle disease in cormorants in Michigan: pathology and virus characterization. *Avian Dis.* **38:**873–878.
5. **Barr, B. C., D. A. Jessup, D. E. Docherty, and L. J. Lowenstine.** 1992. Epithelial intracytoplasmic herpes viral inclusions associated with an outbreak of duck virus enteritis. *Avian Dis.* **36:**164–168.
6. **Buller, B. S., M. Arens, S. P. Hmiel, C. D. Paddock, J. W. Sumner, Y. Rikihisa, A. Unver, M. Gaudreault-Keener, G. A. Manian, A. M. Liddell, N. Schmulewitz, and G. A. Storch.** 1999. *Ehrlichia ewingii,* a newly recognized agent of human ehrlichiosis. *N. Engl. J. Med.* **341:**148–155.
7. **Burkholder, J. M., E. J. Noga, C. W. Hobbs, H. B. Glasgow, Jr., and S. A. Smith.** 1992. New "phantom" dinoflagellate is the causative agent of major estuarine fish kills. *Nature* **358:**407–410.
8. **Carpenter, M. S., M. J. Appel, M. E. Roelke-Parker, L. Munson, H. Hofer, M. East, and S. J. O'Brien.** 1998. Genetic characterization of canine distemper virus in Serengeti carnivores. *Vet. Immunol. Immunopathol.* **65:**259–266.
9. **Centers for Disease Control.** 1998. Rift Valley Fever—East Africa, 1997–1998. *Morb. Mortal. Wkly. Rep.* **47:**261–264.
10. **Crespo, R., H. L. Shivaprasad, P. R. Woolcock, R. P. Chin, D. Davidson-York, and R. Tarbell.** 1999. Exotic Newcastle disease in a game chicken flock. *Avian Dis.* **43:**349–355.
11. **Darwin, C.** 1958. *The Origin of Species by Means of Natural Selection, or the Preservation of Favoured Races in the Struggle for Life.* Penguin, New York, N.Y. [Reprint of 1859 edition.]
12. **Duignan, P. J., J. T. Saliki, D. J. St. Aubin, G. Early, S. Sadove, J. A. House, K. Kovacs, and J. R. Geraci.** 1995. Epizootiology of morbillivirus infection in North American harbor

seals *(Phoca vitulina)* and gray seals (*Halichoerus grypus*). *J. Wildl. Dis.* **31:**491–501.

13. **Duignan, P. J., O. Nielsen, C. House, K. M. Kovacs, N. Duffy, G. Early, S. Sadove, D. J. St. Aubin, B. K. Rima, and J. R. Geraci.** 1997. Epizootiology of morbillivirus infection in harp, hooded, and ringed seals from the Canadian Arctic and western Atlantic. *J. Wildl. Dis.* **33:**7–19.
14. **Elbers, A. R. W., A. Stegeman, H. Moser, H. M. Ekker, J. A. Smak, and F. H. Pluimers.** 1999. The classical swine fever epidemic 1997–1998 in the Netherlands: descriptive epidemiology. *Prev. Vet. Med.* **42:**157–184.
15. **Fischer, J. R., D. E. Stallknecht, M. P. Luttrell, A. A. Dhondt, and K. A. Converse.** 1997. Mycoplasmal conjunctivitis in wild songbirds: the spread of a new contagious disease in a mobile host population. *Emerg. Infect. Dis.* **3:**69–72.
16. **Grattan, L. M., D. Oldach, T. M. Perl, M. H. Lowitt, D. L. Matuszak, C. Dickson, C. Parrott, R. C. Shoemaker, C. L. Kauffman, M. P. Wasserman, J. R. Hebel, P. Charache, and J. G. Morris, Jr.** 1998. Learning and memory difficulties after environmental exposure to waterways containing toxin-producing *Pfiesteria* or *Pfiesteria*-like dinoflagellates. *Lancet* **352:**532–539.
17. **Hedrick, R. P., M. El-Matbouli, M. A. Adkinson, and E. MacConnell.** 1998. Whirling disease: re-emergence among wild trout. *Immunol. Rev.* **166:**365–376.
18. **Kennedy, S.** 1998. Morbillivirus infections in aquatic animals. *J. Comp. Pathol.* **119:**201–225.
19. **Lasley, F. A., S. D. Short, and W. L. Henson.** 1985. *Economic Assessment of the 1983–84 Avian Influenza Eradication Program.* ERS Staff Report AGES841212. National Economics Division, Economic Research Service, U.S. Department of Agriculture, Washington, D.C.
20. **Linthicum, K. J., A. Anyamba, D. J. Tucker, P. W. Kelley, M. F. Myers, and C. J. Peters.** 1999. Climate and satellite indicators to forecast Rift Valley fever epidemics in Kenya. *Science* **285:**397–400.
21. **McNamara, T. S.** 1998. Emerging infections in captive wildlife, p. 381–385. *In* A. M. Nelson and C. R. Horsburgh, Jr. (ed.), *Pathology of Emerging Infections 2.* ASM Press, Washington, D.C.
22. **Meng, X. J., P. G. Halbur, M. S. Shapiro, S. Govindarajan, J. D. Bruna, I. K. Mushahwar, R. H. Purcell, and S. U. Emerson.** 1998. Genetic and experimental evidence for cross-species infection by swine hepatitis E virus. *J. Virol.* **72:**9714–9721.
23. **Meuwissen, M. P. M., S. H. Horst, R. B. M. Huirne, and A. A. Dijknuizen.** 1999. A model to estimate the financial consequences of classical swine fever outbreaks: principles and outcomes. *Prev. Vet. Med.* **42:**249–270.
24. **Patience, C., Y. Takeuchi, and R. A. Weiss.** 1997. Infection of human cells by an endogenous retrovirus of pigs. *Nat. Med.* **3:**282–286.
25. **Perera, R., S. Johnson, M. Collins, et al.** 1994. *Streptococcus iniae* associated with mortality of *Tilapia nilotica* and *T. aurea* hybrids. *J. Aquat. Anim. Health* **6:**335–340.
26. **Pollard, D.** 1999. Newcastle disease outbreak in Australia. *World Poultry* **15:**36–37.
27. **Tanaka, R.** 1993. Foot-and-mouth disease in Italy, p. 8–9. *In Foreign Animal Disease Report,* vol. 21, issue 213. Animal and Plant Health Inspection Service, U.S. Department of Agriculture, Washington, D.C.
28. **Weinstein, M. R., M. Litt, D. A. Kertesz, P. Syper, D. Rose, M. Coulter, A. McGeer, R. Facklam, C. Ostach, B. M. Willey, A. Borczyk, and D. E. Low.** 1997. Invasive infections due to a fish pathogen, *Streptococcus iniae. N. Engl. J. Med.* **337:**589–594.
29. **Wilson, T. M., and C. Tuszynski.** 1997. Foot-and-mouth disease in Taiwan—1997 overview, p. 114–124. *In Proceedings of the 101st Annual Meeting of the U.S. Animal Health Association.* U.S. Animal Health Association, Richmond, Va.
30. **Zlotkin, A., H. Hershko, and A. Eldar.** 1998. Possible transmission of *Streptococcus iniae* from wild fish to cultured marine fish. *Appl. Environ. Microbiol.* **64:**4065–4067.

Emerging Diseases of Animals
Edited by C. Brown and C. Bolin

Chapter 2

Biosafety Classification of Livestock and Poultry Animal Pathogens

J. Scott Rusk

The purpose of this chapter is to introduce a proposed scheme for use as a reference in placing animal pathogens into biosafety classifications. Annual revenues for the U.S. livestock industry are staggering. As an example, the value of all cattle and calves in the United States for 1999 was estimated to be $58.5 billion (14). Producers worldwide depend on good animal health to remain competitive in both national and international trade. The U.S. Department of Agriculture (USDA) Animal and Plant Health Inspection Service (APHIS) is the primary agency responsible for monitoring and controlling animal disease in the United States. APHIS regulations control the movement of animals, animal products, and animal pathogens in order to protect animal health (11). The Veterinary Equivalency and Regionalization Agreements proposed by APHIS also apply to biosafety and biosecurity risk assessment (3, 7). The General Agreement on Trade and Tariffs and the North American Free Trade Agreement regulate international trade and incorporate stipulations about introducing diseases into the United States (9, 18). Additionally, the International Office of Epizootics (OIE) has international scope in world trade and has established criteria for assigning risk levels to certain animal pathogens (10, 16). The importance of an accepted animal disease classification is apparent; however, the universally accepted guidelines for placing pathogens that only infect animals into biosafety risk categories are poorly defined compared with the guidelines for human pathogens.

J. Scott Rusk • National Animal Disease Center, Agricultural Research Service, U.S. Department of Agriculture, Ames, IA 50010.

Many institutions and organizations in the United States must make decisions regarding the placement of animal pathogens into proper biosecurity rankings and categories of biosafety. The Centers for Disease Control and Prevention (CDC) and the National Institutes of Health (NIH) publish guidelines that are universally accepted as the basic standard for biosafety (15). Although this system is appropriate and accepted for agents of human and zoonotic diseases, it cannot be universally applied to biosecurity for pathogens that infect only animals. For example, using human pathogen biosafety guidelines, foot-and-mouth disease virus could be classified as biosafety level 1, which requires minimal containment. Estimated costs associated with an outbreak of foot-and-mouth disease virus, however, are high enough to have a major impact on the U.S. economy. There is no accepted U.S. system for assigning biosecurity classifications to strictly animal pathogens.

Definitions of biosafety and biosecurity cross boundaries and are frequently used in place of each other. For the purposes of this report, biosafety is defined as prevention of exposure to hazardous disease agents or biological products that are capable of producing illness in human beings. Biosecurity is a broader term, referring to controlling the spread of disease agents or hazardous biological products to susceptible hosts or outside to the environment or preventing inconsistent scientific results.

A biosafety classification scheme for animal pathogens is not intended to place undue burden on researchers but to enhance the quality of research and efforts to improve animal health. The use of prudent and reasonable practices to protect against release and spread of animal diseases under experimental study is an important responsibility. The need for proper risk assessment and evaluation of true and perceived risks is on the rise. Many animal diseases that used to ravage the United States have been eradicated and no longer pose daily threats to animal health and economics. Other animal diseases are the subjects of current eradication programs. Reintroduction of eradicated or nearly eradicated animal diseases would have a significant impact on animal health, damaging the farm livestock economy and, in many cases, the ability to freely trade animal products on the global market.

As a starting point, it is necessary to review some of the well-established and universally recognized infectious-agent biosafety classification systems. The overall general intent of these systems is to address biosafety for human and zoonotic pathogens. Guidelines other than those described here exist, but an extensive summation of all worldwide classification systems is beyond the scope of this discussion.

The CDC-NIH publication *Biosafety in Microbiological and Biomedical Laboratories* contains the primary guidelines used for biosafety in the United States (15) (Table 1). Biosafety levels 1 to 4 are used to assign increasing levels of risk and safety. The biosecurity measures needed to work with animal pathogens that do not infect human beings are not included.

The World Health Organization (WHO) *Laboratory Biosafety Manual* includes a section on classification of infective organisms (17). Risk groups I to IV rank an increasing order of risk, from low to high. The WHO classification scheme incorporates both human and animal disease into the risk groups.

Health Canada's *Laboratory Biosafety Guidelines* include criteria for classification of biological agents by risk group (8). Risk groups 1 to 4 indicate increasing risk. The approach is very similar to the WHO guide and includes both human and animal disease agents.

These well-established biosafety guidelines are extremely valuable for application to disease agents that affect the health of human beings. With additional considerations, the principles can also be applied to biosecurity for animal pathogens to prevent contamination or spread of disease agents outside to the environment and to susceptible hosts. A wide variety of animal species are used in animal disease research, which presents unique challenges in biocontainment. These additional considerations must be well thought out and addressed in order to ensure proper containment and control of animal pathogens.

Table 1. CDC-NIH biosafety levels for infectious agents

Biosafety level	Definition
1	Agents not known to consistently cause disease in healthy adult humans, minimal potential hazard to laboratory personnel and environment
2	Indigenous moderate-risk agents present in the community and associated with human disease of various severities, moderate potential hazard to personnel and environment
3	Indigenous or exotic agents with a potential for respiratory transmission (exposure by inhalation) and serious and potentially lethal infection
4	Dangerous and exotic agents that pose high individual risk of life-threatening disease, which may be transmitted via the aerosol route and for which there is no vaccine or therapy

RISK ASSESSMENT

The natural starting point for proper biosecurity classification of strict animal pathogens is risk assessment (13). Risk classification of microorganisms can be based on an abundance of highly variable criteria. In contrast to risk classification of human pathogens, risk classification of animal pathogens considers economic impact, variables of geographic regions, and susceptibility of multiple animal species. These factors and others make development of universal criteria difficult. The risk assessment process is very important, and a standardized system is needed. Because of the factors of economic impact and disease status between countries and regions within countries, it is not possible to devise a single global ranking of animal pathogens.

The assignment of disease-causing agents to specific biosafety levels also depends on how the agent will be used (i.e., concentration or infection of animals). The variation in use and associated risk assessment may result in a single infectious agent's being classified in different risk groups for different experiments.

APPROACHES TO A BIOSAFETY CLASSIFICATION MATRIX FOR ANIMAL PATHOGENS

In 1983, a group of animal disease experts ranked 25 foreign animal diseases in an attempt to create a universal risk evaluation system for strict animal pathogens (12). Each disease was rated according to the criteria of economic risk; vulnerability to introduction, establishment, and spread; epidemiological risks of transmission and persistence; availability of tests and reagents; and research needs.

In 1991, an international veterinary biosafety group began addressing issues of veterinary biosafety, including classification of animal pathogens for biosafety and needs for biocontainment (1). At a second workshop, the following was proposed as a beginning for future considerations for biosafety classification of pathogens strictly infecting animals (5). Note that classification may vary between countries based on assessments of risk.

- The scheme that is developed must be compatible with systems used for human pathogens.
- The scheme should reflect the OIE's list A and list B classification (10).
- The scheme should reflect handling practices based on risk of infection and transmission.

- The scheme should reflect the disease status of the country (endemic or exotic).
- The scheme should be simple and understood by many.
- The scheme should include provisions for large-scale and experimental infections in animals.

A consensus on animal pathogen risk groups was reached at the 5th International Veterinary Biosafety Workshop in 1996 (2) (Table 2). The emphasis of biosafety risk classification was on pathogens that infect only animals (not affecting human health). Low, moderate, high, and severe risk groups were developed, using key words and concepts in each group to help define levels of risk.

The OIE's *Animal Health Code* incorporated a risk classification of animal pathogen matrix (10). The matrix utilizes comprehensive risk factors, and animal pathogens are placed into groups 1 to 4 of increasing risk. Numerous elements are included as part of definitions that assist with risk assessment.

REVISED SYSTEM OF BIOSAFETY CLASSIFICATION FOR STRICT ANIMAL PATHOGENS

There is certainly scientific need to use and work with animal pathogens (some exotic) to improve treatment, prevention, and detection of animal diseases. The potential for introduction or reintroduction of animal pathogens into the United States adds to the need to step up research and control programs. Scientists need better sources for guidance to help determine the proper facilities and procedures for safely and responsibly performing research on animal pathogens.

APHIS issued 3,997 permits to import organisms, vectors, biological materials, and animal products and by-products in fiscal year 1998 (6). Proper handling of these materials ensures reduced risks to the animal

Table 2. International veterinary biosafety classification of animal pathogens

Pathogenicity group	Definition
Low	Unlikely to spread to susceptible hosts, causes mild animal disease
Moderate	Limited spread, causes moderate animal disease
High	Can spread readily, causes serious animal disease
Severe	Will spread readily, causes severe animal disease

health and agricultural economy of the United States. The ability to obtain APHIS permits is directly related to having adequate facilities and procedures in place for proper biosecurity.

A resulting biosafety matrix has been compiled in review of all of the other systems identified in this discussion (Table 3). The CDC-NIH, WHO, Health Canada, International Veterinary Biosafety Group, and OIE guidelines and recommendations have been considered. Appropriate components of each are combined into a single matrix for use in assigning biosecurity classifications to strict animal pathogens. Provisions are included for laboratory biosecurity, animal biosecurity (large and small species), and large-scale biosecurity. No risk group 5 or biosafety level 4 is identified in this matrix. Class 5 is not a biosecurity term but is a special classification for animal pathogens based on foreignness to the United States. The use of class 5 animal pathogens requires special permits from the APHIS Import-Export Offices. Biosafety level 4 is not used in the animal pathogen matrix because

Table 3. Biosafety matrix for livestock and poultry pathogens

Risk group	Biosecurity levels[a]				
	Definition	Laboratory	Animals		Large-scale
			Large	Small	
1	Unlikely to spread; may cause mild disease; low risk—enzootic; no official control programs	1	1 or 2	1	1 or 2
2	Limited spread; produces moderate disease; moderate risk—enzootic; possible control programs	2	2 or 3	2	2 or 3
3	Can spread readily; causes serious disease; high risk—enzootic or exotic; control programs in place; treatment and prevention available	3	3 or 3Ag	3	3 or 3Ag
4	Can spread rapidly; causes severe disease; maximum risk—enzootic or exotic; strict control programs in place; treatment and prevention not effective	3Ag	3Ag	3Ag	3Ag

[a]3Ag, biosecurity level 3-agriculture.

it is reserved for pathogens with specific human health risks. Biosafety level 3-agriculture is maximum containment for animal pathogens.

BIOSAFETY RISK GROUP DEFINITIONS OF ANIMAL PATHOGENS NOT INFECTIOUS TO HUMAN BEINGS

Biological containment of disease agents is more difficult in animal (in vivo) studies than in laboratory (in vitro) manipulations.

Laboratory work allows the containment and control of infectious materials during procedures by utilizing standard laboratory practices and techniques and also specialized primary barrier equipment (15). Primary barriers are any type of equipment or engineering controls, such as biological safety cabinets, enclosed containers, sealed centrifuge cups, special ventilation, and personal protective equipment, which offer significant protection against contamination of personnel and the environment.

Animals (livestock and poultry species) infected with disease agents present unique challenges for biosecurity control. Containment of infected animals, shed disease agents, and associated wastes from infected animals is difficult because standard operating procedures and primary barrier strategies are very limited or not available. As a result, the animal room itself must serve as the biological containment barrier. Some animal pathogen studies (in vivo) warrant additional facility features and procedural controls beyond standard biosafety level 3 requirements to ensure containment and experimental control.

The definitions of laboratory and animal biosecurity levels are shown in Table 4.

CONCLUSIONS

A concise, understandable system for assigning risk categories to animal pathogens is necessary. Many variables contribute to judgments on proper degrees of biosecurity. Current engineering technologies and operational experiences, if applied correctly, are known to effectively ensure safe research on strict animal pathogens with limited and acceptable health risks to animal populations. The comprehensive biosafety matrix presented in Table 3 can serve as a means to focus on the risks of working with animal pathogens and provide methods and background as a basis for appropriate biosecurity. Emphasis on disease eradication, world trade, and agricultural bioterrorism will drive the need for a universal approach to handling livestock and poultry pathogens in the near future.

Table 4. Biosecurity definitions for animal pathogens

Biosecurity level	Definition	
	Laboratory	Animal
1	Risk group 1: microbiological agents not known to cause disease in healthy animals and of low risk to the environment; basic microbiology laboratory of conventional design without special engineering features for containment; potential hazards readily controlled by standard microbiological practices.	Risk group 1: microbiological agents not known to cause disease in healthy animals and of low risk to the environment; basic animal facility of conventional design without special engineering features for containment; potential hazards readily controlled by standard operational practices.
2	Risk group 2: microbiological agents of moderate risk to animals or the environment and normally present in the region; access to the laboratory restricted and work that may cause splashes or aerosol generation conducted in primary containment equipment (biological safety cabinets); directional air flow, nonrecirculating ventilation system recommended; plants and animals not related to research project not allowed in the laboratory; laboratory clothing, gloves, and eye protection may be required for wear by personnel who work in the laboratory.	Risk group 2: microbiological agents of moderate risk to animals or the environment and normally present in the community; access to the facility restricted; directional air flow, nonrecirculating ventilation system recommended; plants and animals not related to research project not allowed in the facility; special protective clothing, gloves, eye protection, and procedures (agent dependent) may be required for personnel who work in the facility when infected animals which cannot be maintained in primary barrier systems to control splashes or aerosol risks are under study.

3	Risk group 3: microbiological agents that may cause serious disease in animals from exposure by inhalation; laboratory separated from remainder of building by a personnel change room (shower optional); access to laboratory restricted; biological safety cabinets and personal protective equipment always used when manipulating infectious agents; laboratory constructed to allow it to be sealed for gas decontamination; ventilation system designed to provide directional airflow from uncontaminated to potentially contaminated areas; air not recirculated; filtration of exhausted air and/or central liquid waste sterilization optional.	Risk group 3: microbiological agents which may cause serious disease in animals from exposure by inhalation; personal protective equipment used for all manipulations of agents or infected animals; animal caging systems used to minimize risk of exposure to people; facility separated from remainder of building by a personnel change room and shower; access to the facility restricted; double doors (self-closing) entry/exit from access corridors and handwashing (foot or elbow) sinks in each animal room near the exit required; floor drains equipped with traps; autoclave located inside the facility; facility constructed to allow gas decontamination; ventilation system (nonrecirculating) designed to provide directional airflow from uncontaminated to potentially contaminated areas; filtration of exhausted air and/or central liquid waste sterilization optional.
3-Agriculture (1, 4)	Risk group 4: designed for certain high-risk and exotic microbiological agents that infect livestock or plants (example: hog cholera virus); mandatory building design includes personnel change rooms with showers; personnel and equipment air locks; double-door autoclave; single-pass, directional, and pressure gradient air system; HEPA filtration (or equivalent) on supply and exhaust air with electrical interlocks to prevent pressurization of laboratory during electrical or mechanical breakdowns; central liquid/solid waste sterilization; and sealed interior surfaces; facility designed to protect the environment and requires special testing and certification procedures for commissioning.	Risk group 4: designed for certain high-risk and exotic microbiological agents that infect livestock or plants (example: hog cholera virus); mandatory building design includes personnel change rooms with showers; personnel and equipment air locks; double-door autoclave; single-pass, directional, and pressure gradient air system; HEPA filtration (or equivalent) on supply and exhaust air with electrical interlocks to prevent pressurization of the facility during electrical or mechanical breakdowns; central liquid/solid waste sterilization; and sealed interior surfaces; facility designed to protect the environment and requires special testing and certification procedures for commissioning, such as pressure decay testing to confirm integrity of biosecurity barriers and construction techniques.

REFERENCES

1. **Barbeito, M. S., G. Abraham, M. Best, P. Cairns, P. Langevin, W. G. Sterritt, D. Barr, W. Meulepas, J. M. Sanchez-Vizcaino, M. Saraza, E. Requena, M. Collado, P. Mani, R. Breeze, H. Brunner, C. A. Mebus, R. L. Morgan, S. Rusk, L. M. Siegfried, and L. H. Thompson.** 1995. Recommended biocontainment features for research and diagnostic facilities where animal pathogens are used. *Rev. Sci. Tech. Off. Int. Epizoot.* **14:**873–887.
2. **Best, M.** 1996. Mission statement and goals of the International Veterinary Biosafety Work Group. *In Proceedings of the 5th International Veterinary Biosafety Workshop.*
3. **Bridges, V. E.** 1998. Risk analysis—an introduction and its application in APHIS:VS, p. 242–247. *In Proceedings of the One Hundred and Second Annual Meeting of the United States Animal Health Association.* U.S. Animal Health Association, Richmond, Va.
4. **Brown, C., and M. Kiley.** 1998. Biocontainment—assessments for animal agriculture, p. 280–283. *In Proceedings of the One Hundred and Second Annual Meeting of the United States Animal Health Association.* U.S. Animal Health Association, Richmond, Va.
5. **Della-Porta, A. J.** 1992. Towards a risk classification scheme for animal pathogens, p. 54–62. *In Proceedings of the Second International Veterinary Biosafety Workshop.*
6. **Kahrs, R. F.** 1998. Annual report to United States Animal Health Association, National Center for Import and Export, Animal and Plant Health Inspection Service, Veterinary Services, p. 338–343. *In Proceedings of the One Hundred and Second Annual Meeting of the United States Animal Health Association.* U.S. Animal Health Association, Richmond, Va.
7. **Kahrs, R. F.** 1998. Applications of the concept of regionalization to domestic animal disease control and global market expansion, p. 207–211. *In Proceedings of the One Hundredth Annual Meeting of the United States Animal Health Association.* U.S. Animal Health Association, Little Rock, Ark.
8. **Kennedy, M. E.** 1996. *Laboratory Biosafety Guidelines.* Laboratory Centre for Disease Control, Health Canada, Ottawa, Ontario, Canada.
9. **North American Free Trade Agreement.** 1992. *Agreement between the Government of the United States of America, the Government of Canada, and the Government of the United Mexican States.* U.S. Government Printing Office, Washington, D.C.
10. **Office International des Epizooties.** 1999. International animal health code for mammals, birds and bees. Office International des Epizooties, Paris, France.
11. **Office of the Federal Register.** 1999. Animal and animal products. *In Code of Federal Regulations* 1999-9 CFR, parts 1–199. U.S. Government Printing Office, Washington, D.C.
12. **Pilchard, E. I., and H. A. McDaniel.** 1983. Foreign diseases and arthropod pests of livestock and poultry, p. 11–23. *In Proceedings of the Eighty-Seventh Annual Meeting of the United States Animal Health Association.* U.S. Animal Health Association, Richmond, Va.
13. **Salman, M.** 1998. Risk analysis methodology: what do we need?, p. 276–278. *In Proceedings of the One Hundred and Second Annual Meeting of the United States Animal Health Association.* U.S. Animal Health Association, Richmond, Va.
14. **U.S. Department of Agriculture.** 1999. *Agricultural Statistics.* U.S. Government Printing Office, Washington, D.C.
15. **U.S. Department of Health & Human Services.** 1999. *Biosafety in Microbiological and Biomedical Laboratories,* 4th ed. U.S. Government Printing Office, Washington, D.C.
16. **Willis, N. G.** 1998. What is the role of the OIE in international trade?, p. 40–44. *In Proceedings of the One Hundred and Second Annual Meeting of the United States Animal Health Association.* U.S. Animal Health Association, Richmond, Va.
17. **World Health Organization.** 1993. *Laboratory Biosafety Manual,* 2nd ed. World Health Organization, London, England.
18. **World Trade Agreement.** 1994. *General Agreement on Tariffs and Trade.* World Trade Organization, Geneva, Switzerland.

Emerging Diseases of Animals
Edited by C. Brown and C. Bolin

Chapter 3

Agroterrorism, Biological Crimes, and Biological Warfare Targeting Animal Agriculture

Terrance M. Wilson, Linda Logan-Henfrey, Richard Weller, and Barry Kellman

There is a rising level of concern that agriculture might be targeted for economic sabotage by terrorists. Knowledge gathered about the Soviet Union biological weapons program following its dissolution in 1991, and about Iraq following the Gulf War, confirmed that animals and agricultural crops were targets of bioweapon development and weaponization. The American public has been exposed to a blitz of news stories about weapons of mass destruction with a renewed emphasis on biological weapons. There have been numerous newspaper articles, books, television specials, radio shows, and new websites, all devoted to the issue of bioweapons and bioterrorism. A number of new books have been published providing some of the history of the development of biological weapons during the past century and describing some of the major players involved in their development (2, 27, 47, 55, 59, 73, 87). More than a dozen countries are suspected of maintaining ongoing offensive biological weapons research programs (20, 59, 64). These revelations are particularly disturbing in light of the fact that most of these countries are States Parties to the Biological and Toxin Weapons Convention that entered into force in 1975.

Terrance M. Wilson • Emergency Programs, Veterinary Services, Animal and Plant Health Inspection Service, U.S. Department of Agriculture, 4700 River Road, Riverdale, MD 20737-1231. **Linda Logan-Henfrey** • Agricultural Research Service, U.S. Department of Agriculture, 5601 Sunnyside Ave., Beltsville, MD 20705-5138. **Richard Weller** • Molecular Biosciences Department, Pacific Northwest National Laboratory, Richland, WA 99352. **Barry Kellman** • International Criminal Justice and Weapons Control Center, DePaul University College of Law, Chicago, IL 60604.

The potential for misusing biotechnology to create more-virulent pathogens and the lack of international means to detect unethical uses of new technologies to create destructive bioweapons are of increasing concern. Foreign animal disease outbreaks, whether naturally occurring or intentionally introduced, involving agricultural pathogens that destroy livestock would have a profound impact on a country's infrastructure, economy, and export markets. In the United States, such outbreaks would also erode consumer confidence in the safety of our food supply. This chapter deals with the history of agroterrorism, biological crimes, and biological warfare directed toward animal agriculture, specifically horses, cattle (both beef and dairy), swine, sheep, goats, and poultry. There are surprisingly few reviews that highlight antilivestock biological weapons and even fewer that provide any credible details as to what pathogens might pose the greatest threats to animal agriculture (2, 23, 25, 39, 40, 51, 64, 84). The threat of biological weapons directed towards crops was recently reviewed (77; http://www.scisoc.org/lecture/Biosecurity/Top.html). The criminal and terrorist threat to processed food has also recently been reviewed (46).

DEFINING TERRORISM

"Local and regional conflicts, famine, economic disparity, mass movements of refugees, and brutal and corrupt regimes contribute to instability and fuel a frustration and a desperation that increasingly finds expression in acts of terror" (70). Simon contends that any attempts to define terrorism are futile and only add to the sense of confusion that people have about what really constitutes a terrorist act (79). He further adds that most definitions of terrorism are either too broad or too narrow. Between 1936 and 1980 there were over one hundred definitions coined for terrorism, and that was long before the current interest on terrorism emerged. Seth Carus, in his detailed scholarly work "Bioterrorism Biocrimes," defines a terrorist as "a non-state actor who uses violence on behalf of a political cause without reference to the moral or political justice of the cause. This includes non-state actors who operate in organized military units (as with guerillas) if the use of biological agents is undertaken covertly using improvised delivery systems" (23). We will adhere to his definition in this chapter. The Department of Defense, the Federal Bureau of Investigation (FBI), and other agencies may have their own definitions of terrorism. A criminal is any perpetrator who uses violence for purposes not related in any way to political, ideological, religious, or social causes (83). Motivation for criminal and terrorist activity may overlap.

Carus (23) defines a biological weapon agent as organisms or toxins produced by organisms that can be used against people, animals, or crops. Chemical agents used as weapons are man-made poisonous substances that kill or incapacitate. Again, the definition of a biological or chemical weapon may vary depending on which agency or institution is defining it. Terrorism has become commonplace throughout the world, is growing in activity, and is driven by psychological, ideological, economic, religious, and environmental factors (23, 54, 78, 79, 84). Terrorism is a battle in which anyone can become a victim and where any individual from any place or position can serve as a perpetrator (69, 79). In its simplest terms, bioterrorism is nothing more than criminal violence. Many nonpolitical factors motivate criminals. Biocrimes targeting livestock may be motivated by revenge or desire for personal financial gain (23). Recently, the term "agroterrorism" as a form of economic attack has emerged in some news articles in reference to biological weapons targeting animal and crop agriculture (42).

WHO MIGHT PRESENT A BIOTERRORIST THREAT?

Traditionally, our primary concerns have centered on state-sponsored terrorism or military programs as the primary drivers of bioweapon development and use. Recently, however, rogue states have been added to the list of potential government-sponsored terrorist groups (20). The Cold War has given way to regional armed conflicts rooted in ethnic clashes and enormous inequalities of income and opportunities for many groups of people (http://www.undp.org.hdro.E5.html). National and international boundaries have been blurred by the virtually free flow of refugees, drugs, weapons, and laundered money. Organized crime syndicates are estimated to gross $1.5 trillion a year (http://www.undp.org.hdro.E5.html). In this environment, there are a number of new actors who have used or might resort to use of bioterrorism. On the rise in the world today are groups driven by strong religious zeal, hatred, or conflict and groups seeking new methods to enact ethnic cleansing. Other groups who might utilize bioterrorism would be organized crime groups such as those that deal in the international narcotics trade. Militia groups, cult sects, and Armageddon believers all have threatened to use or have used terrorism in recent years. Disgruntled individuals and "copy cat" individuals often look for ways to gain public recognition (23, 81). Less traditional groups now resorting to terrorist tactics include antiabortionists, who send letters filled with various powders purported to be anthrax spores. Fortunately, all of these incidences have turned out to be hoaxes. Animal rights groups are renowned for releasing experimental animals from facilities and could pose a new threat of releasing experimental animals infected with dangerous pathogens that could pose a threat to humans and domestic and wild animals. Advocacy groups opposed to genetically modified foodstuffs might inadvertently spread disease in their zeal to destroy genetically modified animals and plants.

The conflict between First World prosperity and Third World poverty continues to be a volatile issue. It is projected that by 2050, 8 billion of the world's 9.5 billion people will live in developing countries, with consequent continually rising social and environmental pressures. The richest fifth of the world, which includes the United States, Canada, and western Europe, consumes 86% of all the goods and services of the world, while the poorest third consumes just 1.3% (http://www.undp.org/hdro/ES.html). Some individuals wishing to correct the economic imbalances of the world might very easily resort to economic terrorism as a means of making a statement. Many peoples of the world resent the global dominance of Western cultural norms, the dominance of multinational companies, and the United States as the sole military superpower now remaining. All of these factors focus attention on U.S. citizens and on the U.S. economy. New forms of warfare that do not require sophisticated equipment or delivery systems are something that pose an emerging threat to the tranquil lifestyle to which we have become accustomed. Future wars may be conducted in a sphere not dominated by military actions but instead may resort to two new methods based on bioterrorism or targeting of other major infrastructure, including communications, computer systems, and commercial sabotage.

HISTORY OF AGROTERRORISM, BIOLOGICAL CRIMES, AND BIOLOGICAL WARFARE TARGETING ANIMAL AGRICULTURE

The literature is replete with information on biological warfare and biocrimes targeting American civilian and military personnel, some of which is factual, balanced, and informative (15, 23, 24, 26, 35, 73, 75). Publicly available scientific literature pertaining to agroterrorism, biological crimes, and biological warfare targeting livestock and poultry dates back nearly 90 years (1, 2, 43, 48, 82–85). These publications discuss, among other things, the criteria for the selection of the biological agent, methods of dissemination, ideal livestock and poultry geographical targeting areas, lists of selected animal and poultry viruses, and negative economic impact figures. Brief descriptions of emergency animal disease response procedures in the event of a biowarfare attack have been presented (1, 7, 25, 33, 37, 43, 44, 80, 82). Livestock and poultry as targets for agroterrorism, biocrimes, and biowarfare development are therefore not a new topic but one that deserves a contemporary review in light of the new economic and military realignment of today's world. Publications dating back to 1952 conclude that targeting livestock and poultry would be easy and would require only a low level of scientific and technical skills to perpetrate (37, 43, 82).

One discovers that the use of disease agents in military conflict is as old as human conflict (87). Documentable incidents from the 20th century involving biological warfare against animals are listed in Table 1. It should be remembered that the first 20th-century allegations of the use of biological warfare agents were against livestock and not humans. These substantially supported allegations were made against Germany, who employed anthrax and glanders against cattle, sheep, horses, and reindeer in Romania, Spain, France, Norway, Argentina, and the United States during World War I (Table 1). A recent investigative report concludes that it was an ambitious and well-planned program, conducted on three continents, but that the success of the attacks was questionable (85). Other authors report that this antilivestock program was successful, killing or incapacitating hundreds of animals (48, 76). Regardless of the success of the operation, it is a historically documented account of a carefully planned and executed clandestine operation in which livestock were the primary target for use of two biological weapons, and both agents continued to be the subject of intensive research by several countries for use against animals and humans. Wheelis (85) concluded, and we concur, that the German World War I antilivestock program was of special significance because (i) it was the first national biological warfare (BW) program, (ii) it was the first BW program with a scientific foundation, (iii) it was one of two confirmed events of the use of BW agents in wartime, (iv) it was the first and perhaps only extensive use of BW agents by secret agents, and (v) these were the first antilivestock BW events which were well documented.

Recent public reports implicate Japan as the inspiration for the modern biological arms race (47, 73). In 1938, the Japanese set up a research and development facility, Unit 731, at Ping Fan, south of Harbin in occupied China. The Japanese, employing a variety of biological agents, reportedly killed at least 850 human subjects in their quest to develop bioweapons and to understand the effects of pathogens on human organ systems (47, 73).

The United Kingdom was the first Western nation to organize an offensive biological weapons program in late 1941 (22). Between the late 1940s and the 1960s, the United States and Canada joined in and developed both

Table 1. Incidences in which animals were targets for biological terrorism (BT), biological crime (BC), or biological warfare (BW) during the last century

Date(s)	Location(s)	Perpetrator	Targets	Biological agent(s) used	Type of incident	Results	Reference(s)
1915–1916	Maryland, Virginia, New York	German agents	Horses, mules	Anthrax, glanders	BW	Unknown, possibly successful	23, 48, 56, 76, 85
1915–1916	Romania	German agents	Horses, sheep, livestock (not specified)	Glanders	BW	Unknown, possibly successful	23, 48, 56, 76, 85
1915–1918	Spain	German agents	Horses, cattle	Anthrax, glanders	BW	Unknown	23, 48, 56, 76, 85
1916	Norway	German agents	Reindeer, cattle	Anthrax, glanders?	BW	Unknown	23, 48, 56, 76, 85
1916–1917	Argentina	German agents	Horses, mules, cattle?	Anthrax, glanders	BW	Unknown, possibly successful	23, 48, 56, 76, 85
1917	France	German agents	Horses	Glanders	BW	Unknown	23, 48, 56, 76, 85
1952	Kenya	Mau Mau freedom fighters	Cattle	Plant toxin (African milk bush plant)	BT	Successful	23
1978–1980	Rhodesia	Rhodesian security forces	Cattle	Anthrax	BW	Successful	17, 28, 59, 63, 71
1982–1984	Afghanistan	Soviet military	Horses	Glanders	BW	Unknown	2, 59
1997	New Zealand	New Zealand farmers	Wild rabbits	Viral hemorrhagic disease of rabbits	BC	Successful	23, 41, 58

offensive and defensive biological weapons programs, building on the knowledge they uncovered from the Japanese biowarfare program (22, 27, 47). Researchers experimented with aerosol dispersion of a number of pathogens that would infect both humans and animals. One significant British antilivestock program was entitled Operation Vegetarian and included the stockpiling of 5,000,000 cattle anthrax cakes which were to be delivered by parachute flares over grazing cattle in Germany. Some of the events that were part of the U.S. offensive biological weapons program are well documented (27, 73). These accounts provide some insight into the U.S. offensive weapons program.

During World War II, the United States and Canada had a highly secret animal disease research program conducted on Grosse Isle in the St. Lawrence River. This first War Disease Quarantine Station was developed to study rinderpest virus, a highly fatal disease of cattle. The possibility of the introduction of rinderpest into the Western Hemisphere had long been a matter of serious concern. As a result of both increased air travel and suspected enemy interest in biological warfare, this possibility became alarming in the latter part of 1941. Following several conferences, a Joint United States-Canadian Commission was appointed by the United States Secretary of War and the Canadian Minister of National Defense. A staff of seven scientists were assigned the task of developing means to protect North American livestock from biowarfare agents targeting livestock and poultry. The staff were given two responsibilities. The first was to prepare a tissue vaccine according to methods previously developed, in order to provide a rapid means of surrounding an epidemic outbreak of disease by using a ring immunization approach. The second task was to investigate the possibility of developing an efficient vaccine which would be produced more economically and without requiring the use of large numbers of animals. The program was successful in developing a rinderpest vaccine grown in chicken eggs (45). Many of the individuals involved in this program went on to have brilliant careers, such as deans of veterinary schools, professors, assistant surgeon general, director of the National Institutes of Health, and chief of the Pathological Division of the Bureau of Animal Industry. This experience undoubtedly influenced H. W. Schoening, who during his tenure at the Bureau of Animal Industry helped support the establishment of the Plum Island Animal Disease Center, a laboratory devoted to study of exotic animal disease. By law, in 1948, Congress mandated that studies of rinderpest and foot-and-mouth disease be conducted on an island setting. Schoening, in anticipation of this important program, initiated a new program to send U.S. veterinarians and microbiologists overseas to work in a number of laboratories conducting research on highly infectious diseases of livestock.

After World War II and during the Cold War years, the U.S. military field-tested two antilivestock and antipoultry viruses. In 1951, hog cholera was weaponized in the E-73 biological bomb, a modification of the M-16 type propaganda leaflet cluster. Two hog cholera bombs were dropped over experimental open pig pens on Eglin Air Force Base, Florida, and exploded at 1,500 ft. The hog cholera virus (Ames, Iowa, strain 1694A) was mixed with turkey feathers, which floated down into the open pig pens. The experimental bombing was a success, judging by the 93 of 115 pigs that developed hog cholera (Department of Defense Special Report 159, 1952). Similarly, biological warfare trials were carried out against poultry using feathers laced with Newcastle disease virus and dispersed by cluster bomb (30, 31). Although the U.S. offensive biological weapons program ended in 1969 when President Richard Nixon abolished the program, Cuba has continued to accuse the United States of using microorganisms and insects as BW agents against its human, animal, and plant populations. On 12 occasions between 1964 and 1967, Cuba has alleged that the United States has intentionally used human, animal, and plant viruses and insects to harm and disrupt the Cuban economy (88). They have specifically accused the United States of intentionally introducing African swine fever in 1971, 1979, and 1981, infectious bronchitis of poultry in 1985, herpes mammillitis in dairy cattle in 1985, and an unidentified cattle skin disease in 1981. Add to those the accusation that the United States deliberately introduced an unidentified sea turtle virus in 1962 and an unnamed rabbit virus in 1993. According to an excellent review by Zilinskas (88), all 12 Cuban allegations lacked scientific merit. In April 1997, Cuba registered a formal complaint with the United Nations (UN) Secretary-General accusing the United States of a potential violation of the BW Convention (BWC) by willfully introducing a plant pest, *Thrips palmi,* to its territory by dispersion from an overflying aircraft. The states parties to the BWC held a formal consultation meeting in August 1997 in Geneva, Switzerland, mediated by the United Kingdom in its role as a BWC depository, to consider the Cuban claim that the United States had violated the BWC. The importance of the allegation lay not only in its seriousness, but also because it was the first test of procedures adopted at the Second Review Conference of States Parties to BWC for consultation and clarification. Both countries gave their statements of evidence to those states parties in attendance. The assembled states did not reach a conclusion, and because of the complexity of the issue and the passage of time, no further investigation has been undertaken.

The unusually large anthrax epidemic in both cattle and humans in Rhodesia (now Zimbabwe) during the war of independence in the 1970s deserves special mention. Prior to the war, anthrax in both animals and humans, while endemic, was rare and well controlled (28, 52, 53, 59, 63). During the war, with the collapse of the government administration, including veterinary services, anthrax became a major disease in cattle, with thousands of cases and deaths (28, 53). There was a corresponding major epidemic in humans, accounting for approximately 10,000 cases and hundreds of deaths. Close examination of the epidemiology of this outbreak indicates that the introduction of anthrax was very likely intentional. The introduction of a biological warfare agent such as anthrax was used to undermine the morale and food supply of those who sought independence. The long-lasting effects on Zimbabwe have continued to impact livestock, wildlife, and humans to the present day.

The biological warfare program of the Republic of South Africa has been partially exposed through the Truth and Reconciliation Commission (14, 59). Preliminary public disclosures strongly suggest that their BW program encompassed antihuman, antianimal, and anticrop activities. Full disclosure relevant to the BW program will most likely never occur because such disclosure would jeopardize government relations with countries that might have cooperated with South Africa in the program but with whom South Africa still maintains diplomatic relations. Research, development, and weaponization of antilivestock agents are reported to have included African swine fever virus and foot-and-mouth disease virus (25, 59).

In a very recent excellent investigative report on the introduction of rabbit calicivirus disease into New Zealand, Steve Goldstein offers a serious, thorough review of agroterrorism and biological crimes (41). The introduction of this virus into New Zealand was a very well-planned, organized criminal event. The participants were boastful and arrogant about their criminal acts and showed little remorse (58). The perpetrators were able to circumvent one of the best and most secure airport and port biosecurity systems in the world. Once the virus had entered the country, it required only a simple, low-tech procedure for the farmers to prepare purees for distribution around their farms. Liver, spleen, and lungs were placed in a kitchen blender and blended into a very potent "kitchen whizz" that was mixed with an appropriate vegetable or food source. The virus entered New Zealand in three ways: mailed into the country in a vial, imported in a vial placed in an air traveler's sock, and imported by an air traveler on a handkerchief that had been drenched in blood and tissues from an infected animal (44). Rabbit calicivirus was handled with impunity by New Zealand farmers because it was not a zoonotic agent.

In 1992 Russia admitted that the former Soviet Union, despite being a codepository of the BWC, had continued a massive offensive biological weapons program which was conducted illegally for years (2, 20, 55, 68,

74). With the breakup of the Soviet Union in 1991, government funding for research decreased dramatically, and scientists who were working in the biological weapons program found themselves unemployed or without funds to conduct research. Many came to the United States and Great Britain and other European countries, but a few were attracted to jobs in rogue states. One early defector, Vladimir Pasechnik, provided significant intelligence information on BW programs. In addition, other sources of information, including satellite imagery, indicated that a superpower BW arms race was on (59). The Sverdlovsk anthrax accident provided additional information regarding the Russian BW program. Another defector was Kanatjan Alibekov, now more commonly known as Ken Alibek, who arrived in the United States in 1992. Alibek had served as the first deputy chief of research for Biopreparat, the umbrella organization that orchestrated the Soviet Union's biological weapons program directed against humans. Biopreparat, established in 1973, was designed to support research and production of bioweapons targeting humans. The program was housed in over 50 laboratory locations and involved tens of thousands of employees (2). This bioweapons program was conducted under the umbrella of multiple ministries. The groups who helped orchestrate this massive program were the Soviet Academy of Sciences, Communist Party Central Committee, the KGB, and the Soviet military. Alibek has stated that the Soviet Union had three bioweapons programs: one in the Ministry of Health, one in the Ministry of Agriculture, and one in the Ministry of Defense.

More is being revealed about the anticrop and antilivestock program, code named "Ecology," of the former Soviet Union through the efforts of new government and nongovernment programs that promote nonproliferation in the former Soviet Union. At its peak, the Soviet Union agriculture BW program is reported to have employed ten thousand staff in an offensive biological weapons program targeting livestock, poultry, and crops. This program was conducted at eight known locations, including the Institute of Veterinary Virology and Microbiology in Pokrof (2, 74). Despite repeated U.S. attempts to obtain permission to visit these animal-agriculture facilities, access, in most cases, continues to be denied. The former Soviet Union program focused on weaponizing a number of antilivestock agents, including foot-and-mouth disease (FMD) virus, rinderpest virus, classical swine fever and African swine fever viruses, sheep and goat pox viruses, *Chlamydia psittaci,* and anticrop agents including wheat rust, rice blast, karnal bunt, and *Fusarium* spp.

Much has been learned about the bioweapons program of Iraq as the result of the UN Special Commission (UNSCOM) (20, 90). The UN sent inspection teams in to examine several Iraqi laboratories and production facilities. Evidence of the production of chemical gases, biological agents, and biological toxins was discovered. It is now well documented that even prior to the Gulf War, the Iraq government used chemical weapons against rebellious Kurdish populations in the northern part of the country and against Iranian soldiers during border conflicts. During one Kurdish uprising in Halabja, Iraqi forces used a combination of four toxic chemical weapons, mustard gas, sarin, tabun, and VX, causing numerous deaths in the village (20). Iraq had an active program to develop biological weapons, but there is no public evidence that Iraq actually utilized bioweapons during the Gulf War. There is, however, clear evidence that Iraq had plans to use biological weapons. The biological weapons identified which were under development were *Bacillus anthracis, Clostridium perfringens,* Congo-Crimean hemorrhagic virus, yellow fever virus, enterovirus 17, human rotavirus, camelpox, FMD virus, and at least four toxins: aflatoxin, botulinum toxin, ricin, and trichothecenes (20, 90; http://www.cnn.com [18 March 1999]).

SELECTION CRITERIA FOR ANTILIVESTOCK AND ANTIPOULTRY BIOLOGICAL WEAPONS

Selection criteria for the most dangerous antilivestock and antipoultry biologic agents have been published (7, 37, 43, 82). The selected animal agent should have the following characteristics:

1. Highly infectious and contagious
2. Good ability to survive in the environment
3. Predictable clinical disease pattern, including morbidity and mortality
4. Pathogenic for livestock or poultry
5. Available and easy to acquire or produce
6. Attributable to a natural outbreak, ensuring plausible deniability
7. Not harmful to perpetrator
8. Easily disseminated

The selection of the animal virus and means of dissemination are key factors in antilivestock biological weapons. Previous reports have indicated that the following agents pose the greatest potential threats (1, 16, 20, 37, 49, 57, 80, 82):

1. Foot-and-mouth disease virus
2. Classical swine fever virus
3. African swine fever virus
4. Rinderpest virus
5. Rift Valley fever virus
6. Avian influenza virus
7. Velogenic viscerotropic Newcastle disease (VVND) virus
8. Venezuelan equine encephalomyelitis virus
9. Bluetongue virus
10. Sheep and goat pox viruses
11. Pseudorabies virus (Aujeszky's disease)
12. Vesicular stomatitis virus
13. Teschen disease virus (porcine enterovirus 1)
14. Porcine enterovirus type 9
15. Lyssaviruses and rabies viruses
16. Lumpy skin disease virus
17. Porcine reproductive and respiratory syndrome virus
18. African horse sickness virus
19. *Bacillus anthracis* (anthrax)
20. *Chlamydia psittaci* (ornithosis, psittacosis, chlamydiosis)
21. *Cowdria ruminantium* (heartwater, cowdriosis)
22. Screwworm (myiasis)

While viruses pathogenic for humans might make excellent agents for biocrimes or bioterrorist attacks, they are poorly adapted for strategic deployment, with the exception of smallpox, because of their lack of stability in the environment. This animal pathogen list has very few agents in common with the list often given as high-risk antihuman bioweapons agents (15, 20). In fact, with the exception of Venezuelan equine encephalitis virus, Rift Valley fever virus, *C. psittaci,* and *B. anthracis,* the agent list is completely different. Often, the public and even professionally trained scientists assume that the potential BW agent list for humans would be the same as for livestock and poultry. However, the list of livestock and poultry pathogens is based on economic trade impact and ease of transmissibility, while that for humans is based more on high mortality or fear. Anthrax appeared on the list of potential animal bioweapons as early as 1952 (82). However, thereafter anthrax was assumably dropped from the antianimal biological weapons agent lists because an effective vaccine had been developed.

The animal pathogens and the species affected that are most often mentioned as being the most important as potential agroterrorism attack agents are, for cattle, foot-and-mouth disease virus and rinderpest virus; for swine, foot-and-mouth disease virus, classical swine fever virus, and African swine fever virus; and for poultry, avian influenza virus and Newcastle disease virus.

It should be noted that all of these viruses spread quite readily on their own and would not require any microbiological manipulation or weaponization to make them effective agents to engender an outbreak.

THE OFFICE INTERNATIONAL DES EPIZOOTIES

The Office International des Epizooties (OIE) (http://www.oie.int/info) serves as the world animal health organization and is headquartered in Paris, France. This important organization was founded in 1924 and presently has 151 member countries. The OIE is an independent organization and is funded by member countries. The main objectives of the OIE are to inform governments of the occurrence and course of animal diseases throughout the world and of ways to control these diseases; coordinate, at the international level, studies devoted to the surveillance and control of animal diseases; and harmonize regulations for trade in animals and animal products among member countries. Through international cooperation, the OIE develops procedures, guidelines, and information sharing to help prevent the spread and introduction of important infectious diseases of livestock and poultry. The OIE maintains two levels of animal disease lists. List A animal diseases characteristically are highly infectious, spread rapidly irrespective of international borders, often cause catastrophic economic losses, and have major socioeconomic impact on affected countries. These list A diseases are rigorously monitored on a global basis by the OIE, and international agreements require member countries to report outbreaks of these diseases to the OIE within 24 h of laboratory confirmation. Trading partners impose severe trade restrictions on members who have outbreaks of list A diseases. The 15 list A animal diseases consist of 14 viral diseases (foot-and-mouth disease, swine vesicular disease, classical swine fever, African swine fever, rinderpest, peste des petits ruminants, vesicular stomatitis, bluetongue, Rift Valley fever, lumpy skin disease, sheep and goat pox, African horse sickness, highly pathogenic avian influenza, and Newcastle disease) and one bacterial disease (contagious bovine pleuropneumonia). After the original diagnosis of a list A disease, the country must periodically report specific information on the epidemiology, control, and eradication progress to OIE. All this information, including the original diagnosis, is immediately electronically communicated to all OIE member countries. Upon receiving this information, member countries may enact trade embargoes against the infected country. This consequence is the major negative economic impact when a list A disease is naturally or intentionally introduced into a country. For example, when Taiwan diagnosed FMD on 17 March 1997 and communicated this to OIE, this diagnosis was immediately communicated to all OIE member countries. Upon receipt of this information, all Taiwan pork export products were immediately embargoed. Taiwan's once very lucrative billion-dollar pork export market literally disappeared overnight following the diagnosis of FMD. Over the years this will amount to approximately a $15 billion loss to the Taiwan pork industry and related industries.

List B diseases are grouped by animal species and are considered of socioeconomic or public health importance within countries. There are some misconceptions that list B diseases are less important than list A diseases and do not have as much of an effect on international trade. This, however, is not necessarily true. Take for example the emergence of bovine spongiform encephalopathy (BSE; "mad cow disease") in Great Britain. This disease has caused significant and long-standing barriers to trade of meat and meat products from Great Britain to other countries, including the United States. Despite the enormous economic impact of this disease, it is still classified as list B and not a list A disease. This classification is based on the fact that although it is transmissible, it is not highly contagious. A few other examples of important OIE list B diseases that are closely monitored in domestic and international trade are brucellosis and tuberculosis in cattle, porcine reproductive and respiratory syndrome and trichinellosis of swine, and Marek's disease and infectious laryngotracheitis of poultry. Natural or intentional outbreaks of these diseases do have significant economic impact.

RESPONSE TO INCURSION OF A FOREIGN ANIMAL DISEASE IN THE UNITED STATES

The emergency program (EP) of the Animal and Plant Health Inspection Service (APHIS) Veterinary Service (VS) of the U.S. Department of Agriculture (USDA) is well positioned to respond to an intentionally or naturally introduced foreign animal disease (19). EP, as the "911" of the U.S. livestock industry, would coordinate the entire emergency response plan in close coordination with state veterinary officials, veterinary colleges, industry officials, the Department of Defense, the Federal Emergency Management Agency, the American Veterinary Medical Association, private veterinary practitioners, and livestock and poultry producers. Critical to this response would be the approximately 350 federal, state, military, university, and private veterinarians specially trained as foreign animal disease diagnosticians, who are strategically located across the United States, available to respond within 24 h to a suspected foreign animal disease outbreak. Supporting this infrastructure is the Regional Emergency Animal Disease Eradication Organization (READEO), of which EP maintains two, the eastern and western READEOs. The READEO is a 38-person team available for immediate call-up to control and eradicate a foreign animal disease. Smaller teams, three or four persons, called emergency response teams, may be called up in smaller outbreaks. The READEO trains regularly with field and tabletop computer exercises. The most recent full-scale field exercise was conducted on 6 November 1998, when the USDA simulated the intentional release of a virus similar to foot-and-mouth disease virus. A simulated terrorist group entered three livestock auctions or sale barns in California, Minnesota, and Florida simultaneously and released the viral agent. The eastern and western READEOs were fully mobilized to the field and exercised with local and regional veterinary and legal personnel, including the FBI, for 1 week. The Canadian veterinary emergency response team also participated because the simulated exercise included having the disease outbreak spread into Canada.

Integral to the READEO is the legal unit (67). This unit would have special forensic and investigative responsibilities in an outbreak that was intentionally introduced. The Office of Inspector General (OIG) staff and other local, regional, and national law enforcement personnel, including the FBI, will staff the legal unit in the READEO. They will spearhead the investigation and collection of forensic evidence and material should there be evidence of an intentional or criminal act. This OIG unit would coordinate with other law enforcement organizations throughout the United States, including local, regional, and national police staff and the FBI.

In an intentional or naturally introduced foreign animal disease, the READEOs will operate from field headquarters selected at the onset of the outbreak, very likely an established emergency or disaster response center in the state involved. The headquarters national emergency response team will operate from an emergency management operation center, the Emergency Response Center, which is an integral part of the APHIS emergency management system. This modern, fully equipped 8,000-ft^2 facility is located in the USDA APHIS headquarters building in Riverdale, Md.

CONTEMPORARY ANTILIVESTOCK AND ANTIPOULTRY AGROTERRORISM, BIOLOGICAL CRIMES, AND BIOLOGICAL WARFARE ISSUES

An agroterrorism attack on the U.S. livestock industry would cause major economic losses for the livestock industry and related spinoff agroindustries. An intentional introduction of a foreign animal disease, especially a zoonotic agent, would cause panic and social instability and certainly result in a decrease in public confidence in the local, regional, and national safety of food. It is believed that the bovine spongiform encephalopathy epidemic and resultant public outcry were partly responsible for the fall of the U.K. Tory government under John Major during the mid-1990s.

The intentional introduction of a highly transmissible, economically costly livestock or poultry disease would be an inexpensive operation. One is reminded of the millions of dollars required to build, maintain, and operate a modern laboratory for human pathogens for use in bioterrorism; such an infrastructure is not needed in the case of animals. Antilivestock programs are not manpower or cash intensive. Animal agents are readily available from clinical specimens that can be collected throughout the world where the diseases are endemic and where many of the list A diseases occur on a frighteningly regular basis. For example, FMD viruses have been reported by the OIE to occur in 25 countries in the last 18 months, permitting ready access to field specimens. In a recent important New Zealand veterinary publication, it was stated that the terrorist or criminal introduction of FMD represented a most likely source of introduction over the next 20 years (38).

VALUE OF U.S. AGRICULTURE

The economies of many countries around the world are inextricably linked to agriculture, and the United States is no exception. In the United States, agriculture is the largest single sector in our economy, making up approximately 13.3% of the gross domestic product (GDP). Over 24 million Americans are employed in some aspect of agriculture, making it the single largest employer. In 1999, agricultural cash receipts from crops and livestock and poultry were $190.7 billion; livestock and poultry accounted for almost half of this total ($95 billion).

The United States is the world's leading exporter of agricultural products, with growing annual income from exports of $53.7 billion in 1998. U.S. agricultural exports account for about 15% of all global agricultural exports. The share of U.S. agricultural production exported is more than double that of other major U.S. industries and therefore contributes as a positive factor toward the U.S. balance of trade. In 1998, the United States exported $10.9 billion of animals and animal products. The cost of food to the U.S. consumer is less than 11% of per capita disposable income, giving Americans the least expensive and the highest-quality food in the world.

Farm gate receipts represent but a small portion of the overall value of agriculture to the U.S. economy. There are a number of allied industries, such as supplies, transporters, distributors, food processors, exporters, fast food chains, and restaurants that are all part of the greater web of the U.S. food industry. Consequently, the downstream effect of undermining U.S. agriculture through intentional introduction of animal or plant disease pathogens with the malicious intent of undermining the U.S. economy would create a tidal wave effect felt by all of these sectors of our economy, including, ultimately, the American consumer. Any country that has agriculture as a central theme in its economy, as does the United States, is equally vulnerable to the negative effects of catastrophic animal disease outbreaks that would negatively impact agricultural production.

ECONOMIC IMPACT OF AN INTENTIONALLY INTRODUCED LIVESTOCK DISEASE

Many factors are of importance in determining the negative economic impact of an intentional introduction or natural incursion of a foreign animal disease. The total economic impact will be very much the same for a foreign animal disease, whether it is a natural incursion or an intentional introduction. Certainly the type of disease agent and the potential range of species susceptible to that agent would be critical. For example, FMD virus affects all cloven-hoofed animals, including approximately 60 species of wildlife and zoo animals. FMD would therefore, have a much greater economic impact than a virus that caused disease in only one species (60). The ease of transmission on fomites and the possibility of airborne viral dissemination would also be important, as would the presence of subclinical or carrier animals. Our trading partners would very likely enact similar trade embargoes or restrictions on the U.S. livestock industry whether the disease was a natural incursion or an intentional introduction. A review of the financial losses associated with natural foreign animal disease outbreaks in other countries might provide an idea of the costs involved. In order to determine such economic losses, several factors must be considered, and in all cases such information is not factored in. Indeed, losses due to a foreign animal disease might not be fully known for 3 to 5 years after the end of the outbreak. The costs of a foreign animal disease are (i) costs of diagnosis and surveillance; (ii) direct costs of depopulation, cleaning and disinfection, and quarantine; (iii) direct, indirect, and induced losses in the economy of the country or state; and (iv) losses due to trade restrictions (62). Rarely are all these factors considered when determining reported losses in a country. A study published in 1994 indicated that if African swine fever were established in the U.S. swine population, the cost over a 10-year period would be approximately $5.4 billion. Today, that figure would likely be three to five times higher. A recent FMD outbreak in Italy in 1993 cost $8.6 million to eradicate and $3.2 million for cleaning, disinfection, and carcass disposal. Indirect trade losses were estimated at $120 million (21). A recent study from California presented eight different scenarios associated with a theoretical FMD outbreak. The suggested economic losses in each of these situations were staggering. Depending on the duration of the outbreak and geographical spread, losses ranged from $6 billion to $13 billion in just a few weeks (34). Delay in the eradication or control of FMD might cost California $1 billion per day in trade sanctions (57).

Included here are some economic figures representing the significant losses caused by foreign animal diseases. Diagnosis, control, surveillance, and eradication programs for BSE have cost the United Kingdom $7 billion (see chapter 7). The costs of this frightening outbreak to the United Kingdom trade and export markets have not yet been published but will certainly be in the billions of dollars. FMD was diagnosed in Taiwan in March 1997 and is still ravaging the economy of Taiwan over 3 years later. The costs for diagnosis, surveillance, depopulation, cleaning, disinfection, and related eradication programs have been set at approximately $4 billion (86). Despite this tremendous expenditure of funds, FMD has not been eradicated from Taiwan. Indirect economic losses due to trade embargoes will not be realized for many years but will be near $15 billion. Australia has experienced several small, localized VVND outbreaks in the past 10 years. The 1999 VVND epidemic costs, including diagnosis, quarantine, cleaning and disinfection, and depopulation, have been estimated to be about $15 million. Trade and export losses will certainly push that figure much higher. The economic costs of the hog cholera outbreak in the Netherlands in 1997–1998 will approach $2.3 billion (61). From February to April 1999, Malaysia experienced a devastating epidemic of a viral disease in both swine and humans. The virus has been termed Nipah virus.

To date, many hundreds of humans have been infected, and 109 have died. Over a million pigs were depopulated to control and eradicate the disease. The once economically viable swine industry in Malaysia has been totally destroyed, amounting to millions of U.S. dollars. The full negative impact of the epidemic has not yet been determined (M. Bunning, personal communication).

Many of the OIE list A diseases can infect free-ranging and zoo-contained wild animals. Should an intentionally introduced list A disease cross over to wildlife in the United States, the negative economic consequences could be catastrophic. People in the United States spend approximately $100 billion dollars annually on hunting, fishing, and nonconsumption wildlife activities. These activities would be seriously jeopardized by such an outbreak (V. Nettles, personal communication).

POTENTIAL DISSEMINATION OF AN INTENTIONALLY INTRODUCED ANTILIVESTOCK AGENT

Livestock movement around and across the United States is a very big business and would greatly facilitate the hitchhiking of a highly contagious viral disease such as FMD. A pound of meat travels about 1,000 miles on the hoof before it reaches the dinner table. The volume, length, and rapidity of movement vary from state to state and reflect many market factors. In the Southeast, 2,000 to 3,000 cattle and feeder calves move daily from many auction or sale barns to various locations in the East and Midwest, some moving as far as 1,500 miles in 3 days. These animals may be shipped to slaughter, feed lots, and farms and ranches. During this movement the animals come in contact with a variety of fomites that can readily harbor and transmit the virus. People, trucks, animal crates, clothing, and other fomites are a well-recognized source for transmitting the viruses responsible for FMD, hog cholera, and other infectious diseases.

In the western United States, economic forces within the various cattle industries increasingly contribute to wider and more rapid movement of livestock both within California and to other states. Dissemination within California is increasing, particularly in the dairy industry. One example is the move toward contract rearing of replacement heifers by large calf-raising operations, which may typically manage 10,000 to 40,000 calves from up to 80 or more dairies. Calves are transported daily from the dairies to the calf-raising site, and each week weaned calves are hauled back to dairies of other facilities that raise heifers (34). The low operating margin for dairy and beef enterprises dictates that marginally productive cattle need to be culled quickly and that replacement stock be obtained quickly from whatever source can offer a competitive price that will be expected to maximize the marginal return. Consequently, cattle are moved rapidly into and out of the state according to the market for replacement cattle, stockers, feeders, etc. In a recent survey of some representative sales yards, it was estimated that 20 to 30% of cattle were consigned to a nonslaughter destination more than 30 to 40 km from the sales yard location (M. Thurmond, personal communication). Cattle can readily cross several states within 36 to 48 h after leaving the sales yard, depending on the route, the number of drivers, and the number of stops made.

These auction and sale barns are ideal locations to intentionally introduce a highly contagious virus because there is minimal security at such auctions and the sale barn contains ample numbers of disseminators, the animal themselves, that will soon be shipped to many different locations, carrying the virus with them. In the late 1950s and 1960s, U.S. officials staged several test exercises at several auction and sale barns in the United States to practice the intentional introduction of a virus such as FMD. The men, posing as cattle buyers, disseminated a water spray (mimicking FMD virus release) from small hand-held spray cans into cattle pens, alleyways, and loading docks (30, 31). They were able to do this in several locations across the United States without interception. We believe that today a similar event could be duplicated without interception. Indeed, today, it might be easier, as some auction and sale barns have catwalks above the pens, making intentional introduction from above the animals much easier. Selecting a disease which has a slightly longer incubation period and a disease that may have silent carrier animals (that may show no signs of the disease but will be shedding the virus) would greatly facilitate the rapid spread of the virus. FMD virus is an example of such a virus.

ON-FARM BIOSECURITY

On-farm and livestock auction and stockyard security is an important aspect in preventing the intentional or natural introduction of an exotic or foreign animal disease. The top 30 cattle feeders in the United States have the capacity to maintain from 50,000 to 500,000 head of cattle at any one time in their feedlots. Meat packers process approximately 30,000 head of cattle a year, with four companies processing approximately 80% of the carcasses sent to slaughter. Over 5 million head of these cattle pass through public markets and auctions. Thus, on average, almost

100,000 head of cattle are slaughtered per day in the United States. We will not include a comprehensive review of the literature on this subject matter for cattle (dairy or beef), swine, sheep, and poultry but will glean a few relevant issues for presentation.

The six largest pork producers in the United States maintain from 100,000 to 700,000 sows per year. Biosecurity on swine farms includes such activities as (i) animal quarantine procedures, (ii) blood testing of new arrivals, (iii) control of human movement onto the farm, (iv) control of vermin, (v) farm locations, and (vi) movement of vehicles onto the farm. A survey of pork producers has indicated discouraging and dangerous trends in overall basic elements of on-farm biosecurity activities (5). For example, half the swine farms receiving new breeding stock did not isolate or quarantine new animal purchases. The 1995 survey showed that 60% did not control or limit the entrance of visitors, technical representatives, or feed delivery trucks and personnel (5, 34). In California and other states, it is common to cull diseased animals for slaughter, and many culled sows from California are shipped to the Midwest for slaughter, providing easy movement of a highly contagious virus (3, 13).

In California, the large operations closely observe and practice basic on-farm biosecurity measures, but medium and small producers pay limited attention to biosecurity. Renderers pick up most dead swine on the farm, providing an easy avenue for disease movement. In some instances, lots of finishing hogs may be diverted from slaughter and sent to another finishing operation, allowing easy spread of a foreign animal disease.

Some large U.S. dairies now have over 10,000 milking cows. Biosecurity on dairy and beef farms in the United States is less well documented (11, 12). Approximately 70% of producers do not isolate or quarantine calves and young heifers upon arrival. The percentage of farms that do not isolate dry calves, lactating cows, weaned heifers, and bulls is much higher. On nearly 33% of dairy farms, young calves have contact with different age groups, allowing easy disease spread. Biosecurity practices on dairy and beef operations require significant improvement (6, 8–10).

In California, the dairy industry requires large quantities of services, support, and inputs because of its vast economic size. Milk may be collected up to three times per day on a large farm. Milk trucks may visit as many dairies as necessary to completely fill their tanks, providing easy movement of a highly contagious viral animal disease. Animal buyers in California and other states may visit up to 25 dairies per day. Movements of people and vehicles, sharing of equipment, and other high-risk biosecurity breaches indicate that biosecurity on the farm in the United States is low.

PORT SECURITY PROCEDURES RELATED TO FOREIGN ANIMAL DISEASE EXCLUSION

APHIS Plant Protection and Quarantine (PPQ) staff maintain inspection sites and activities in the United States at 82 international airports, 71 maritime seaports, and 25 land border crossings. In a cooperative effort, PPQ has the important responsibility of ensuring compliance with agricultural product rules and regulations, while the customs authority is charged with the responsibility of enforcing all rules and regulations relevant to the entrance into the United States of agricultural products.

The international movement and transport of people, cargo ships, airplanes, and cars with their corresponding cargo and baggage is a major economic industry and is rapidly growing. Statistics from the customs fiscal year 1999 indicate that 584,004 commercial and 141,120 private international aircraft entered the United States. Transportation traffic into the United States during customs fiscal year 1999 included 200,000 ships, 125,000,000 privately operated vehicles, 463,000 buses, and 39,000 trains. PPQ and customs officials inspected approximately 400,000,000 international travelers who entered the United States via air, car, boat, truck, train, and bus and as pedestrians. In 1998 almost 15 million poultry, hatching eggs, embryos, semen samples, and livestock animals were imported into the United States. In addition, international mail volume into the United States in 1997 was 2.2 million air parcels and letters and 1.9 million sea parcels and letters. PPQ and customs officials in 1998 did intercept numerous illegal meat, poultry, and animal byproduct contraband. These products were intercepted at seaports, airports, and land border ports and through inspection of international mail. Approximately 300,000 intercepts were made for meat, poultry, and animal byproducts.

Despite the most stringent rules and regulations and the most vigilant PPQ and customs staff, including sophisticated X-ray detector equipment and 58 detector dog teams at 21 locations, the sheer volume of people and material at times overwhelms the system. At these particular times, the system is very susceptible to the intentional or criminal introduction of a serious animal disease pathogens.

DETERRENCE UNDER THE BIOLOGICAL AND TOXIN WEAPONS CONVENTION

In response to biological and chemical attacks during World War I, the Geneva Protocol for the Prohibition of the Use of Asphyxiating, Poisonous and Other Gases, and of Bacteriological Methods of Warfare entered into force in 1925. The treaty banned the use of biological, toxin, and chemical weapons but not their development or storage and contained no provisions for determining compliance. The Soviet Union, France, and Great Britain signed the treaty, but the United States did not ratify it until 1975. During World War II, new and more toxic nerve agents were developed, and research and development was begun on biological weapons. For example, throughout the 1930s and 1940s Japan developed a biological weapons project, Unit 731, in occupied China and used those weapons in China and Manchuria. Fortunately, neither the Allies nor the Axis Powers deployed such weapons in combat during World War II. Several decades passed before the international community augmented the Geneva Protocol with a more comprehensive prohibition against the production and possession of biological weapons. The Biological and Toxin Weapons Convention (BWC) opened for signature on 10 April 1972. Article I of this treaty mandates that

> Each State Party to this Convention undertakes never in any circumstances to develop, produce, stockpile or otherwise acquire or retain: (1) Microbial or other biological agents, or toxins whatever their origin or method of production, of types and quantities that have no justification for prophylactic, protective or other peaceful purposes; (2) Weapons, equipment or means of delivery designed to use such agents or toxins for hostile purposes in armed conflict.

The BWC entered into force on 26 March 1975 and has been ratified by 141 countries and signed by an additional 18. However, this arms control agreement lacks what many believe must be a fundamental component of any such accord—a means to detect noncompliance. Negotiators did not include verification measures in the BWC because at that time it was believed that biological weapons lacked military utility. The revelation that one of the depository nations for the BWC, the Soviet Union, maintained an active BW program in violation of the convention until at least 1992; the finding by UNSCOM that Iraq produced and stockpiled biological munitions prior to the 1991 Persian Gulf War (18); intelligence information indicating that several other countries are actively pursuing a BW capability (20); and advances in biotechnology over the last 25 years have significantly altered that view.

Deterring inappropriate dual use of the biological research and development infrastructure and biotechnological production facilities through legally binding and enforceable compliance to the BWC is critical. The BWC is nearly 28 years old, yet the BWC, which seeks to reduce the threat of biological warfare, lacks a legally binding protocol to monitor compliance. This weakness has limited the BWC's ability to deter biological weapons programs, even though it prohibits virtually all activities that could be part of a biological weapons program. It did not prevent Iraq from developing biological weapons (18), nor did it deter continuance of the Soviet Union biological weapons programs that employed over tens of thousands of people and produced tens of tons of biological weapons.

Since 1995, the Ad Hoc Group of State Parties to the BWC has met periodically to negotiate a compliance regimen that will strengthen the treaty. To date, the Ad Hoc Group has achieved only modest progress. It is considering a variety of measures to strengthen the BWC, including annual declarations of activities of concern, investigations, and other on-site activities (50, 89). Key to any protocol is the provision for filing annual declarations of activities or capabilities that might be relevant to the BWC. Activities currently being considered include biological defense, presence of listed agents, work with listed agents, production capacity, vaccine production, biopesticide research and production, biological containment, and aerobiology. These are referred to as declaration triggers. All of these activities are inherent in conduct of research and development by the United States government, industry, and academic institutions.

Ad Hoc Group negotiators are considering whether to require declarations from institutions possessing select plant, animal, and human pathogens and toxins or whether to focus the compliance regimen on specific activities with pathogens—such as large-scale production, work with listed agents and toxins, and aerobiology—that could be relevant to the development of biological weapons. For example, the term "work with listed agents and toxins" means any manipulation with a listed agent or toxin and covers research and development, production and recovery, and diagnosis. This includes studying the properties of those agents and toxins, detection and identification methods, genetic modification, aerobiology, prophylaxis, treatment methods, and maintenance of culture collections. The inclusion of microorganisms and toxins capable of affecting plants and animals in the lists of agents and toxins (Table 2) means that U.S. agencies and facilities working with those materials could be subject to declaration under a legally binding compliance protocol.

Although animal and plant pathogens have figured prominently in the history of biological warfare, highlighted by Germany's ambitious biological warfare program during World War I aimed at its enemies' livestock, they have received scant attention compared to pathogens capable of causing human disease (26). Among the first to rediscover the importance of animal and plant pathogens as potential biological warfare agents was the Australia Group. Formed in 1984, the Australia

Table 2. Animal and plant pathogens currently being considered by the Ad Hoc Group of State Parties to the BWC

Zoonotic pathogens
Rift Valley fever virus
Monkeypox virus
Eastern, Western, and Venezuelan equine encephalitis viruses
Bacillus anthracis
Brucella melitensis, Brucella suis
Burkholderia mallei, Burkholderia pseudomallei
Francisella tularensis
Yersinia pestis
Animal pathogens
African horse sickness virus
African swine fever virus
Avian influenza virus
Hog cholera virus
Bluetongue virus
Foot-and-mouth disease virus
Lumpy skin disease virus
Newcastle disease virus
Peste des petits ruminants virus
Porcine enterovirus type 1
Rinderpest virus
Vesicular stomatitis virus
Plant pathogens
Colletotrichum coffeanum var. *virulans*
Dothistroma pini
Erwinia amylovora
Ralstonia solanacearum
Puccinia graminis
Sugarcane Fiji disease virus
Tilletia indica
Xanthomonas albilineans
Xanthomonas campestris pv. citri
Sclerotinia sclerotiorum
Peronospora hyoscyami de Bary f. sp. *Tabacina (Adam) skalicky*
Claviceps purpurea

Group is an informal and voluntary consortium of nations whose goal is the limitation of chemical and biological weapon proliferation. Members meet annually to review national export controls to try to limit the transfer of materials and equipment that could be used to create chemical and biological weapons. The group has created lists of items whose export should be controlled, as well as warning lists of items whose purchase could be indicative of proliferation activities. Their list of animal and plant pathogens of concern is very similar to Table 2. With U.S. government funding and attention focused on countering biological weapons targeted at humans, agencies and groups are just now becoming aware of the threat posed by agroterrorism and biological warfare directed against our nation's animals and crops (42). The threat of agroterrorism was given credence in recent years when information revealed the size and extent of the Soviet Union's antilivestock and anticrop biological weapons program. At the moment, the United States is considered highly vulnerable to such a threat. The U.S. government has launched a national effort to counter the threat posed by bioterrorism, and the agricultural sector is now included as a key component of that effort. The importance attached to the role of agriculture in the global economy is underscored by the emphasis given to animal and plant pathogens by the Ad Hoc Group of State Parties to the BWC. Agroterrorism should be viewed as an act of economic warfare. In this age of global agribusiness, any country that has its livestock or crops infected by endemic or exotic pathogens, either naturally or intentionally, is rapidly barred from export markets. Developing countries in particular recognize the dire consequences that an agroterrorism attack would have on the political, social, and economic sectors of their society and potentially on national survival itself. We need to view the threat posed by biological warfare and the use of biological weapons by terrorists as multidimensional and invest our resources accordingly.

LEGAL PROHIBITION OF AGROTERRORISM

Agroterrorism presents serious challenges to a legal system's ability to protect agroindustries from serious threats. America has laws to prosecute and punish agroterrorists. Unfortunately, however, agroterrorism demonstrates the limited ability of legal prohibitions to reduce prevailing risks.

Punishment for Committing an Act of Agroterrorism

Punishing agroterrorists, once they are apprehended and enough evidence is obtained for a conviction, is straightforward. Under 18 U.S.C. §43, anyone who intentionally causes physical disruption to the functioning of an animal enterprise by causing the loss of animals or property (in excess of $10,000) will be fined or imprisoned for up to 1 year. For agroterrorists who cause serious bodily injury to another person, the maximum jail sentence extends up to 10 years; if death is caused, the perpetrator can be imprisoned for any term of years. Moreover, the guilty agroterrorist will have to pay restitution for the loss of production or farm income reasonably attributable to the offense.

The law provides stiffer penalties for terrorists who use weapons of mass destruction, including a biological agent, toxin, vector, or disease organism. Under 18 U.S.C. §2332, the punishment is severe, up to life imprisonment; if death results, the perpetrator may be executed. The term biological agent includes a biological product that may be engineered or naturally occurring. However, the application of §2332 to agroterrorism is problematic because, under that section, the use of weapons of mass destruction has to be against a person or against the property of the United States. An attack, even using biological agents, against only agricultural land or livestock would not qualify as weapons-of-mass-destruction terrorism. Proposals are currently under consideration in Congress that would either raise the penalties under 18 U.S.C. §43 or amend the definition of weapons-of-mass-destruction terrorism to include attacks against private property. Moreover, these proposals would also establish a National Animal Terrorism and Ecoterrorism Incident Clearinghouse for information on incidents for purposes of cross-referencing and assisting investigations.

Possessing Animal Pathogens

The problem with U.S. laws pertaining to agroterrorism is that they operate after the fact. Only a terrorist who has actually committed an act of agroterrorism and who has been caught can be prosecuted. Meanwhile, of course, the damage would be done, and prosecuting the perpetrator may mean little to the sector that has been devastated. Moreover, in view of the delayed nature of agroterrorism, a perpetrator may have considerable opportunity to commit the crime and then flee beyond detection and apprehension.

For this reason, many commentators have urged that criminal penalties should be imposed for various activities before an attack. Most often discussed is a prohibition against the possession of a pathogen (human or animal), even if that possessor has not yet used the material to commit a terrorist act. Under the Virus, Serum, and Toxin Act (21 U.S.C. §151 et seq.), it is illegal to prepare and sell or to import any worthless or harmful virus, serum, or toxin for domestic animals. However, possession by itself is not prohibited by this provision.

More generally, 18 U.S.C. §175 prohibits knowing development or possession of any biological agent, toxin, or delivery system for use as a weapon. A terrorist who develops or possesses any biological agent for use as a weapon can be fined or imprisoned. Another section, 18 U.S.C. §177, establishes an affirmative defense when the development or possession is for a prophylactic, protective, or other peaceful purpose. The problem from a law enforcement perspective is for a prosecutor to produce evidence which proves that an accused's possession of deadly agents is "for use as a weapon" and not for a legally acceptable purpose. The Justice Department (like virtually every commentator) has recommended that unlicensed possession of pathogens be criminalized without regard to purpose or intent or to whether those pathogens have in fact been used to deadly effect. According to this proposal, only licensed facilities could legally develop or possess pathogens. The licensing process would be the proper venue to determine the applicant's purpose and intent, and necessarily, any possessor of pathogens who does not successfully obtain a license would be subject to prosecution.

Detecting Terrorist Conspiracies

Even laws that prohibit possession of pathogens may be inadequate (or too late) to prevent an agroterrorist from inflicting a catastrophe. Therefore, law enforcement officials such as the FBI and state and local police should be authorized to conduct investigations of suspicious activity. It is in this context, however, that federal law is most troubling.

The current laws that authorize investigations of potential terrorism rest on assumptions that may be misaligned for preventing and punishing agroterrorism. These assumptions are that an act of terrorism is likely to be carried out by an identifiable group whose distinct political agenda is overtly hostile to the American system or that is linked to a foreign group which opposes U.S. policy. Our laws reflect the belief that tracking groups who may be suspected of harboring such viewpoints is the best way to prevent an attack. Accordingly, law enforcement officials are authorized to monitor those groups and the people who support them, and strict punishments apply to apprehended perpetrators. Yet many experts believe that an act of bioterrorism, or more specifically of agroterrorism, is not likely to be undertaken by an organized political organization, and therefore monitoring their activities is not likely to be a productive focus. This criticism is enhanced by the fact that agroterrorism has few telltale markings or characteristics that law enforcement officials in the field can trace to detect preparations for an attack. Thus, the problem, in legal terms, is not that there are no laws to address the problem, but that the mechanisms adopted by those laws may be less than optimally effective.

By focusing on groups that are known to be ideologically antagonistic to the United States government, law enforcement officials can trace money and, in some cases, group leaders who are affiliated with international terrorist networks. If one of these known groups is planning an agroterrorist attack, it is possible that it will be detected. However, if the attack is planned by some person or group who is not known, detection is extremely unlikely. There is another problem with this approach: civil liberties advocates object to focusing on groups who have a hostile ideology lest that become a way to stifle unwelcome speech and political association. Many commentators assert that law enforcement should focus on the means of conducting terrorism and not on the ideology of the potential terrorist. This makes a great deal of sense in connection with nuclear terrorism as well as some forms of chemical and biological terrorism where the technology, equipment, and materials that would be necessary for an attack are sufficiently specialized to be distinguished from commonly obtainable items. Yet it is unlikely that the same holds true for agroterrorism, which can be undertaken using only naturally occurring disease agents.

The conclusion must be that our laws can punish an agroterrorist who has been caught after committing a crime. This prospect of punishment may act as a deterrent. However, for potential terrorists who either believe they can escape detection or are willing to accept punishment (perhaps even seek punishment to become martyrs), the deterrent value of punishment is insignificant. For our country to be truly secure, our laws should enable the FBI and others to detect terrorism before it happens without overstepping constitutional protections. Currently, this is less a reality than an aspiration.

CONCLUSIONS

The U.S. livestock and agriculture-related industries are highly vulnerable to the introduction of an exotic or foreign animal disease which would have significant economic and public health consequences. The biological agents, both viral and bacterial, are readily available throughout the world, and dissemination is easily accomplished. Acquisition and dissemination of the animal biological agents are unsophisticated activities, requiring only minimal scientific or laboratory skills and education, and therefore the ability exists for intentional introduction. Many countries and states and nonstate groups possess the ability to obtain these agents and to disseminate them. The key problem is to measure the intention of countries, states, and nonstates to conduct acts of intentional introduction. A sound intelligence system that is cognizant of the needs of agriculture is essential.

To date, there have been no known biological attacks on U.S. livestock or poultry industries (32, 36, 83). We must remain alert to the possibility, however, because the United States and agroindustries of other countries have been the subject of criminal chemical attacks which have caused significant economic damage. In Wisconsin in 1996, chlordane was introduced into animal feed byproducts. The motive was to destroy a business competitor. This criminal event was successful, and as a result approximately 4,000 farmers in four Midwestern states, Wisconsin, Michigan, Illinois, and Minnesota, received contaminated feed. A repeat criminal chemical introduction took place at the rendering company. On 14 September 1999, Brian W. (Skip) Lea was indicted on two counts of agricultural product tampering. He faces a possible 3-year prison term and a fine of $250,000 for each count (83). Between 1977 and 1979 but most likely in February 1978, a terrorist group contaminated Israeli citrus produce (oranges, lemons, and grapefruit) with mercury. Contaminated produce was discovered in the Netherlands, Germany, Belgium, Sweden, and the United Kingdom. The citrus export market of Israel immediately dropped by approximately 40%, causing a significant economic impact on that country's economy. These biocrimal acts and the consequent economic impact clearly demonstrate that agriculture can be the target of an economically devastating terrorist or criminal attack (23, 65, 72).

Acknowledgments. Many people provided encouragement and assistance with the collection of material and references used to write this chapter. We valued the input from ARS and APHIS scientists who had valuable first-hand experience with the most important list A diseases: D. Swayne of the Southeast Poultry Research Laboratory, Athens, GA; P. Mason of the Plum Island Animal Disease Center, Plum Island, N.Y.; S. Bolin and W. Mengeling of the National Animal Disease Center, Ames, Iowa; and D. Gregg of the Foreign Animal Disease Diagnostic Laboratory, Greenport, N.Y. Staff from the USDA APHIS PPQ, J. Gray, J. Cougil, R. Sponaugle, and J. Smith, provided valuable data concerning animal, people, and product movement through borders, ports, and airports. C. Tuszynski, V. Bridges, E. Bush, and K. Forsyth of the Center for Epidemiology and Animal Health, APHIS, VS, Ft. Collins, Colo., provided timely information on the economic impact of animal diseases. We thank D. Freeman, AFMIC; G. W. Christopher, USAMRID; M. Thurmond, University of California, Davis; W. S. Carus, National Defense University; and J. B. Tucker, director, and Jason Pate, manager, CBRN Terrorism Database of the Monterey Institute of International Studies, for their assistance in gathering needed information and references and for review of this chapter. T. Facer, M. Kauppi, and D. Huckle of the Joint Military Intelligence College, DIAC, always encouraged our pursuit of this topic and provided valuable resources. We thank P. Chalk at Rand for valuable discussions. N. Robinson provided needed information on U.S. auction markets. B. Lautner and D. Pyburn at the National Pork Producers' Council provided needed information on farm biosecurity. A special thanks goes to Troy McCullough, *Ames Tribune*,

Ames, Iowa, for his invaluable assistance. J. Eifling, Chief Librarian, and staff at NADC, Ames, Iowa, as well as T. Allen, National Agriculture Library, College Park, Md., provided material references at a moment's notice. We extend sincere thanks to M. Tran-Knechtges, D. Shifflett, and C. Denner Payne of the Collection and Information Requirements Branch, AFMIC, for their support in obtaining needed reference materials. J. Cullinan and M. Brennan, *Annals of the New York Academy of Sciences,* and R. Zilinskas provided invaluable reference material in a very timely fashion. We could not have written this chapter without the timely library support received from the above groups. We thank Ronald Atlas, University of Louisville, and Wallace Deen for valuable comments and review. We extend a very special thanks to James Kvach, Chief Scientist, AFMIC, who was of constant support and encouragement, and M. Harrison, University of Georgia, Athens, who stuck with us through the typing and numerous revisions of this chapter. Her willingness to help and valuable suggestions are greatly appreciated.

REFERENCES

1. **Agricultural Research Service.** 1961. A leader's guide to agriculture's defense against biological warfare and other outbreaks. ARS Special Report ARS 22–75. U.S. Department of Agriculture, Washington, D.C.
2. **Alibek, K., and S. Handelman.** 1999. *Biohazard.* Random House, New York, N.Y.
3. **Amass, S. F., and L. K. Clark.** 1999. Biosecurity consideration for pork production units. *Swine Health Prod.* **7:**217–228.
4. **Animal and Plant Health Inspection Service.** 1999. Regional Emergency Animal Disease Eradication Organization (READEO). Manual Emergency Program. APHIS. U.S. Department of Agriculture, Washington, D.C.
5. **Animal and Plant Health Inspection Service.** 1999. National Animal Health Monitoring System (NAHMS). Part III. The U.S. Pork Industry 1990–1995. U.S. Department of Agriculture, Washington, D.C. October 1997.
6. **Animal and Plant Health Inspection Service.** 1998. NAHMS. Part IV. Changes in the U.S. beef cow-calf industry, 1993–1997. May 1998. U.S. Department of Agriculture, Washington, D.C.
7. **Animal and Plant Health Inspection Service.** 1998. The threat of biological terrorism to U.S. agriculture. U.S. Department of Agriculture, Washington, D.C.
8. **Animal and Plant Health Inspection Service.** 1996. NAHMS. Part II. Changes in the U.S. dairy industry; 1991–1996. Sept. 1996. U.S. Department of Agriculture, Washington, D.C.
9. **Animal and Plant Health Inspection Service.** 1996. NAHMS. Information sheet: biosecurity practices on US dairy farms. May 1996. U.S. Department of Agriculture, Washington, D.C.
10. **Animal and Plant Health Inspection Service.** 1993. NAHMS. Dairy herd management practices focusing on preweaned heifers. July 1993. U.S. Department of Agriculture, Washington, D.C.
11. **Animal and Plant Health Inspection Service.** 1993. NAHMS. Biosecurity measures in dairy herds. March 1993. U.S. Department of Agriculture, Washington, D.C.
12. **Animal and Plant Health Inspection Service.** 1992. NAHMS. Highlights of the national swine survey. March 1992. U.S. Department of Agriculture, Washington, D.C.
13. **Anonymous.** 1999. National symposium on medical and public health response to bioterrorism. *Emerg. Infect. Dis.* **5:**1–602.
14. **Anonymous.** 1999. Chemical and biological warfare. *In South Africa Truth and Reconciliation Commission,* vol. 2. Grove's Dictionaries Inc., Capetown, Rebublic of South Africa.
15. **Anonymous.** 1997. Review of public health and biological terrorism. *JAMA* **278:** 347–446.

16. **Anonymous.** The biological and chemical warfare threat, p. 1–54. U.S. Government Document.
17. **Barnaby, W.** 1997. *The Plague Makers: The Secret World of Biological Warfare.* Bath Press, London, United Kingdom.
18. **Barton, R.** 1998. The application of the UNSCOM experience to international biological arms control. United National Special Commission. *Crit. Rev. Microbiol.* **24:**219–233.
19. **Bowman, Q. P., and J. M. Arnoldi.** 1999. Management of animal health emergencies in North America: prevention, preparedness, response and recovery. *Rev. Sci. Tech. Off. Int. Epizoot.* **18:**76–103.
20. **British Medical Association.** 1999. *Biotechnology Weapons and Humanity.* Harwood Academic Publishers, Australia.
21. **Brown, C. C. and B. D. Slenning.** 1996. Impact and risk of foreign animal disease. *J. Am. Vet. Med. Assoc.* **208:**1038–1040.
22. **Bryden, J.** 1989. *Deadly Allies.* McClelland & Steward Inc., Toronto, Ontario, Canada.
23. **Carus, W. S.** 1998. *Bioterrorism and Biocrimes: the Illicit Use of Biological Agents in the 20th Century* (September 1998 revision). Center for Counterproliferation Research, National Defense University, Washington, D.C.
24. **Caudle, L. C.** 1997. The biological warfare threat, p. 451–466. *In* N. R. Zajtchuk and R. S. Bellamy (ed.), F. R. Sidell, E. T. Takafuji, and D. R. Franz (special ed.), *Medical Aspects of Chemical and Biological Warfare.* Office of Surgeon General, Border Institute, Walter Reed Army Medical Center, Washington, D.C.
25. **Chalk, P.** 1999. *The Political Terrorist Threat to Agriculture and Livestock.* RAND Corporation. DRR-2187-OSD. National Defense Research Institute, Washington, D.C.
26. **Christopher, G. W., T. J. Cieslak, J. A. Pavlin, and E. M. Eitzen, Jr.** 1997. Biological warfare: a historical perceptive. *JAMA* **278:**412–417.
27. **Cole, L. A.** 1997. *The Eleventh Plague: the Politics of Biological and Chemical Warfare.* W. H. Freeman and Company, New York, N.Y.
28. **Davies, J. C. A.** 1985. A major epidemic of anthrax in Zimbabwe: the experience at the Beatrice Road Infections Hospital, Harare. *Cent. Afr. J. Med.* **31:**176–180.
29. **Department of Agriculture.** 1998. Office of Inspector General 5-year summary of investigative activity. U.S. Department of Agriculture, Washington, D.C.
30. **Department of Defense.** 1999. Biological threat to livestock. Foot and mouth disease—a foreign threat to U.S. livestock. Film Production no. P6566. U.S. Army Materiel Command. Fort Detrick Productions, Frederick, Md.
31. **Department of Defense.** 1999. Livestock as military targets. Film, production no. 4903. U.S. Army Materiel Command. Fort Detrick Productions, Frederick, Md.
32. **Department of State.** 1999. Patterns of global terrorism. Department of State Publication 10610, April 1999. Department of State, Washington, D.C.
33. **Douglas, J. D., Jr., and N. C. Livingstone.** 1987. *America the Vulnerable: the Threat of Chemical and Biological Warfare.* Lexington Books, D. Cheath and Co., Lexington, Mass.
34. **Ekboir, J. M.** 1999. *Potential Impact of Foot-and-Mouth Disease in California.* Agricultural Issues Center, University of California, Davis.
35. **Falkenrath, R. A., R. D. Newman, and B. A. Thayer.** 1998. *America's Achilles Heel.* MIT Press, Cambridge, Mass.
36. **Federal Bureau of Investigation.** 1997. *Terrorism in the United States.* Counterterrorism Threat Assessment and Warning Unit, National Security Division, U.S. Department of Justice, Federal Bureau of Investigation, Washington, D.C.
37. **Fothergill, D. L.** 1961. Biology warfare and its effects on foods. *J. Am. Dietetic Assoc.* **38:**249–252.
38. **Forbes, R. N., R. L. Sanson, and R. S. Morris.** 1994. Application of subjective methods to

the determination of the likelihood and consequences of the entry of foot-and-mouth disease in New Zealand. *N. Z. Vet. J.* **42:**81–88.

39. **Frazier, T. W., and D. C. Richardson.** 1999. *Food and Agriculture Security. Ann. N. Y. Acad. Sci.* **894:**1–233.
40. **Gillespie, J. R.** 2000. The underlying interrelated issues of biosecurity. *J. Am. Vet. Med. Assoc.* **24:**622–623.
41. **Goldstein, S.** 2000. Rabbit response: rabbit hemorrhagic virus New Zealand. 13 February 2000, p. 8–16, 22, 23. *Philadelphia Enquirer,* Philadelphia, Pa.
42. **Goldstein, S.** 1999. U.S. could face new terror tact: Ag warfare, p. 1. 22 June 1999. *Philadelphia Inquirer,* Philadelphia, Pa.
43. **Gordon, J. C., and S. B. Nielsen.** 1986. Biological terrorism: a direct threat to our livestock industry. *Mil. Med.* **151:**357–363.
44. **Gorman, S.** 1999. Bioterrorism down on the farm. Be afraid, be moderately afraid. Special Report. *Natl. J.* **31:**812–813.
45. **Hale, M. W., and R. V. L. Walker.** 1946. Rinderpest. XIII. The production of rinderpest vaccine from an attenuated strain of virus. *Am. J. Vet. Res.* **7:**199–211.
46. **Hall, S. D.** 1999. *U.S. Food Vulnerability to Intentional Contamination (Bioterrorism): History, Perspectives and Prevention.* M.S. thesis, University of Texas-Houston Health Science School of Public Health, Houston, Tex.
47. **Harris, S. H.** 1994. *Factories of Death: Japanese Biological Warfare, 1934–45, and the American Coverup.* Routledge, London, United Kingdom.
48. **Hugh-Jones, M.** 1992. Wickham Steed and German biological warfare research. *Intelligence Natl. Security* **7:**379–402.
49. **Huxsoll, D. L., W. C. Patrick III, and C. D. Parrott.** 1987. Veterinary services in biological disasters. *J. Am. Vet. Med. Assoc.* **190:**714–722.
50. **Kadlec, R. P., A. P. Zelicoff, and A. M. Vrtis.** 1997. Biological weapons control: prospects and implications for the future. *JAMA* **278:**351–356.
51. **Kadlec, R. P.** 1995. B. R. Schneider and L. E. Grinter (ed.), Biological Weapons for Waging Economic Warfare, p. 251–266. Air University Press, *Battlefield of the Future.* Maxwell Air Force Base, Ala.
52. **Kobuch, W. E., J. Davis, K. Fleischer, M. Isaacson, and P. C. B. Turnbull.** 1990. Clinical and epidemiological study of 621 patients with anthrax in Western Zimbabwe. Proceedings of the International Workshop on Anthrax. *Salisbury Med. Bull.* **68 (Suppl.):**34–38.
53. **Lawrence, J. A., C. M. Foggin, and R. A. I. Norval.** 1980. The effects of the war on the control of diseases of livestock in Rhodesia (Zimbabwe). *Vet. Rec.* **107:**82–85.
54. **Laquer, W.** 1999. *The New Terrorism: Fanaticism and the Arms of Mass Destruction.* Oxford University Press, New York, N.Y.
55. **Lederberg, J.** 1999. Introduction, *In* J. Lederberg (ed.), *Biological Weapons: Limiting the Threat.* The MIT Press, Cambridge, Mass.
56. **Leitenberg, M.** 1971. The problems of chemical and biological warfare. Stockholm Int. Peace Res. Inst. **1:**214–230.
57. **Livestock Conservation Institute.** 1999. Animal Health Emergency Management update: vet to the rescue of animal health in Oregon. Livestock Conservation Institute summer/fall newsletter.
58. **MacKereth, G.** 1997. Rabbit haemorrhagic disease virus (RHDV)—the New Zealand experience. Foreign Animal Disease Training Course, Foreign Animal Disease Diagnostic Laboratory, Plum Island Animal Disease Center, New York, N.Y.
59. **Mangold, T., and J. Goldberg.** 1999. *Plague Wars: a True Story of Biological Warfare.* St. Martin's Press, New York, N.Y.

60. **McCauley, E. H., J. C. New, Jr., W. B. Sundquist, N. A. Aulaqi, and W. M. Miller.** 1979. *A Study of the Potential Economic Impact of Foot-and-Mouth Disease in the United States.* University of Minnesota, under a cooperative agreement with the Animal and Plant Health Inspection Service. U. S. Government Printing Office, Washington, D.C.
61. **Meuwissen, M. P. H., S. H. Horst, R. B. M. Huirue, and A. A. Dijkhuizen.** 1999. A model to estimate the financial consequences of classical swine fever outbreaks: principles and outbreaks. *Prev. Vet. Med.* **42:**249–270.
62. **Murray, G., and P. M. Thornber.** 1999. Management of animal health emergencies. *OIE Revue Sci. Tech.* **18:**1–287.
63. **Nass, M.** 1992. Anthrax epizootic in Zimbabwe, 1978–1980: due to deliberate spread? A new etiology. *PSR Q.* **2:**198–209.
64. **National Academy of Sciences.** 1999. *Chemical and Biological Terrorism: Research and Development to Improve Civilian Medical Response.* Institute of Medicine, National Research Council, National Academy of Sciences. National Academy Press, Washington, D.C.
65. **Neher, N.J.** 1996. *Food Terrorism: The Need for a Coordinated Response.* The Wisconsin Experience, Agricultural Resource Management Division, Wisconsin Department of Agriculture, Madison, Wis.
66. **Noah, D., A. L. Sobel, S. M. Ostroff, and J. A. Kildew.** 1998. Biological warfare training: infectious disease outbreak differentiate criteria. *Mil. Med.* **163:**198–201.
67. **Office of the Inspector General, USDA.** 1998. Five-year summary of investigative activity. U.S. Department of Agriculture, Washington, D.C.
68. **Preston, R.** 1998. The bioweaponeers. *The New Yorker,* March 9, p. 52–65.
69. **Preston, R.** 1997. *The Cobra Event.* Random House, New York, N.Y.
70. **Probst, P. S.** 1999. Terrorism overview. *Ann. N. Y. Acad. Sci.* **894:**154–158.
71. **Pugh, A. O., and J. C. A. Davies.** 1990. Human anthrax in Zimbabwe. Proceedings of the International Workshop on Anthrax. *Salisbury Med. Bull.* **68:**32–33.
72. **Purver, R.** 1995. *Chemical and Biological Terrorism: the Threat According to the Open Literature.* Canadian Security Intelligence Service, Canadian Security Intelligence Agency, Ottawa, Canada.
73. **Regis, E.** 1999. *The Biology of Doom: the History of America's Secret Germ Warfare Project.* Henry Holt and Company, New York, N.Y.
74. **Rimmington, A.** 1999. *Anti-Livestock and Anti-Crop Offensive Biological Warfare Programs in Russia and the Newly Independent Republic.* Center for Russian and Eastern European Studies, University of Birmingham, Edgbaston, Birmingham, United Kingdom.
75. **Roberts, B. (ed.).** 1997. *Terrorism with Chemical and Biological Weapons: Collaborative Risks and Responses.* Chemical and Biological Arms Control Institute, Alexandria, Va.
76. **Robertson, A. G., and L. J. Robertson.** 1995. From ASPS to allegations: biological warfare in history. *Mil. Med.* **160:**369–373.
77. **Rodgers, P., S. Whitby, and M. Dando.** 1999. Biological warfare against crops. *Sci. Am.* **280:**70–75.
78. **Simon, J. D.** 1997. Biological terrorism: preparing to meet the threat. *JAMA* **278:**428–430.
79. **Simon, J. D.** 1994. *The Terrorist Trap: America's Experience with Terrorism.* Indiana University Press, Bloomington.
80. **Stalheim, O. H. V.** 1987. Veterinary services in emergencies: food safety and inspection. *J. Am. Vet. Med. Assoc.* **190:**723–732.
81. **Stern, J.** 1999. The prospect of domestic bioterrorism. *Emerg. Infect. Dis.* **5:**517–522.
82. **Todd, F. A.** 1952. Biological warfare against our livestock. *N. Am. Vet.* **33:**689–691.
83. **Tucker, J., and J. Pate.** 2000. *Chemical Biological Radiological Nuclear Database.* Monterey Institute of International Studies, Monterey, Calif.

84. **Tucker, J.** 1999. Historical trends related to bioterrorism: an empirical analysis. *Emerg. Infect. Dis.* **5:**498–504.
85. **Wheelis, M.** 1999. Biological sabotage in World War I, p. 35–62. *In* E. Geissler and J. E. van Courtland Moon (ed.), *Biological and Toxic Weapons: Research Development, and Use from the Middle Ages to 1945.* International and Peace Research Institute, Stockholm, Sweden.
86. **Yang, P. C., R. M. Chu, W. B. Chung, and H. T. Sung.** 1999. Epidemiological characteristics and financial costs of the 1997 foot-and-mouth disease epidemic in Taiwan. *Vet. Rec.* **145:**731–734.
87. **Zajtchuk, R., F. R. Sidell, E. T., Takafuji, and D. R. Franz.** 1997. Medical aspects of chemical and biological warfare, p. 451–466. *In Textbook of Military Medicine, Part I: Warfare, Weaponry, and the Casualty.* Borden Institute, Walter Reed Army Medical Center, Washington, D.C.
88. **Zilinskas, R.** 1999. Cuban allegations of biological warfare by the United States: assessing the evidence. *Crit. Rev. Microbiol.* **25:**173–227.
89. **Zilinskas, R. A.** 1998. Verifying compliance to the Biological and Toxin Weapons Convention. *Crit. Rev. Microbiol.* **24:**195–218.
90. **Zilinskas, R. A.** 1997. Iraq's biological weapons: the past as future? *JAMA* **278:**418–424.

Emerging Diseases of Animals
Edited by C. Brown and C. Bolin

Chapter 4

Xenotransplantation

Laurie G. O'Rourke

No forum on emerging diseases is charged with more emotion than that dealing with xenotransplantation. The issues are many—scientific, ethical, legal—but underlying all is the unrelenting fact that there is no way to accurately predict the unknown. Journals, books, government agencies, professional organizations, magazines, and newspapers have published a great deal of information on this subject. The history and possibilities of xenotransplantation were recently reviewed (48, 84, 85, 125). This chapter will review the microbiological risks of xenotransplantation as they currently exist, with the caveat that the technical tools of molecular biology are rapidly adding to our knowledge in leaps and bounds.

Xenotransplantation derives from (i) the Greek word *xenos,* which denotes a relationship to foreign material, (ii) the Latin prefix *trans,* meaning through, across, or beyond, and (iii) the Latin verb *plantare,* literally to plant, meaning the insertion or application of tissue in or on a living body (45). Xenotransplantation is defined as the transplantation of living cells, tissues, or organs from one species into or on another.

Organ transplantation, an experimental procedure about 30 years ago, has become a widely accepted treatment for organ failure. Successful outcomes for whole-organ recipients are produced by the life-long suppression and/or modulation of the host's immune system. As the medical community has improved its ability to transplant organs and the pharmaceutical and biotechnical industries have created novel and specific immunomodulating drugs that have more predictable and fewer adverse effects, the medical indications for organ transplantation have expanded. Tragically, the numbers of people in need of a transplant have increased while the donor rate has remained static, producing a hopeless imbalance in organ supply

---

Laurie G. O'Rourke • Preclinical Safety—USA, Department of Toxicology and Pathology, Novartis Pharmaceuticals Corporation, 59 Route 10, 406/185, East Hanover, NJ 07936-1080.

and demand and, more horrifically, generating black market sales of organs (125). Within the United States alone, approximately 10 people with end-stage disease of vital organs will die each day while awaiting transplantation (United Network for Organ Sharing, http://www.unos.org). These waiting lists are not inclusive. They include only those people who "qualify," based on age, cause of organ failure, history of previous organ transplantation, co-morbid diseases, and immune status. One solution to this imbalance in supply and demand is xenotransplantation (84, 86, 140).

During the same 30-odd years, explosions in molecular biology provided scientists with the ability to identify and manipulate DNA. The use of these tools in the search for ways to overcome the formidable immunological barriers to achieving successful engraftment of animal transplants in humans has recently been reviewed (18, 28, 92, 106). Recent xenotransplantation clinical trials have used animal-derived cells to treat Parkinson's disease (25, 43) and diabetes mellitus (48, 70, 93). Devices containing pig hepatocytes have been developed for extracorporeal treatment of fulminant liver failure (46, 130), and encapsulated pancreatic islet cells are being proposed for treatment of diabetes mellitus (142). Source animals have ranged from clinically healthy pigs of unknown pedigree and minimal medical history to transgenic pigs with complete historical records.

The ethical issues that surround the debate on xenotransplantation extend far beyond the scope of this chapter, and many reviews and discussions are available for more in-depth reading (2, 41, 60, 78, 107, 114, 162, 169). The ethical issue for xenotransplantation from the microbiological aspect is the risk of introducing an unknown infectious agent into a naïve human population versus the individual benefit of a life-giving organ transplantation and improved quality of life.

Arguments favoring the selection of closely related primate species (concordant xenotransplantation) point out the homology in genetics, anatomy, and physiology. However, concordance increases the potential risk of transmission of microbiological agents (15). The debate on whether or not primates should be used as organ donors has taken place in open forums and journal pages (10, 12, 13, 14, 15; Developing U.S. public health policy in xenotransplantation [transcript], www.fda.gov/ohrms/dockets/dockets/96m0311/tr00001.pdf and tr00002.pdf; Cross-species infectivity and pathogenesis [workshop], www.niaid.nih.gov/dait/cross-species/contents.htm). The current position taken by the Center for Biologics Evaluation and Research of the U.S. Food and Drug Administration is presented in *Guidance for Industry: Public Health Issues Posed by the Use of Nonhuman Primate Xenografts* (161). This document states that the use of nonhuman primate xenografts "raises substantial public health concerns" and poses significant infectious disease risk, concluding that there is not "sufficient information to adequately assess and potentially to reduce the risks." Because this is the dominant global opinion, the microbiological risks of xenotransplantation will be limited to the current species of choice, the pig (38, 72, 135, 172).

There is a large database of information on pigs, from basic anatomy and physiology to specifics of their biochemistry and genome. Production of pigs in barrier facilities operated under very high standards of animal husbandry can consistently yield qualified-pathogen-free animals (73). Moreover, genetic engineering has made it possible to produce pigs transgenic for human proteins that may protect a transplanted swine organ from hyperacute rejection in a human.

THE RISKS OF CROSS-SPECIES INFECTION, REAL AND POTENTIAL

A number of advisory boards and individuals have reviewed the risks of xenotransplantation (2, 11, 12, 13, 14, 15, 27, 36, 37, 57, 58, 59, 78, 101, 102, 103, 118, 121, 123, 159, 160, 161, 169; Biological Response Modifiers Advisory Committee [transcript], www.fda.gov/ohrms/dockets/ac/97/transcpt/3365t1.pdf and www.fda.gov/ohrms/dockets/ac/99/transcpt/3517t1.rtf; Proceedings of the OECD Conference on Xenotransplantation; Onions et al., submitted for publication; Proceedings of the 4th National Symposium on Biosafety 1996, www.cdc.gov/od/ohs/symposium/symp 15.htm, -symp27.htm, and -symp31.htm; xenotransplantation: guidance on infectious disease prevention and management, www.who.int/emc-documents/zoonoses/docs/whoemczoo981.pdf; K. Yamanouchi, personal communication). To determine risk, the following questions must be asked about each known or potential agent: What is the risk of exposure? What is the likelihood that the agent will establish itself in the new host? What is the risk that the agent will disseminate in the host and produce disease? What is the risk of transmission to the public at large? Finally, what level of risk is acceptable? While the health risk to the individual is important, the obligation to protect the public is primary. Could xenotransplantation unleash an infectious agent with the potential to adapt to humans and wreak havoc on our species, or is this just science fiction? As Robin Weiss (166) cautioned,

> It took more than 20 years for HIV-1 to spread out of Africa, and it is only after 55 years of individual benefit from antibiotics that we are facing the public health threat of

multidrug-resistant microbes. The ethical and technical problems of maintaining vigilance over xenotransplantation should not be underestimated.

Genetics and Infectious Agents

The phylogentic distance between pigs and humans extrapolates to immediate organ rejection, termed hyperacute rejection (HAR). HAR is antibody and complement dependent and cannot be prevented by immunosuppressive drugs (40, 106). Strategies to overcome HAR include modulation of the immune response and generation of pigs transgenic for overexpression of fucosyltransferase or human complement regulatory proteins, such as decay-accelerating factor (DAF or CD55), membrane cofactor protein (MCP or CD46), or CD59, or lacking α1, 3-galactosyltransferase activity, all entities involved in acute immunological response at the cell membrane (79, 106).

Viruses and bacteria use membrane receptors to gain entry into the host cell. Measles virus (family *Paramyxoviridae*) binds specifically to human CD46, a cellular complement regulatory protein, which results in significant inhibition of interleukin-12 production by monocytes (44, 81, 112, 166a). Human CD46 also serves as a receptor for M protein of group A streptococci (115). CD55 is a receptor for echoviruses (164) and coxsackieviruses (22) and is linked with the external membrane of human T-cell leukemia virus type 1 (HTLV-1) and human cytomegalovirus (CMV) (156). Additionally, during the budding phase, human immunodeficiency virus type 1 (HIV-1) selectively acquires host-encoded proteins, one of which is CD55 (156). Finally, molecular mimicry to CD59 by herpesvirus saimiri, *Entamoeba* spp., and *Schistosoma* spp. has been identified as a self-preserving means used by these agents to inhibit complement (171). The cellular expression of these receptor molecules by the transgenic xenograft may help mitigate HAR, but they may also set the stage for cross-species transmission and emergence of a new disease.

Could transgenic expression of human proteins on the xenograft make it more vulnerable to human pathogens, particularly in a pharmacologically immunized recipient? Viral adaptation in a new host can lead to evolutionary changes that include increased virulence and a broader range of susceptible hosts (34, 157). It has been suggested that overexpression of human complement regulatory proteins on the xenograft cells could enhance the risk of cross-species infection (165). Could human viruses adapt to replication in transgenic pig tissues, modify, and return to the environment as new infectious agents of domestic and wild pigs?

The role of the carbohydrate galactose α-1,3 galactose (Gal) epitope on transmembrane glycoproteins in HAR stands as a major obstacle to graft survival and has recently been reviewed (79). Anti-Gal antibody is one of the natural antibodies to carbohydrates that develop after birth, perhaps through sensitizing events associated with microbial colonization of the gastrointestinal tract, and is the most abundant natural antibody identified in human serum. Also present in apes and Old World monkeys but not in the serum of any other mammal, including New World monkeys, anti-Gal is directed against the terminal disaccharide Gal (61, 62). Selective binding of anti-Gal to pig organs perfused with human plasma, triggering complement-mediated lysis of xenogenic cells, was demonstrated, and the sugar component was identified (39, 69). Efforts to overcome this antibody-mediated rejection include depletion of anti-Gal antibodies, immunosuppression of B cells, and modification of Gal expression or fucosyltransferase on pig organs by genetic engineering (16). However, blocking anti-Gal antibodies to prevent xenograft rejection could have a deleterious effect on the recipient.

Anti-Gal antibodies are known to have some protective roles against various pathogens, including bacteria, protozoa (*Leishmania, Trypanosoma,* and *Plasmodium* spp.), and enveloped viruses that carry Gal on their surfaces (17, 39, 63, 70a, 134, 135, 136, 145, 146). Furthermore, anti-Gal may protect against the horizontal transmission of retroviruses from species expressing Gal, since the glycoproteins of the host cell membrane become incorporated in the envelope. This suggests that production of virus particles from xenografts lacking the Gal epitope may go unnoticed by the host's immune system, undergo adaptation, and produce infection (133, 146, 165).

Microbiological Risk of Cross-Species Infection

The risk of potential adaptation and cross-species transmission of infectious agents in a xenotransplant recipient is enhanced by several factors, which typically include pharmacological immunosuppression, direct contact with the recipient's cells, and the potential for chronic exposure. Although there is far greater control over the microbiological status of an organ obtained from a qualified-pathogen-free pig, the potential risk of transfer of an infectious agent remains, along with the consequences for public health.

History past and present has taught us the importance of vigilance with respect to cross-species infection and emerging diseases (109, 110, 157). Examples of agents, infectious or benign in their natural host, that have jumped species remind us that virulence or host predilection can be altered by a single point mutation, as was the case with canine parvovirus (34).

The microbiological risks that a donor pig organ presents to an immunosuppressed recipient can be divided into three categories: (i) known or suspect pathogens and zoonotics, (ii) opportunistic pathogens, and (iii) potential pathogens. Arguably, the risks of known and opportunistic agents are similar to those of allotransplantation, but because these agents can wreak havoc in naïve or immunocompromised individuals, surveillance diagnostics are critical to demonstrate their absence. An extensive review of agents that should be excluded from the organ-source pigs and from the donor organs has recently been prepared (117a). Bacterial risks have been considered minimal from the public health perspective because these organisms are easier to eliminate with barrier animal facilities, and when they do occur, they are generally amenable to antimicrobial therapy. In contrast, the risks posed by viruses represent real and potential hazards.

There is naïve security in presuming that exclusion of known viruses from the herd by rigorous animal husbandry practices will provide a safe biological product. Negative testing does not ensure a safe product, and it is only possible to test for known agents. Validated, definitive diagnostic tests, i.e., direct detection of the agent as with a PCR assay, rather than indirect evidence, as with serology, are not available for all known agents identified as potential zoonotics. Yet-to-be-recognized microbial agents could gain access to the animals and become established if the barrier were to fail; therefore, testing for agents associated with barrier breakdown must be done on a routine basis to ensure the high health status of the herd. Recent examples of newly identified paramyxoviruses in pigs in Asia and Australia with zoonotic mortality (31, 32, 124) serve as reminders that surveillance lists are dynamic and must be continually updated and supported with diagnostic tests.

Finally, potential and unknown pathogens present the greatest risk in xenotransplantation because little is known about these agents or their significance to swine health or their ability to infect or adapt to a human host. The best animal husbandry, barrier facilities, and diagnostic tools cannot eliminate pathogens that are capable of infecting fetuses congenitally or that can be transmitted in the germ line.

POTENTIAL AND RECENTLY IDENTIFIED ZOONOTIC PATHOGENS: TRANSMISSIBLE SPONGIFORM ENCEPHALOPATHIES

Prions have been identified as the agents responsible for transmissible spongiform encephalopathies (TSE) (127). TSE have now been identified in a variety of species. In cattle and cats, the disease has been attributed to the recycling of animal protein in feed. In humans, TSE occur sporadically, genetically, or via infectious transmission.

The incidence of sporadic TSE in pigs, as in other mammals, is considered extremely rare, and no cases have been reported. Pigs appear to have a high resistance to experimental TSE. No clinical or histopathological disease was detected 52 to 76 months after intracerebral inoculation of brain tissue from kuru patients (67). Transmission of bovine spongiform encephalopathy required coexposure by intracerebral, intravenous, and intraperitoneal inoculation of infected cattle brain (42).

Strict breeding practices and feed that is void of any animal-derived protein combined with the extremely rare incidence of spontaneous TSE render the possibility of prion transmission via xenograft remote. Furthermore, if TSE transmission did occur, the disease risk would be limited to the recipient; there would be no public health risk.

Porcine Circoviruses

Porcine circovirus (PCV) was first detected in 1974 as a persistent contaminant of porcine kidney cell line PK-15 (148, 149). In the 1980s and early '90s, serological surveys in herds, slaughter pigs, laboratory pigs, and wild boars in Germany, Canada, Britain, and the United States suggested that PCV was a common global virus in all swine populations (47, 53, 76, 150). In vitro investigations with PCV demonstrated that the virus could replicate in pig and bovine leukocyte cell cultures (4). Furthermore, accumulation of cytoplasmic viral surface antigen was found to be independent of DNA synthesis and caused cell destruction (152). However, experimental inoculation with PCV failed to produce clinical disease or pathological changes, although detection of antibodies to the virus and recovery of virus from nasal swabs and fecal samples (5, 150) confirmed infection.

Toward the end of the 1990s, a new variant of PCV was associated with severe multisystemic wasting disease of swine in Canada, the United States, Europe, the British Isles, and Japan (6, 8, 55, 71, 83, 96, 108, 119). Inoculation of gnotobiotic or colostrum-deprived piglets with the new variant of PCV produced clinical signs and lesions compatible with postweaning multisystemic wasting syndrome (PMWS), confirming Koch's postulate (7, 8, 19, 56, 82).

Nuclear sequencing made it possible to further characterize PCV, making a case for phylogenetic alignment with plant circo-like viruses (nanoviruses) through the identification of replication initiator protein (Rep), which allows it to replicate in a rolling-circle mechanism (104, 105).

It has been proposed that circoviruses may be the result of a recombination event that took place in a vertebrate between a calicivirus and a plant nanovirus, perhaps through exposure to nanovirus-infected sap (68). Molecular biophysical characterization of transfusion-transmitted virus (TTV), a nonenveloped, single-stranded DNA virus of humans (113), suggests that this virus is also circular and similar to circoviruses in its genomic organization (104). However, other researchers believe there are enough differences to warrant classification into a new virus family, tentatively named *Circinoviridae* (163).

The prevalence of TTV in the general population suggests vertical transmission (113, 126, 132, 138, 173). TTV DNA sequences have been detected in several species of nonhuman primates (116, 163), and the detection of human TTV viremia in a chimpanzee used in transmission experiments with human hepatitis virus (116) indicates that cross-species infection is possible.

The zoonotic potential for PCV and other circo-like viruses under natural conditions is unknown. Antibodies reacting with PCV have been detected in serum samples collected from humans, mice, and cattle, but the possibility of a cross reactivity to TTV in the human subjects has not been ruled out (151). Circoviruses appear to be more recently evolved agents, possibly resulting from a recombination of plant and vertebrate viruses, and as such, they may have a greater propensity for change and adaptation.

Porcine Hepatitis E

Hepatitis E virus (HEV), currently classified as a calicivirus, is endemic in many developing countries and associated with epidemics of high mortality, particularly in young adults (20). Transmission is fecal-oral, most frequently associated with contaminated water. New strains have been isolated from individuals living exclusively in countries where human HEV is not considered a public health risk, such as isolates US-1 and US-2 identified in the United States (137). However, serological surveys of humans in nonepidemic populations suggests a much higher rate of exposure than can be explained, even when travel and asymptomatic infections are factored in. Since many of the caliciviruses demonstrate broad host range and diverse mechanisms of transmission, the zoonotic potential of HEV was raised. The range of tropism that caliciviruses may display is attributed to their primitive RNA viral status and associated replicative mechanisms that can be expected to generate numerous mutants (139). Pigs are considered the natural host for several caliciviruses and are susceptible to several others, including human HEV (21, 139).

In 1997, a novel swine HEV that cross-reacted with antibody directed against the capsid antigen of human HEV was identified in North America (97). The virus was found to be ubiquitous in the majority of pigs older than 3 months of age. No clinical disease was detected in either naturally infected piglets or intravenously inoculated specific-pathogen-free (SPF) pigs. However, 2 weeks postinoculation, virus was recovered on nasal and rectal swab samples collected from the SPF pigs, with continued shedding for another 4 to 8 weeks (98). Viremia was detected in the same pigs 4 to 6 weeks postinoculation, lasting 1 to 3 weeks. All pigs seroconverted, as did an uninoculated sentinel pig. Successful infection in pigs with human HEV is strain dependent: infection was produced in pigs inoculated with the US-2 isolate, but no infection occurred with inocula of human strains Sar-55 and Mex-14 (98, 99). The zoonotic potential of swine HEV was further demonstrated by producing infection in rhesus monkeys and chimpanzees, which, as surrogate human models, underscore the risk of infection to humans by swine HEV (99).

These data and the high seroprevalence of anti-HEV antibodies in humans with no corroborating history of exposure support the hypothesis that human HEV is a zoonotic infection. Since the identification of swine HEV, more emphasis has been placed on the pig as the natural reservoir for human HEV. Swine HEV has been detected in domestic and wild-caught pigs in Australia, but further studies will be necessary to determine if there is a risk associated with human HEV in that country (33). A serological survey of herds from two countries with endemic human HEV and two that were not suggested that swine HEV is enzootic in pigs regardless of whether human HEV is endemic in the respective human population (100). However, data from a study in Taiwan detected swine HEV RNA in 1.3% of the animals surveyed, suggesting that swine may experience subclinical HEV infection and serve as a reservoir (170). Searching for a ubiquitous reservoir that would be widely scattered and cross rural and urban boundaries, investigators trapped wild rats from disparate regions of the United States (80). Seropositivity ranged from 44 to 90%, and the prevalence of anti-HEV antibodies increased in parallel with the estimated age of the rats.

New Members of the *Paramyxoviridae*: Menangle, Nipah, and Hendra Viruses

Recent emerging viruses identified as members of the family *Paramyxoviridae* have been found to remain quiescent or produce subclinical illness in their natural hosts but are able to exhibit increased pathogenicity when transferred to a new host. These viral diseases are covered thoroughly in another chapter in this volume (see chapter 5). What are the infectious disease risks in xenotransplantation with respect to newly identified zoonotic pathogens such as those that have recently been identified in Australia and Malaysia? Foremost they stand as prime examples of the unexpected, underscoring the need for constant vigilance and reminding us that there are viruses waiting to be discovered. These viruses appear to be transmitted through contact with bats or their droppings, urine, and saliva, but the results of experimental disease indicate that this is not straightforward. The circumstances necessary to set the stage for such outbreaks have not all been identified. Donor or sentinel animals may well need to be tested for these viruses to support claims of a secure barrier facility.

Porcine Cytomegalovirus

Porcine cytomegalovirus (PCMV), a member of the *Betaherpesvirinae*, is a slow-growing, highly species-specific virus. Attempts to cultivate PCMV in cells from other species, including rabbits, mice, hamsters, chicken embryos, cattle (52), and humans (D. E. Onions, personal communication, 1996), have failed. PCMV is frequently associated with clinically silent infections and can induce a latent infection, with virus being excreted in the presence of circulating antibody (49). Of greatest concern from the perspective of raising pigs as potential organ donors is transplacental passage and fetal infection by PCMV (50, 51). Virus-free herds might possibly be established via hysterotomy, separation of newborns from mothers, and frequent testing for the presence of viral DNA.

The concerns regarding PCMV as a potential pathogen arise from the virulent and aggressive behavior of human CMV in allotransplantation. Human CMV is similar to PCMV in healthy individuals but can become highly pathogenic in immunocompromised or naïve individuals and is the most important single cause of posttransplantation infections (77). Treatment with antiviral agents such as gancyclovir can arrest viral proliferation, but as with all herpesvirus infections, there is no cure. The fear of PCMV behaving as a zoonotic agent comes from parallels drawn between human CMV and PCMV. However, there is no solid evidence to date that PCMV will infect cells of any species other than pig.

Porcine Gammaherpesviruses

Using a modified consensus PCR assay targeting the highly conserved herpesviral DNA polymerase, investigators recently identified the existence of two porcine gammaherpesviruses (54). Data suggest that these viruses are lymphotropic and ubiquitous in commercial pig herds. Phylogenetic analysis aligns these viruses most closely with bovine gammaherpesviruses. These two viruses have been named porcine lymphotropic herpesvirus type 1 (PLHV-1) and PLHV-2 and firmly classified in the subfamily *Gammaherpesvirinae* (158). Efforts to find a natural reservoir have suggested the transfer of virus between feral and domestic pig populations (158). PLHV appears to be apathogenic, although investigators have suggested that the usually short life span of a commercial pig could prevent recognition of a PLHV-associated disease (54).

The zoonotic potential for PLHV is unknown. As an informational backdrop, gammaherpesviruses include Epstein-Barr virus (EBV), human herpesvirus 8 (HHV-8), herpesvirus saimiri (primates), and retroperitoneal fibromatosis-associated herpesvirus of rhesus macaques (RFHVMm), all of which can remain latent for years. Posttransplantation lymphoproliferative disorder (PTLD) is believed to be caused by reactivation of latent EBV secondary to immunosuppression. The lesion may present as lymphocytic hyperplasia (primarily B cells) or may progress to unrelenting malignancy. Inbred mice chronically infected with murine gammaherpesvirus develop lymphoproliferative disease that parallels PTLD, the incidence of which can be greatly increased by immunosuppressive therapy (143). HHV-8 and RFHVMm have been implicated in the pathogenesis of Kaposi's sarcoma (KS) and KS-like neoplasias in humans and primates, respectively, and have a strong association with lentivirus infection or immunosuppression (35, 131).

Could a latent porcine gammaherpesvirus in the donor organ reactivate under the heavy and constant immunosuppression necessary to prevent graft rejection? Recent work suggests that HHV-8 can be transmitted by donor organs and produce KS in patients who become seropositive and are severely immunosuppressed (129). Since members of this family of herpesviruses can produce cross-species infection (e.g., malignant catarrhal fever [MCF]), the potential for host infection must be considered. Much more research is needed to characterize these viruses and study their biology and pathogenic potential. The danger to the transplant recipient lies in developing a malignant disease; the danger to society may be the emergence of a new virus.

Porcine Endogenous Retrovirus

Vertebrate DNA contains thousands of retroviral elements (proviruses) that are inherited in a Mendelian fashion after infection of germ cells during evolution (24). Most of these endogenous retroviruses have been part of their hosts' genome. The RNA viruses have more plasticity than their DNA cousins, resulting in a greater ability to change (mutate) and adapt. Endogenous retroviruses have been part of their hosts' genomes for tens of thousands of years and are not associated with disease. Simian immunodeficiency virus is not pathogenic in the African green monkey and is highly productive but nonpathogenic in sooty mangabeys, the proposed reservoir species for HIV-2 (9, 64, 65, 66, 111). The gibbon ape leukemia virus (GALV) behaves as an exogenous retrovirus, but DNA evidence points to the endogenous retrovirus of *Mus caroli* as the origin (154, 155). Experimental work has also demonstrated the ability of retroviruses to infect across species. Three of eight immunosuppressed macaques inoculated intravascularly with endogenous murine leukemia virus (MuLV) developed lymphomas (128). Acquisition of endogenous murine retroviral sequences by human lymphoblastic cells transplanted into nude and SCID mice models demonstrated xenotropism (29, 91). These findings highlight the potential risk for infection when there is direct cell contact between donor and recipient cells against an immunosuppressed backdrop. Further in vitro work has suggested that coinfection with endogenous MuLV and HIV-1 may alter the biological properties of HIV-1 (91).

Another retrovirus, simian foamy virus, has been detected in humans who have occupational contact with primates (30, 74, 75). Although none of the individuals testing positive has developed any evidence of clinical disease and transmission to spouses has not been detected, testing has demonstrated persistent infection. Serum and, where available, postmortem tissue samples collected from human recipients of baboon livers have demonstrated simian retroviral sequences and microchimerism (15). Consequently, once seen as evolutionary curiosities, these endogenous retroviral elements are the focus of intense investigation, no longer shelved as fossil remnants. There is evidence that the conservation of endogenous retroviruses in vertebrates supports an active role in the host. Endogenous retroviruses are thought to confer resistance to infection by viruses that use the same cellular receptor for entry (23). Human endogenous retroviruses, shown to be transcriptionally active especially in placenta, embryonic tissue, and human cell lines (87), are believed to play a role in regulation of gene expression and have been implicated in autoimmune disease (86, 87). Certainly, the expanding interest in pigs as organ donors has placed the subject of endogenous retroviruses, in particular the porcine endogenous retrovirus (PERV), front and center in the microbiological risk analysis of xenotransplantation.

PERV is a member of the gammaretrovirus (C-type retrovirus) family (26, 89, 105, 143, 153, 168). Early work with PERV, first identified as spontaneously produced virus in PK-15 and PFT porcine oviduct cell lines and later associated with a porcine lymphoma, suggested that the virus was unable to infect cells from other species. These observations were overturned when studies demonstrated productive infection of human cell lines by virus obtained from PK-15 cells and later from stimulated peripheral blood lymphocytes (118, 122, 147, 167). Many in the scientific community and the population at large are quick to compare PERV with HIV, but the differences between the complex leukemia viruses (HTLV) or the lentiviruses (HIV) and the gammaretroviruses suggest otherwise. The complex retroviruses are more capable of mutagenicity and adaptability, rendering them more efficient pathogens. Feline leukemia virus (FeLV) has been intensely studied in natural and experimental disease (117) and has been used as the prototype pathogen to best guess how PERV might behave in a susceptible host (cross-species infectivity and pathogenesis website cited above).

Transmission and disease induction by gammaretroviruses require active viral replication, persistent plasma viremia, and viral shedding. Adult animals are difficult to infect, the infectious dose level is moderately high, and the majority of infected individuals eliminate the virus and provirus, recovering fully. In contrast, lentiviruses are considered highly infectious, and the majority of exposed individuals are infected for life. Retroviral infection induces immunosuppression, but the frequency and severity are far greater with lentiviral infections. The gammaretroviruses produce disease by integrating next to an oncogene, producing hematopoietic neoplasias.

As shown by looking at the *pol* gene, PERV is phylogenetically related to the leukemia viruses GALV, FeLV, and MuLV (Fig. 1). In their natural hosts, these simple C-type oncoviruses cause leukemia and lymphomas plus degenerative conditions such as immunosuppression and anemia. Consistent within the family *Retroviridae,* these viruses contain two copies of an RNA genome that is surrounded by a protein coat or capsid, which is in turn surrounded by an envelope derived from the plasma membrane of the infected host cell. Full nucleotide sequences of a PERV-B subgroup derived from virions released from PK-15 cells (Q-One Biotech Ltd. & Imutran Ltd., international patent application WO 97/40167, 1997) and of a proviral clone of PERV-C derived from miniature swine (3) have been completed. While the *gag* and *pol* genes remain relatively consistent among the PERV variants, their *env* genes are distinct and code for three variants, designated PERV-A, PERV-B, and PERV-C (88). Correlation of the differences between variants in their SU surface glycoprotein coding regions with production, infection, and replication of PERV in cell culture demonstrated three functional classes with different host ranges (147).

Takeuchi et al. (147) screened 32 cell lines of various origins for expression of PERV receptors and defining host range. Not surprisingly, all three variants are able to infect and replicate in pig cells. Interference

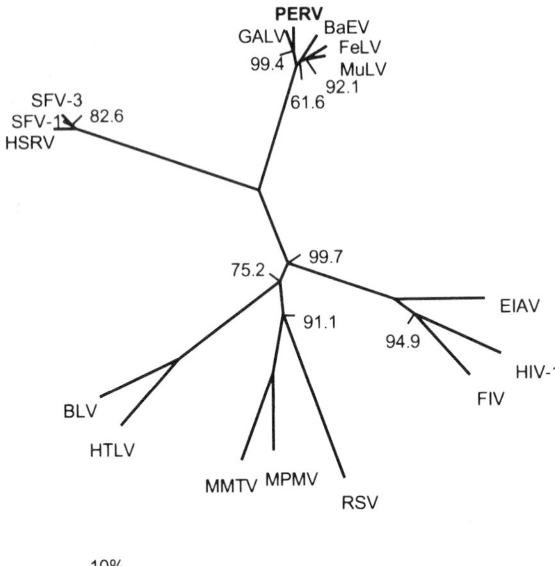

Figure 1. *pol* phylogenetic tree. An unrooted phylogenetic tree of retroviruses based on alignment of Pol amino acid sequences is shown. The tree was constructed from an alignment of 16 representative retrovirus sequences, corresponding to amino acids 178 to 740 of the predicted PERV-A Pol polyprotein. The values given at the branch forks are the bootstrap percentages for 2,000 samplings, with missing values all being 100%. Sequences were aligned using the CLUSTAL W program, and the PHYLIP package was used for protein distance calculation and tree construction. The solid line below the diagram represents 10% divergence. Diagram courtesy of D. E. Onions, Q-One Biotech, Glasgow, United Kingdom. HSRV, human respiratory syncytial virus; BaEV, baboon endogenous virus; SFV, simian foamy virus; EIAV, equine infectious anemia virus; FIV, feline immunodeficiency virus; RSV, Rous sarcoma virus; MPMV, Mason-Pfizer monkey virus; MMTV, mouse mammary tumor virus; BLV, bovine leukemia virus.

studies demonstrated that all three variants use receptors distinct from each other and from those used by other mammalian C-type retroviruses (e.g., GALV, FeLV-B, and MuLV). However, variable susceptibility of cell lines to infection suggests that PERV receptors are not ubiquitously expressed. PERV-A and PERV-B are capable of infecting several human cell lines and releasing infectious C-type retrovirus (122, 167), while PERV-C is more restricted (147). PERV-C is probably the most recently evolved variant, since expression of transcripts has not been detected in swine testes (ST)-IOWA or PK-15 cells, it is limited in its host range and cell tropism, and copy numbers vary from very few in Large White and Landrace × Duroc pigs to many in the minipig (118, 147). Negative success in vector transduction on nonhuman primate cells and cell lines has led some investigators to doubt the utility of primate animal models for study of PERV zoonosis (147). However, other scientists have identified rhesus and chimpanzee cell lines that are susceptible to PERV infection (Biological Response Modifiers Advisory Committee website, cited above). The possibility of using a susceptible nonprimate species for in vivo infectivity and pathogenicity studies (species whose cell lines are susceptible to PERV-A and/or PERV-B include mink, rat, mouse, rabbit, cat, and dog [147]) has been considered. However, producing infection, disease, and transmission in another animal species will not answer the question of how PERV might behave in an immunosuppressed human being. Nor will a failure to establish persistent infection in immunosuppressed animals provide a solid safety net against PERV replication, viremia, and viral shedding in human xenograft recipients.

The guinea pig holds potential as a small-animal model for experimental study of cross-species PERV infectivity studies. Recently, immunocompetent guinea pigs were infected with PERV-B as a result of a routine procedure to raise antiserum (D. E. Onions, personal communication, 2000). The animals were injected with a cell-free preparation harvested from infected 293 cells. PERV proviral DNA and genomic RNA were detected in spleen samples, and serum antibodies were detected against both *gag* and *env* proteins. It is hypothesized that the animals experienced a transient productive infection which transitioned to a latent infection. None of the animals developed clinical disease, and serological testing gave no evidence of transmission to control animals. This first report of in vivo transmission of PERV is likely to be followed by additional animal models.

Research efforts to characterize PERV in transgenic pigs and their primate recipients are ongoing. PERV mRNA has been identified at variable levels in all tissues examined, indicating that viral expression is not restricted (Biological Response Modifiers Advisory Committee website, cited above). Transmission electron microscopy of tissues was negative, although virus-like particles (VLP) were detected in serum. Although the VLP did not bind antibodies against recombinant *gag* or whole virus, product-enhanced reverse transcriptase analysis produced positive results. Human 293 cells cultured in the presence of the same porcine serum for 10 passages were negative for PERV infection by PCR, suggesting that the VLP and reverse transcriptase activity did not necessarily indicate the presence of infectious virions. However, other investigators have suggested that there is a low-level viremia in some pigs (117a). Using real-time PCR assays, terminal samples collected from baboon and cynomolgus monkey xenograft recipients showed no evidence of PERV infection, nor was antibody against PERV detected by Western blot (Biological Response Modifiers Advisory Committee website, cited above). Similar diagnostic techniques used in other laboratories have produced similar results, i.e., no evidence of PERV infection or exposure in their xenografted primate models.

A surprisingly large number of human patients have been exposed to living porcine tissue via extracorporeal splenic, hepatic, or renal perfusion, implantation of pancreatic islet cells or fetal neuronal cells, or full-thickness skin grafts. A retrospective study of these patients detected no viremia, although persistent microchimerism was detected in 23 patients for up to 8.5 years posttreatment (120). Antibody to recombinant *gag* and whole virus was detected in two individuals. Similar assays used to detect HIV or HTLV look for antibody against at least two viral antigens to rule out the possibility of cross-reactive antibody; whole virus is not used. The technologies used to detect PERV DNA, mRNA, reverse transcriptase activity, and anti-PERV represent the most sensitive methods available. Validation of assays and interpretation of the data and their significance are challenging because there is no patient population. Although the sensitivity and specificity of the PCR-based assays can be determined mathematically and are based on Poissonian distribution, the significance of a false-positive or false-negative result cannot be measured against the incidence of disease in the population.

Will the combination of immunosuppression and host response result in an environment that favors selection of PERV variants able to efficiently infect both the recipient and the human population? Normally, human complement can lyse nonhuman retroviruses, but under the circumstances of xenotransplantation, this ability might be lessened or inactivated by the presence of transgenic cell membranes allowing the virus to masquerade as self. It has been suggested that since most endogenous xenotropic retroviruses are nonpathogenic, transmission of virus would have little consequence for the recipient or for society (141). However, it has also been suggested that PERV may be able to recombine with human retroviral elements, generating a whole new viral agent (166).

The risk of PERV infection may be acceptable to a xenotransplant candidate who is faced with probable death due to the shortage of available donor organs. However, the risk of such a patient serving as a mixing vessel for a novel viral agent that threatens public health cannot be ignored.

UNKNOWN PATHOGENS

It is not possible to design diagnostic tools with the specificity to detect an unrecognized etiological agent. Xenograft sponsors will need to rise to the challenge of monitoring patients and donor pig herds proactively with newly validated diagnostics as new disease entities are recognized and identified as potential infectious etiological agents. Clinicians will need to work hand in hand with sponsors and regulatory authorities, being vigilant in their frequent monitoring of patients. They will need to be on the alert for infections that are inconsistent with the usual pattern of opportunistic infections likely to emerge at various postoperative stages related to changes in immune deficiencies over time. Any evidence of unexplained disease symptoms or laboratory test changes that might suggest the presence of a potential pathogen or pathogen not yet discovered will require further investigation. Clinicians will also need to be sensitive to other circumstances that might suggest transmission to the recipient's intimate contacts. It will be a formidable task to differentiate between alterations that are (i) pharmacological and secondary to immunosuppression, (ii) the result of a known infectious agent, or (iii) signs of an emerging disease. The clinician will need to carefully consider the identity of any zoonotic agent detected posttransplantation in a xenograft recipient before incriminating the organ as the pathogen's source.

As has been mentioned several times above, the risk of the unknown is not easily defined. The recent identification of two porcine gammaherpesviruses and the virulent and zoonotic Nipah virus in Malaysia demonstrates clearly that our current knowledge of potential porcine and zoonotic pathogens is incomplete. Screening for viruses that have not yet been identified is obviously not possible, although eliminating known viruses that have various modes of transmission should reduce the risk of disease transmission to the recipient by the xenograft.

Many viral genes that code for immune-modulating proteins have been identified, and many "orphan" proteins will probably prove to have roles in evading host response. The impact, if any, of the recipient's pharmacologically altered cytokine milieu on microbiological evolution remains to be determined. The strategies used to fool the immune system into accepting a xenotransplant may well set the stage for old viruses to acquire new tricks, including enhanced virulence and broader host range. Detailed contingency plans dedicated to life-long surveillance of recipients and their close contacts will need to be designed. These plans must coordinate reporting efforts of primary physicians, public health officials, animal health officials, and pharmaceutical sponsors and must be well rehearsed prior to xenotransplantation. Explicit attention will need to be given to the diagnostic tools used to ensure that assays are valid and appropriately sensitive. However, only retrospectively will it be possible to determine whether surveillance methods have been adequate to monitor and protect the public.

Acknowledgments. I extend special thanks to D. Onions and the expert panel assembled by Novartis Pharmaceuticals to evaluate the zoonotic risk of xenotransplantation for sharing their expertise in the complexities of cross-species infections and emerging diseases.

REFERENCES

1. **Abe, K., T. Inami, K. Ishikawa, S. Nakamura, and S. Goto.** 2000. TT virus infection in nonhuman primates and characterization of the viral genome: identification of simian TT virus isolates. *J. Virol.* **74:**1549–1553.
2. **Advisory Group of the Ethics of Xenotransplantation.** 1997. *Animal Tissues into Humans.* Her Majesty's Stationery Office, London, United Kingdom.
3. **Akiyoshi, D. E., M. Denaro, H. Zhu, J. L. Greenstein, P. Banerjee, and J. A. Fishman.** 1998. Identification of a full-length cDNA for an endogenous retrovirus of miniature swine. *J. Virol.* **72:**4503–4507.
4. **Allan, G. M., F. McNeilly, J. C. Foster, and B. M. Adair.** 1994. Infection of leucocyte cell cultures derived from different species with pig circovirus. *Vet. Microbiol.* **41:**267–279.
5. **Allan, G. M., F. McNeilly, J. P. Cassidy, G. A. C. Reilly, B. Adair, W. A. Ellis, and M. S. McNulty.** 1995. Pathogenesis of porcine circovirus, experimental infections of colostrum deprived piglets and examination of pig fetal material. *Vet. Microbiol.* **44:**49–64.
6. **Allan, G. M., F. McNeilly, S. Kennedy, B. Daft, E. G. Clarke, J. A. Ellis, D. M. Haines, B. M. Meehan, and B. M. Adair.** 1998. Isolation of porcine circovirus-like viruses from pigs with a wasting disease in the USA and Europe. *J. Vet. Diagn. Investig.* **10:**3–10.
7. **Allan, G. M., S. Kennedy, F. McNeilly, J. C. Foster, J. A. Ellis, S. J. Krakowka, B. M. Meehan, and B. M. Adair.** 1999. Experimental reproduction of severe wasting disease by co-infection of pigs with porcine circovirus and porcine parvovirus. *J. Comp. Pathol.* **121:**1–11.
8. **Allan, G. M., F. McNeilly, B. M. Meehan, S. Kennedy, D. P. Mackie, J. A. Ellis, E. G. Clark, E. Espuna, N. Saubi, P. Riera, A. Botner, and C. E. Charreyre.** 1999. Isolation and characterization of circoviruses from pigs with wasting syndromes in Spain, Denmark and Northern Ireland. *Vet. Microbiol.* **66:**115–123.
9. **Allan, J. S., M. Short, M. E. Taylor, S. Su, V. M. Hirsch, P. R. Johnson, G. M. Shaw, and B. H. Hahn.** 1991. Species-specific diversity among simian immunodeficiency viruses from African green monkeys. *J. Virol.* **65:**2816–2828.
10. **Allan, J. S.** 1994. Primates and new viruses. *Science* **265:**1345–1346.
11. **Allan, J. S.** 1995. Xenograft transplantation and the infectious disease conundrum. *ILAR J.* **37:**37–48.
12. **Allan, J. S.** 1996. Xenotransplantation and possible emerging infectious diseases. *Mol. Diagn.* **1:**1–18.
13. **Allan, J. S.** 1996. Xenotransplantation at a crossroads: prevention versus progress. *Nat. Med.* **2:**18–21.
14. **Allan, J. S.** 1997. Silk purse or sow's ear. *Nat. Med.* **3:**275–276.
15. **Allan, J. S.** 1998. The risk of using baboons as transplant donors. Exogenous and endogenous viruses. *Ann. N. Y. Acad. Sci.* **862:**87–99.
16. **Alwayn, I. P. J., M. Basker, L. Buhler, and D. K. C. Cooper.** 1999. The problem of anti-pig antibodies in pig-to-primate xenografting: current and novel methods of depletion and/or suppression of production of anti-pig antibodies. *Xenotransplantation* **6:**157–168.
17. **Avila, J. L., M. Rojas, and U. Galili.** 1989. Immunogenic Gal alpha 1-3Gal carbohydrate epitopes are present on pathogenic American trypanosoma and leishmania. *J. Immunol.* **142:**2828–2834.
18. **Bach, F.** 1998. Xenotransplantation: problems and prospects. *Annu. Rev. Med.* **49:**301–310.
19. **Balasch, M., J. Segales, C. Rosell, M. Domingo, A. Mankertz, A. Urniza, and J. Plana-Duran.** 1999. Experimental inoculation of conventional pigs with tissue homogenates from pigs with post-weaning multisystemic wasting syndrome. *J. Comp. Pathol.* **121:**139–148.
20. **Balayan, M. S.** 1997. Epidemiology of hepatitis E virus infection. *J. Viral Hepatitis* **4:**155–165.

21. **Balayan, M. S., R. K. Usmanov, D. I. Zamyatina, and F. R. Karas.** 1990. Brief report: experimental hepatitis E infection in domestic pigs. *J. Med. Virol.* **32:**58–59.
22. **Bergelson, J. M., J. G. Mohanty, R. L. Crowell, N. F. St. John, D. M. Lublin, and R. W. Finberg.** 1995. Coxsackievirus B3 adapted to growth in RD cells binds to decay-accelerating factor (CD55). *J. Virol.* **69:**1903–1906.
23. **Best, L., P. R. Le Tissier, and J. P. Stoye.** 1997. Endogenous retroviruses and the evolution of resistance to retroviral infection. *Trends Microbiol.* **5:**313–318.
24. **Boeke, J. D., and J. P. Stoye.** 1997. Retrotransposons, endogenous retroviruses, and the evolution of retroelements, p. 343–435. *In* J. M. Coffin, S. H. Hughes, and H. E. Varmus (ed.), *Retroviruses.* Cold Spring Harbor Laboratory Press, Cold Spring Harbor, N. Y.
25. **Borlongan, C. V., D. W. Cahill, T. Freeman, and P. R. Sanberg.** 1994. Recent advances in neural transplantation: relevance to neurodegenerative disorders. *J. Fla. Med. Assoc.* **81:**689–694.
26. **Bouillant, A. M., and A. S. Greig.** 1975. Type C virus production by a continuous line of pig oviduct cells. *J. Gen. Virol.* **27:**173–180.
27. **Brown, J., A. L. Matthews, P. A. Sanstrom, and L. E. Chapman.** 1998. Xenotransplantation and the risk of retroviral zoonosis. *Trends Microbiol.* **6:**411–415.
28. **Buhler, L., T. Friedman, J. Iacomini, and D. K. C. Cooper.** 1999. Xenotransplantation—state of the art—update 1999. *Front. Biosci.* **4:**D416–D432.
29. **Cavallo, R., G. Valente, C. Jemma, G. Gribaudo, S. Landolfo, and G. Cavallo.** 1994. Presence of murine retroviral sequences in human cell line transplanted in immunosuppressed SCID mice. *Microbiologica* **17:**195–202.
30. **Centers for Disease Control and Prevention.** 1997. Nonhuman primate spumavirus infections among persons with occupational exposure—United States, 1996. *Morb. Mortal. Wkly. Rep.* **46:**129–131.
31. **Centers for Disease Control and Prevention.** 1999. Outbreak of Hendra-like virus—Malaysia and Singapore, 1998–1999. *Morb. Mortal. Wkly. Rep.* **48:**265–269. (Erratum, **48:**339.)
32. **Centers for Disease Control and Prevention.** 1999. Update: outbreak of Nipah virus—Malaysia and Singapore, 1999. *Morb. Mortal. Wkly. Rep.* **48:**335–337.
33. **Chandler, J. D., M. A. Riddell, F. Ki, R. J. Love, and D. A. Anderson.** 1999. Serological evidence for swine hepatitis E virus infection in Australian pig herds. *Vet. Microbiol.* **68:**95–105.
34. **Chang, S. F., J. Y. Sgro, and C. R. Parrish.** 1992. Multiple amino acids in the capsid of canine parvovirus coordinately determine the canine host range and specific antigenic and haemagglutination properties. *J. Virol.* **66:**6858–6867.
35. **Chang, Y., E. Cesarman, M. S. Pessin, F. Lee, J. Culpepper, D. M. Knowles, and P. S. Moore.** 1994. Identification of herpesvirus-like-DNA sequences in AIDS-associated Kaposi's sarcoma. *Science* **266:**1865–1869.
36. **Chapman, L. E., T. M. Folks, D. R. Salomon, A. P. Patterson, T. E. Eggerman, and P. D. Noguichi.** 1995. Xenotransplantation and xenogeneic infections. *N. Engl. J. Med.* **333:**1498–1501.
37. **Chapman, L. E.** 1999. Xenogeneic infections and public health. *Clin. Exp. Pharmacol. Physiol.* **26:**1005–1008.
38. **Cooper, D. K. C., Y. Ye, L. L. Rolf, Jr., and N. Zuhdni.** 1991. The pig as a potential organ donor for man, p. 481–500. *In* D. K. C. Cooper, E. Kemp, K. Reemtsma, and D. J. G. White (ed.), *Xenotransplantation,* 1st ed. Springer, Heidelberg, Germany.
39. **Cooper, D. K. C., E. Koren, and R. Oriol.** 1994. Oligosaccharides and discordant xenotransplantation. *Immunol. Rev.* **141:**31–58.
40. **Cozzi, E., and D. J. G. White.** 1995. The generation of transgenic pigs as potential organ donors for humans. *Nat. Med* **1:**964–966.

41. **Daar, A. S., and D. Phil.** 1997. Ethics of xenotransplantation: animal issues, consent, and likely transformation of transplant ethics. *World J. Surg.* **21:**975–982.
42. **Dawson, M., G. A. Wells, B. N. Parker, and A. C. Scott.** 1990. Primary parenteral transmission of bovine spongiform encephalopathy to the pig. *Vet. Rec.* **127:**338.
43. **Deacon, T., J. Schumacher, J. Dinsmore, C. Thomas, P. Palmer, S. Kott, A. Edge, D. Penney, S. Kassissieh, P. Dempsey, and O. Isacson.** 1997. Histological evidence of fetal pig neural cell survival after transplantation into a patient with Parkinson's disease. *Nat. Med.* **3:**350–353.
44. **Dorig, R. E., A. Marcil, A. Chopra, and C. D. Richardson.** 1993. The human CD46 molecule is a receptor for measles virus (Edmonston strain). *Cell* **75:**295–305.
45. ***Dorland's Illustrated Medical Dictionary,* 27th ed.** 1988. W. B. Saunders, Philadelphia, Pa.
46. **Dowling, D. J., and D. J. Mutimer.** 1999. Artificial liver support in acute liver failure. *Eur. J. Gastroenterol. Hepatol.* **11:**991–996.
47. **Dulac, G. C., and A. Ahmad.** 1989. Porcine circovirus antigens in PK-15 cell line (ATCC CCL-33) and evidence of antibodies to circovirus in Canadian pigs. *Can. J. Vet. Res.* **53:**431–433.
48. **Edge, A. S., M. E. Gosse, and J. Dinsmore.** 1998. Xenogeneic cell therapy: current progress and future developments in porcine cell transplantation. *Cell Transplant.* **7:**525–539.
49. **Edington, N., R. G. Watt, and W. Plowright.** 1976. Cytomegalovirus excretion in gnotobiotic pigs. *J. Hyg. (Lond.)* **77:**283–290.
50. **Edington, N., R. G. Watt, and W. Plowright.** 1977. Experimental transplacental transmission of porcine cytomegalovirus. *J. Hyg. (Lond.)* **78:**243–251.
51. **Edington, N., A. E. Wrathall, and J. T. Done.** 1988. Porcine cytomegalovirus (PCMV) in early gestation. *Vet. Microbiol.* **17:**117–128.
52. **Edington, N.** 1992. Cytomegalovirus, p. 250–256. *In* A. D. Leman, B. E. Straw, W. L. Mengeling, S. D'Allaire, and D. J. Taylor (ed.), *Diseases of Swine,* 7th ed. Iowa State University Press, Ames.
53. **Edwards, S., and J. J. Sands.** 1994. Evidence of circovirus infection in British pigs. *Vet. Rec.* **134:**680–681.
54. **Ehlers, B., S. Ulrich, and M. Goltz.** 1999. Detection of two novel porcine herpesviruses with high similarity to gammaherpesviruses. *J. Gen. Virol.* **80:**971–978.
55. **Ellis, J., L. Hassard, E. Clark, J. Harding, G. Allan, P. Willson, J. Strokappe, K. Martin, F. McNeilly, B. Meehan, D. Todd, and D. Haines.** 1998. Isolation of circovirus from lesions of pigs with postweaning multisystemic wasting syndrome. *Can. Vet. J.* **39:**44–51.
56. **Ellis, J., S. Krakowka, M. Lairmore, D. Haines, A. Bratanich, E. Clark, G. Allan, C. Konoby, L. Hassard, B. Meehan, K. Martin, J. Harding, S. Kennedy, and F. McNeilly.** 1999. Reproduction of lesions of postweaning multisystemic wasting syndrome in gnotobiotic piglets. *J. Vet. Diagn. Investig.* **11:**3–14.
57. **Fishman, J., D. Sachs, and R. Shailch.** 1998. *Xenotransplantation: Scientific Frontiers and Public Policy. Ann. N. Y. Acad. Sci.,* vol. 862. New York Academy of Science, New York, N.Y.
58. **Fishman, J. A.** 1997. Xenosis and xenotransplantation: addressing the infectious risks posed by an emerging technology. *Kidney Int. Suppl.* **58:**S41–S45.
59. **Fishman, J. A.** 1999. Infection in xenotransplantation: a clinical approach. *Transplant. Proc.* **31:**2225–2227.
60. **Florencio, P. S., and T. Caulfield.** 1999. Xenotransplantation and public health: identifying the legal issues. *Can. J. Public Health* **90:**282–284.
61. **Galili, U., E. A. Rachmilewitz, A. Peleg, and I. Flechner.** 1984. A unique natural human IgG antibody with anti-alpha-galactosyl specificity. *J. Exp. Med.* **160:**1519–1531.
62. **Galili, U., M. R. Clark, S. B. Shohet, J. Buehler, and B. A. Macher.** 1987. Evolutionary relationship between the natural anti-Gal antibody and the Gal alpha 1→3Gal epitope in primates. *Proc. Natl. Acad. Sci. USA* **84:**1369–1373.

63. **Galili, U., R. E. Mandrell, R. M. Hamadeh, S. B. Shohet, and J. M. Griffis.** 1998. The interaction between the human natural anti-α-galactosyl IgG (anti-Gal) and bacteria of the human flora. *Infect. Immun.* **57:**1730–1737.

64. **Gao, F., L. Yue, A. T. White, P. G. Pappas, J. Barchue, A. P. Hanson, B. M. Greene, P. M. Sharp, G. M. Shaw, and B. H. Hahn.** 1992. Human infection by genetically-diverse SIVsm-related HIV-2 in west Africa. *Nature* **358:**495–499.

65. **Gao, F., L. Yue, D. L. Robertson, S. C. Hill, H. Hui, R. J. Biggar, A. E. Neequaye, T. M. Whelan, D. D. Ho, and G. M. Shaw.** 1994. Genetic diversity of human immunodeficiency virus type 2: evidence for distinct sequence subtypes with differences in virus biology. *J. Virol.* **68:**7433–7447.

66. **Gao, F., E. Bailes, D. L. Robertson, Y. Chen, C. M. Rodenburg, S. F. Michael, L. B. Cummins, L. O. Arthur, M. Peeters, G. M. Shaw, P. M. Sharp, and B. H. Hahn.** 1999. Origin of HIV-1 in the chimpanzee *Pan troglodytes troglodytes. Nature* **397:**436–441.

67. **Gibbs, C. J., Jr., D. C. Gajdusek, and H. Amyx.** 1979. Strain variation in the viruses of Creutzfeldt-Jakob disease and kuru, p. 87–110. *In* S. B. Prusiner and W. J. Hadlow (ed.), *Slow Transmissible Diseases of the Nervous System,* vol. 2. *Pathogenesis, Immunology, Virology, and Molecular Biology of the Spongiform Encephalopathies.* Academic Press, New York, N.Y.

68. **Gibbs, M. J., and G. F. Weiller.** 1999. Evidence that a plant virus switched hosts to infect a vertebrate and then recombined with a vertebrate-infecting virus. *Proc. Natl. Acad. Sci. USA* **96:**8022–8027.

69. **Good, A. H., D. K. C. Cooper, A. J. Malcolm, R. M. Ippolito, E. Koren, F. A. Neethling, Y. Ye, L. R. Zuhdi, and L. R. Lamontagne.** 1992. Identification of carbohydrate structures which bind human anti-porcine antibodies: implications for discordant xenografting in man. *Transplant. Proc.* **24:**559–562.

70. **Groth, C. G., A. Tibell, L. Wennberg, and O. Korsgren.** 1996. Experimental and clinical experience with xenoislet transplantation. *Transplant. Proc.* **28:**3515.

70a.**Hamadeh, R. M., G. A. Jarvis, U. Galili, R. E. Mandrell, P. Zhou, and J. M. Griffiss.** 1992. Human natural anti-Gal IgG regulates alternative complement pathway activation on bacterial surfaces. *J. Clin. Investig.* **89:**1223–1235.

71. **Hamel, A. L., L. L. Lin, and G. P. Nayar.** 1998. Nucleotide sequence of porcine circovirus associated with postweaning multisystemic wasting syndrome in pigs. *J. Virol.* **72:**5262–5267.

72. **Hammer, C.** 1997. Evolution: its complexity and impact on xenotransplantation, p. 716–735. *In* D. K. C. Cooper, E. Kemp, J. L. Platt, and D. J. G. White (ed.), *Xenotransplantation,* 2nd ed. Springer, Heidelberg, Germany.

73. **Harris, D. L., and T. J. L. Alexander.** 1999. Methods of disease control, p. 1077–1110. *In* B. E. Straw, S. D'Allaire, W. L. Mengeling, and D. J. Taylor (ed.), *Diseases of Swine,* 8th ed. Iowa State University Press, Ames.

74. **Heneine, W., N. W. Lerche, T. Woods, T. Spira, J. M. Liff, W. Eley, J. L. Yee, J. E. Kaplan, and R. F. Khabbaz.** 1993. The search for human infection with simian type D retroviruses. *J. Acquir. Immune Defic. Syndr.* **6:**1062–1066.

75. **Heneine, W., W. M. Switzer, P. Sandstrom, J. Brown, S. Vedapuri, C. A. Schable, A. S. Khan, N. W. Lerche, M. Schweizer, D. Neumann-Haefelin, L. E. Chapman, and T. M. Folks.** 1998. Identification of a human population infected with simian foamy viruses. *Nat. Med.* **4:**403–407.

76. **Hines, R. K., and P. D. Lukert.** 1995. Porcine circovirus: a serological survey of swine in the United States. *Swine Health Prod.* **3:**71–73.

77. **Ho, M., and J. S. Drummer.** 1995. Infections in transplant recipients, p. 2709–2717. *In* G. L. Mandell, J. E. Bennett, and R. Dolin (ed.) *Mandell, Douglas and Bennett's Principles and Practice of Infectious Diseases,* 4th ed., vol. 2. Churchill Livingstone, New York, N.Y.

78. **Institute of Medicine.** 1996. *Xenotransplantation: Science, Ethics, and Public Policy.* National Academy Press, Washington, D.C.
79. **Joziasse, D. H., and R. Oriol.** 1999. Xenotransplantation: the importance of the Galα1,3Gal epitope in hyperacute vascular rejection. *Biochim. Biophys. Acta* **1455:**403–418.
80. **Kabrane-Lazizi, Y., J. B. Fine, J. Elm, G. E. Glass, H. Higa, A. Diwan, C. J. Gibbs, Jr., X. J. Meng, S. U. Emerson, and R. H. Purcell.** 1999. Evidence for widespread infection of wild rats with hepatitis E virus in the United States. *Am. J. Trop. Med. Hyg.* **61:**331–335.
81. **Karp, C. L., M. Wysocka, L. M. Wahl, J. M. Ahearn, P. J. Cuomo, B. Sherry, G. Trinchieri, and D. E. Griffin.** 1996. Mechanism of suppression of cell-mediated immunity by measles virus. *Science* **273:**228–231.
82. **Kennedy, S., D. Moffett, F. McNeilly, B. Meehan, J. Ellis, S. Krakowka, and G. M. Allan.** 2000. Reproduction of lesions of postweaning multisystemic wasting syndrome by infection of conventional pigs with porcine circovirus type 2 alone or in combination with porcine parvovirus. *J. Comp. Pathol.* **122:**9–24.
83. **Kiupel, M., G. W. Stevenson, S. K. Mittal, E. G. Clark, and D. M. Haines.** 1998. Circovirus-like viral associated disease in weaned pigs in Indiana. *Vet. Pathol.* **35:**303–307.
84. **Lanza, R. P., D. K. C. Cooper, and W. L. Chick.** 1997. Xenotransplantation. *Sci. Am.* **277:**54–59.
85. **Lanza, R. P., and D. K. C. Cooper.** 1998. Xenotransplantation of cells and tissues: application to a range of diseases, from diabetes to Alzheimer's. *Mol. Med. Today* **4:**39–45.
86. **Larsson, E., and G. Andersson.** 1998. Beneficial role of human endogenous retroviruses: facts and hypotheses. *Scand. J. Immunol.* **48:**329–338.
87. **Leib-Mosch, C., M. Bachmann, R. Brack-Werner, T. Werner, V. Erfle, and R. Hehlmann.** 1992. Expression and biological significance of human endogenous retroviral sequences. *Leukemia* **6:**725–755.
88. **Le Tissier, P., J. P. Stoye, Y. Takeuchi, C. Patience, and R. A. Weiss.** 1997. Two sets of human-tropic pig retrovirus. *Nature* **389:**681–682.
89. **Lieber, M. M., C. J. Sherr, R. E. Benveniste, and G. J. Todaro.** 1975. Biologic and immunologic properties of porcine type C viruses. *Virology* **66:**616–619.
90. **Lu, C. Y., T. A. Khair-el-Din, I. A. Dawidson, T. M. Butler, K. M. Brasky, M. A. Vazquez, and S. C. Sicher.** 1994. Xenotransplantation. *FASEB J.* **8:**1122–1130.
91. **Lusso, P., F. Di Marzo Veronese, B. Ensoli, G. Franchini, C. Jemma, S. DeRocco, V. S. Kalyanaraman, and R. C. Gallo.** 1990. Expanded HIV-1 cellular tropism by phenotypic mixing with murine endogenous retroviruses. *Science* **247:**848–852.
92. **Makrides, S. C.** 1998. Therapeutic inhibition of the complement system. *Pharmacol. Rev.* **50:**59–87.
93. **Mandel, T. E., and N. Koulmanda.** 1994. Xenotransplantation of islets of Langerhans in diabetes mellitus. *Transplant. Proc.* **26:**1110–1112.
94. **Mankertz, A., F. Persson, J. Mankertz, G. Blaess, and H. J. Buhk.** 1997. Mapping and characterization of the origin of DNA replication of porcine circovirus. *J. Virol.* **71:**2562–2566.
95. **Meehan, B. M., J. K. Creelan, M. S. McNulty, and D. Toss.** 1997. Sequence of porcine circovirus DNA: affinities with plant circoviruses. *J. Gen. Virol.* **78:**221–227.
96. **Meehan, B. M., F. McNeilly, D. Todd, S. Kennedy, V. A. Jewhurst, J. A. A. Ellis, L. E. Hassard, E. G. Clark, D. M. Haines, and G. M. Allan.** 1998. Characterization of novel circovirus DNAs associated with wasting syndromes in pigs. *J. Gen. Virol.* **79:**2171–2179.
97. **Meng, X. J., R. H. Purcell, P. G. Halbur, J. R. Lehman, D. M. Webb, T. S. Tsareva, J. S. Haynes, B. J. Thacker, and S. U. Emerson.** 1997. A novel virus in swine is closely related to the human hepatitis E virus. *Proc. Natl. Acad. Sci. USA* **94:**9860–9865.

98. **Meng, X. J., P. G. Halbur, J. S. Haynes, T. S. Tsareva, J. D. Bruna, R. L. Royer, R. H. Purcell, and S. U. Emerson.** 1998. Experimental infection of pigs with the newly identified swine hepatitis E virus (swine HEV), but not with human strains of HEV. *Arch. Virol.* **143:**1405–1415.
99. **Meng, X. J., P. G. Halbur, M. S. Shapiro, S. Govindarajan, J. D. Bruna, I. K. Mushahwar, R. H. Purcell, and S. U. Emerson.** 1998. Genetic and experimental evidence for cross-species infection by swine hepatitis E virus. *J. Virol.* **72:**9714–9721.
100. **Meng, X. J., S. Dea, R. E. Engle, R. Friendship, Y. S. Lyoo, T. Sirinarumitr, K. Urairong, D. Wang, D. Wong, D. Yoo, Y. Zhang, R. H. Purcell, and S. U. Emerson.** 1999. Prevalence of antibodies to the hepatitis E virus in pigs from countries where hepatitis E is common or is rare in the human population. *J. Med. Virol.* **59:**297–302.
101. **Michaels, M. G., and R. L. Simmons.** 1994. Xenotransplantation: concerns aired over potential new infections. *ASM News* **61:**442–443.
102. **Michaels, M. G. and R. L. Simmons.** 1994. Xenotransplantation-associated zoonoses: strategies for prevention. *Transplantation* **57:**1–7.
103. **Michaels, M. G.** 1997. Infectious concerns of cross-species transplantation: xenozoonoses. *World J. Surg.* **21:**968–974.
104. **Miyata, H. H. Tsunoda, A. Kazi, A. Yamada, M. A. Khan, J. Murakami, T. Kamahora, K. Shiraki, and S. Hino.** 1999. Identification of a novel GC-rich 113-nucleotide region to complete the circular, single-stranded DNA genome of TT virus, the first human circovirus. *J. Virol.* **73:**3582–3586.
105. **Moennig, V., H. Frank, G. Hunsmann, P. Ohms, H. Schwarz, and W. Schafer.** 1974. C-type particles produced by a permanent cell line from a leukemic pig. *Virology* **57:**179–188.
106. **Mollnes, T. E., and A. E. Fiane.** 1999. Xenotransplantation: how to overcome the complement obstacle? *Mol. Immunol.* **36:**269–276.
107. **Morgan, F.** 1997. Babe the magnificent organ donor? The perils and promises surrounding xenotransplantation. *J. Contemp. Health Law Policy* **14:**127–160.
108. **Morozov, I., T. Sirinarumitr, S. D. Sorden, P. G. Halbur, M. K. Morgan, K. J. Yoon, and P. S. Paul.** 1998. Detection of a novel strain of porcine circovirus in pigs with postweaning multisystemic wasting syndrome. *J. Clin. Microbiol.* **36:**2535–2541.
109. **Morse, S. S., and A. Schluederberg.** 1990. Emerging viruses: the evolution of viruses and viral diseases. *J. Infect. Dis.* **162:**1–7.
110. **Murphy, F. A.** 1994. New, emerging and reemerging infectious diseases. *Adv. Virus Res.* **4:**1–52.
111. **Myers, G., K. MacInnes, and B. Korber.** 1992. The emergence of simian/human immunodeficiency viruses. *AIDS Res. Hum. Retroviruses* **8:**373–386.
112. **Naniche, D., G. Varior-Krishnan, F. Cervoni, T. F. Wild, B. Rossi, C. Rabourdin-Combe, and D. Gerlier.** 1993. Human membrane cofactor protein (CD46) acts as a cellular receptor for measles virus. *J. Virol.* **67:**6025–6032.
113. **Nishizawa, T., H. Okamoto, K. Konishi, H. Yoshizawa, Y. Miyakawa, and M. Mayumi.** 1997. A novel DNA virus (TTV) associated with elevated transaminase levels in posttransfusion hepatitis of unknown etiology. *Biochem. Biophys. Res. Commun.* **241:**92.
114. **Nuffield Council on Bioethics.** 1996. *Animal-to-human transplants: the Ethics of Xenotransplantation.* Nuffield Council on Bioethics, London, United Kingdom.
115. **Okada, N., M. K. Liszewski, J. P. Atkinson, and M. Caparon.** 1995. Membrane cofactor protein (CD46) is a keratinocyte receptor for the M protein of the group A streptococcus. *Proc. Natl. Acad. Sci. USA* **92:**2489–2493.
116. **Okamoto, H., M. Fukuda, A. Tawara, T. Nishizawa, Y. Itoh, I. Hayasaka, F. Tsuda, T. Tanaka, Y. Miyakawa, and M. Mayumi.** 2000. Species-specific TT viruses and cross-species infection in nonhuman primates. *J. Virol.* **74:**1132–1139.

117. **Onions, D. E.** 1994. Viruses as the aetiological agents of leukaemia and lymphomas, p. 34–71. *In* A. Burnett, J. Armitage, A. Newland, and A. Keating (ed.) *Cambridge Medical Reviews: Haematological Oncology.* Cambridge University Press, Cambridge, United Kingdom.

117a. **Onions, D., D. K. C. Cooper, T. J. L. Alexander, C. Brown, E. Claassen, J. Foweraker, D. L. Harris, B. Mahy, A. D. M. E. Osterhaus, P.-P. Pastoret, K. Yamanouchi, and P. Minor.** An assessment of the risk of xenozoonotic disease in pig-to-human xenotransplantation. *Xenotransplantation,* in press.

118. **Onions, D., D. Hart, C. Mahoney, D. Galbraith, and K. Smith.** 1998. Endogenous retroviruses and the safety of porcine xenotransplantation. *Trends Microbiol.* **11:**430–431.

119. **Onuki, A., K. Abe, K. Togashi, K. Kawashima, A. Taneichi, and H. Tsunemitsu.** 1999. Detection of porcine circovirus from lesions of a pig with wasting disease in Japan. *J. Vet. Med. Sci.* **61:**1119–1123.

120. **Paridis, K., G. Langford, Z. F. Long, W. Heneine, P. Sandstrom, W. M. Switzer, L. E. Chapman, C. Lockey, D. Onions, the Xen 111 Working Group, and E. Otto.** 1999. Search for cross-species transmission of porcine endogenous retrovirus in patients treated with living pig tissue. *Science* **285:**1236–1241.

121. **Pastoret, P. P., S. Escutenaire, and B. Brochier.** 1999. Zoonotic risks linked to xenotransplantation. *Infect. Dis. Rev.* **1:**47–79.

122. **Patience, C., Y. Takeuchi, and R. A. Weiss.** 1997. Infection of human cells by an endogenous retrovirus of pigs. *Nat. Med.* **3:**282–286.

123. **Patience, C., Y. Takeuchi, and R. A. Weiss.** 1998. Zoonosis in xenotransplantation. *Curr. Opin. Immunol.* **10:**539–542.

124. **Philbey, A. W., P. D. Kirkland, A. D. Ross, R. J. Davis, A. B. Gleeson, R. J. Love, P. W. Daniels, A. R. Gould, and A. D. Hyatt.** 1998. An apparently new virus (family *Paramyxoviridae*) infectious for pigs, humans, and fruit bats. *Emerg. Infect. Dis.* **4:**269–271.

125. **Pool, R.** 1998. Saviors. *Discover,* May 1998, p. 53–57.

126. **Prescott, L. E. and P. Simmonds.** 1998. Global distribution of transfusion-transmitted virus. *N. Engl. J. Med.* **339:**776–777.

127. **Prusiner, S. B.** 1982. Novel proteinaceous infectious particles cause scrapie. *Science* **216:**136–144.

128. **Purcell, D. F., C. M. Broscius, E. F. Vanin, C. E. Buckler, A. W. Nienhuis, and M. A. Martin.** 1996. An array of murine leukemia virus-related elements is transmitted and expressed in a primate recipient of retroviral gene transfer. *J. Virol.* **70:**887–897.

129. **Regamey, N., M. Tamm, M. Wernli, A. Witschi, G. Thiel, G. Cathomas, and P. Erb.** 1998. Transmission of human herpesvirus 9 infection from renal-transplant donors to recipients. *N. Engl. J. Med.* **339:**1358–1363.

130. **Riordan, S. M., and R. Williams.** 1999. Extracorporeal support and hepatocyte transplantation in acute liver failure and cirrhosis. *J. Gastroenterol. Hepatol.* **14:**757–770.

131. **Rose, R. M., K. B. Strans, E. R. Schultz, G. Schaefer, G. W. Rankin, Jr., M. E. Thouless, C. C. Tsai, and M. L. Bosch.** 1997. Identification of two homologs of the Kaposi's sarcoma-associated herpesvirus (human herpesvirus 8) in retroperitoneal fibromatosis of different macaque species. *J. Virol.* **71:**4138–4144.

132. **Ross, R. S., S. Viazov, V. Runde, U. W. Schaefer, and M. Roggendorf.** 1999. Detection of TT virus DNA in specimens other than blood. *J. Clin. Virol.* **13:**181–184.

133. **Rother, R. P., W. L. Fodor, J. P. Springhorn, C. W. Birks, E. Setter, M. S. Sandrin, S. P. Squinto, and S. A. Rollins.** 1995a. A novel mechanism of retrovirus inactivation in human serum mediated by anti-α-galactosyl natural antibody. *J. Exp. Med.* **182:**1345–1355.

134. **Rother, R. P., S. P. Squinto, J. M. Mason, and S. A. Rollins.** 1995b. Protection of retrovi-

ral vector particles in human blood through complement inhibition. *Hum. Gene Ther.* **6:**429–435.

135. **Sachs, D. H.** 1994. The pig as a potential xenograft donor. *Vet. Immunol. Immunopathol.* **43:**185–191.
136. **Satapathy, A. K., and B. Ravindram.** 1999. Naturally occurring alpha-galactosyl antibodies in human sera display polyreactivity. *Immunol. Lett.* **69:**347–351.
137. **Schlauder, G. G., G. J. Dawson, J. C. Erker, P. Y. Kwo, M. F. Knigge, D. L. Smalley, J. E. Rosenblatt, S. M. Desai, and I. K. Mushahwar.** 1998. The sequence and phylogenetic analysis of a novel hepatitis E virus isolated from a patient with acute hepatitis reported in the United States. *J. Gen. Virol.* **79:**447–456.
138. **Simmonds, P., F. Davidson, C. Lycett, L. E. Prescott, D. M. MacDonald, J. Ellender, P. L. Yap, C. A. Ludlam, G. H. Haydon, J. Gillon, and J. M. Jarvis.** 1998. Detection of a novel DNA virus (TT virus) in blood donors and blood products. *Lancet* **352:**191.
139. **Smith, A. W., D. E. Skilling, N. Cherry, J. H. Mead, and D. O. Matson.** 1998. Calicivirus emergence from ocean reservoirs: zoonotic and interspecies movements. *Emerg. Infect. Dis.* **4:**13–20.
140. **Steele, D. J., and H. J. Auchincloss.** 1995. Xenotransplantation. *Annu. Rev. Med.* **46:**345–360.
141. **Stoye, J. P.** 1997. Proviruses pose potential problems. *Nature* **386:**126–127.
142. **Sun, Y. L., X. Ma, D. Zhou, I. Vacek, and A. M. Sun.** 1993. Porcine pancreatic islets: isolation, microencapsulation, and xenotransplantation. *Artif. Organs* **17:**723–733.
143. **Sunil-Chandra, N. P., J. Amo, J. Fazakerley, and A. A. Nash.** 1994. Lymphoproliferative disease in mice infected with murine gammaherpesvirus 68. *Am. J. Pathol.* **145:**818–826.
144. **Suzuka, I., K. Sekiguchi, and M. Kodama.** 1985. Some characteristics of a porcine retrovirus from a cell line derived from swine malignant lymphomas. *FEBS Lett.* **183:**124–128.
145. **Takeuchi, Y., C. D. Porter, K. M. Strahan, A. F. Preece, K. Gustafsson, F. L. Cosset, R. A. Weiss, and M. K. Collins.** 1996. Sensitization of cells and retroviruses to human serum by (α1-3) galactosyltransferase. *Nature* **379:**85–88.
146. **Takeuchi, Y., S. H. Liong, P. D. Bieniasz, U. Jager, C. D. Porter, T. Friedman, M. O. McClure, and R. A. Weiss.** 1997. Sensitization of rhabdo-, lenti-, and spumaviruses to human serum by galactosyl (α1-3) galactosylation. *J. Virol.* **71:**6174–6178.
147. **Takeuchi, Y., C. Patience, S. Magre, R. A. Weiss, P. T. Banerjee, P. Le Tissier, and J. P. Stoye.** 1998. Host range and interference studies of three classes of pig endogenous retrovirus. *J. Virol.* **72:**9986–9991.
148. **Tischer, I., R. Rasch, and G. Tochtermann.** 1974. Characterization of papovavirus- and picornavirus-like particles in permanent pig kidney cell lines. *Zentbl. Bakteriol. Mikrobiol. Hyg. Ser. A* **226:**153–167.
149. **Tischer, I., H. Geldblom, W. Vettermann, and M. A. Koch.** 1982. A very small porcine virus with circular single-stranded DNA. *Nature* **295:**64–66.
150. **Tischer, I., W. Mields, D. Wolff, M. Vagt, and W. Griem.** 1986. Studies on epidemiology and pathogenicity of porcine circovirus. *Arch. Virol.* **91:**271–276.
151. **Tischer, I., J. Bode, H. Timm, D. Peters, R. Rasch, S. Pociuli, and E. Gerike.** 1995. Presence of antibodies reacting with porcine circovirus in sera of humans, mice, and cattle. *Arch. Virol.* **140:**1427–1439.
152. **Tischer, I., D. Peters, and S. Pociuli.** 1995. Occurrence and role of an early antigen and evidence for transforming ability of porcine circovirus. *Arch. Virol.* **140:**1799–1816.
153. **Todaro, G. J., R. E. Benveniste, M. M. Lieber, and C. J. Sherr.** 1974. Characteristics of a type C virus released from the porcine cell line PK15. *Virology* **58:**65–74.
154. **Todaro, G. J., R. E. Benveniste, C. J. Sherr, M. M. Lieber, and R. Callahan.** 1975. Infectious primate type C virus group: evidence for an origin from an endogenous virus of the rodent, Mus caroli. *Bibl. Haematol.* **43:**115–120.

155. **Todaro, G. J., M. M. Lieber, R. E. Benveniste, and C. J. Sherr.** 1975. Infectious primate type C viruses: three isolates belonging to a new subgroup from the brains of normal gibbons. *Virology* **67:**335–343.
156. **Tremblay, M. J., J.-F. Fortin, and R. Cantin.** 1998. The acquisition of host-encoded proteins by nascent HIV-1. *Immunol. Today* **19:**346–351.
157. **Truyen, U., C. R. Parrish, T. C. Harder, and O.-R. Kaaden.** 1995. There is nothing permanent except change: the emergence of new virus diseases. *Vet. Microbiol.* **43:**103–122.
158. **Ulrich, S., M. Goltz, and B. Ehlers.** 1999. Characterization of the DNA polymerase loci of the novel porcine lymphotropic herpesviruses 1 and 2 in domestic and feral pigs. *J. Gen. Virol.* **80:**3199–3205.
159. **U.S. Department of Health and Human Services, Food and Drug Administration.** 1996. Draft Public Health Service guidelines on infectious disease issues in xenotransplantation. *Fed. Register* **61:**49929.
160. **U.S. Department of Health and Human Services, Food and Drug Administration.** 1999. Draft guidance for industry: precautionary measures to reduce the possible risk of transmission of zoonoses by blood and blood products from xenotransplantation product recipients and their contacts. *Fed. Register* **64:**73563.
161. **U.S. Department of Health and Human Services, Food and Drug Administration.** 1999. Guidance for industry: public health issues posed by the use of nonhuman primate xenografts. *Fed. Register* **64:**16743.
162. **Vanderpool, H. Y.** 1999. Commentary: a critique of Clark's frightening xenotransplantation scenario. *J. Law Med. Ethics* **27:**153–157.
163. **Verschoor, E. J., S. Langenhuijzen, and J. L. Heeney.** 1999. TT viruses (TTV) of nonhuman primates and their relationship to the human TTV genotypes. *J. Gen. Virol.* **80:**2491–2499.
164. **Ward, T., P. A. Pipkin, N. A. Clarkson, D. M. Stone, P. D. Minor, and J. W. Almond.** 1994. Decay-accelerating factor CD55 is identified as the receptor for echovirus 7 using CELICS, a rapid immuno-focal cloning method. *EMBO J.* **13:**5070–5074.
165. **Weiss, R. A.** 1998. Transgenic pigs and virus adaptation. *Nature* **391:**327–328.
166. **Weiss, R. A.** 1999 Xenografts and retroviruses. *Science* **285:**1221–1222.
166a. **Wild, T. F., D. Naniche, C. Rabourdin-Combe, D. Gerlier, E. Malvoisin, V. Lecouturier, and R. Buckland.** 1995. Mode of entry of morbilliviruses. *Vet. Microbiol.* **44:**267–270.
167. **Wilson, C. A., S. Wong, J. Muller, C. E. Davidson, T. M. Rose, and P. Burd.** 1998. Type C retrovirus released from porcine primary peripheral blood mononuclear cells infects human cells. *J. Virol.* **72:**3082–3087.
168. **Woods, W. A., T. S. Papas, H. Hirumi, and M. A. Chirigos.** 1973. Antigenic and biochemical characterization of the C-type particle of the stable porcine kidney cell line PK15. *J. Virol.* **12:**1184–1186.
169. **World Health Organization.** 1997. Report of WHO consultation on xenotransplantation. World Health Organization, Geneva, Switzerland.
170. **Wu, J. C., C. M. Chen, T. Y. Chiang, I. J. Sheen, J. Y. Chen, W. H. Tsai, Y. H. Huang, and S. D. Lee.** 2000. Clinical and epidemiological implications of swine hepatitis E virus infection. *J. Med. Virol.* **60:**166–171.
171. **Wurzner, R.** 1999. Evasion of pathogens by avoiding recognition or eradication by complement, in part via molecular mimicry. *Mol. Immunol.* **36:**249–260.
172. **Ye, Y., M. Niekrasz, S. Kosanke, R. Welsh, H. E. Jordan, J. C. Fox, W. C. Edwards, C. Maxwell, and D. K. Cooper.** 1994. The pig as a potential organ donor for man: a study of potentially transferable disease from donor pig to recipient man. *Transplantation* **57:**694–703.
173. **Yokozaki, S., Y. Fukuda, I. Nakano, Y. Katano, M. Okamura, and M. Naruse.** 1999. TT virus: a mother-to-child transmitted rather than bloodborne virus. *Blood* **83:**3569–3570.

Emerging Diseases of Animals
Edited by C. Brown and C. Bolin

Chapter 5

New Fruit Bat Viruses Affecting Horses, Pigs, and Humans

Peter T. Hooper

In the autobiographical "Virus Hunter," C. J. Peters of the Centers for Disease Control and Prevention described a hypothetical scenario of a lethal and contagious paramyxovirus in humans emerging in 2012 in Southeast Asia and spreading to at least Europe and Japan (25). The virus he used in his scenario was Hendra virus (HeV). He likened the emergence of this virus in his hypothetical situation to that of another paramyxovirus, measles virus, 5,000 years before. His vision has proven to be remarkably prophetic. A virus closely related to HeV, Nipah virus (NV), broke out in Southeast Asia in 1999, killing more than 100 people. This chapter describes a remarkable series of new zoonotic viruses, including HeV and NV, that have emerged from fruit bats in Australia and Malaysia. Serendipity, the faculty of making desirable but unsought findings, played a big role in their discovery. They may not have ever been found, or at least not been associated with bats, if HeV had not been found first, isolated during an unusual outbreak of fatal disease in humans and horses in 1994.

HENDRA VIRUS

The HeV infections of 1994 might have passed unnoticed except for the severity of the outbreak. In two neighboring stables at Hendra, Brisbane, Australia, 21 horses and two humans became ill, with 14 horses and one human dying in the acute phase. So many horses and their trainer

Peter T. Hooper • CSIRO Australian Animal Health Laboratory, PO Bag 24, Geelong, Victoria 3220, Australia.

fatally affected at the same time meant that the full resources of governments and their laboratories were activated. Within days of receipt of samples at the laboratory, the new paramyxovirus HeV was isolated. Experimental infection of horses soon satisfied Koch's postulates, and HeV was confirmed as the cause of the outbreak (14, 21). Subsequently, two other incidents of disease have been identified. In 1995, a farmer near Mackay, Australia, died of encephalitis (23) caused by HeV that was retrospectively traced to contact with two HeV-infected horses in August 1994 (13, 27). In 1999, a horse died of the disease near Cairns, Australia (6, 12).

In the original outbreak, HeV was isolated from a range of tissues from horses and from the kidney of the deceased horse trainer (21). Cell cultures showed a cytopathic effect with focal syncytia. In addition to exhibiting the pleomorphism and sizes characteristic of the *Paramyxoviridae,* the virus had a characteristic "double-fringed" envelope not observed in other members of the *Paramyxoviridae* (16). Sequences of the matrix and fusion genes and part of the phosphoprotein gene were consistent with the family *Paramyxoviridae* (20). As it was closer to members of the *Morbillivirus* genus than to other genera in the family, it was provisionally called equine morbillivirus. Subsequent studies have shown that HeV has a significantly larger genome than other previously known members of the *Paramyxoviridae* (19, 20). Sequence comparisons have confirmed that although the virus is a member of the *Paramyxoviridae* subfamily, it displays only low-level homology with other members of the genus *Morbillivirus* (33, 39). It also has no or negligible immunological cross-reactivity with a range of antisera to other paramyxoviruses (20, 21). The virus therefore could not be easily classified in any of the existing genera in the family *Paramyxoviridae* and so was renamed HeV (20, 33).

Pathogenesis and Epidemiology

In the initial investigations, two horses were inoculated with specimens from field cases and two more horses were inoculated with the isolated virus in tissue culture supernatant. All four horses developed similar disease. Subsequently, disease was also reproduced in cats and guinea pigs but not in dogs, rabbits, rats, mice, or chickens similarly challenged with isolated virus (35). The incubation period in these and subsequent experiments was 5 to 10 days irrespective of the dose of virus or its route of administration (14, 36). The course of disease in field and experimental cases was typically short, about 2 days, with clinical signs in all horses consistent with severe pneumonia. Less frequently there were neurological signs consistent with either cerebral infarction or meningoencephalitis. The most significant gross lesions in horses were severe pulmonary edema and vascular congestion (14). In many field cases but not the experimental cases, the airways were filled with thick stable foam, which was occasionally blood tinged. The important histological findings in horses were interstitial pneumonia and generalized fibrinoid degeneration of small blood vessels and, characteristically in early cases, endothelial syncytial cells (14, 36). There were strong reactions to immunohistochemical tests that located viral antigen in blood vessels, especially endothelial cells (14). In some cases, the virus induced nonsuppurative encephalitis characterized by perivascular cuffing of lymphocytes, neuronal necrosis and focal gliosis in the human case in Mackay (17, 27) and in experimental horses and guinea pigs (36). The neurotropism was confirmed by immunohistochemical staining of neurons.

In the original Hendra outbreak, even though there were as many as 21 horses infected, the disease did not spread from the original foci even though there was a sizeable equine population in racing stables or associated with the training track in the neighborhood (18, 34). Likewise, there was no spread in the other two outbreaks. In one of these, infection was confined to the index horse, to another horse, and to one person in contact with the body fluids of both infected horses (27). In the third outbreak, there was only the index case. A second horse in the paddock and in-contact people were not infected (6). The minimal level of contagiousness and lack of risk from aerosol infection were consistent with experimental reproductions in which, despite every opportunity for aerosol spread, there was no transmission from HeV-inoculated horses to in-contact horses or cats (36). Furthermore, HeV was not isolated from the nasal cavity or trachea of infected horses, but virus was present in the urine. The experimental disease may have differed slightly from the field cases in that there was no frothy nasal discharge from the experimental horses. In the field, there was froth, indicating the presence of pulmonary exudate in expired air. It was postulated that virus-rich fluid derived from such affected lungs could have been a hazard to close-contact humans and horses (I. C. Douglas, personal communication). Spread by this intimate contact would be different from aerosol spread, which would lead to contagiousness.

There is serological and virological evidence that fruit bats (flying foxes; *Pteropodidae*) are likely wildlife reservoirs of the viruses (10, 38). Recent studies have found an overall antibody prevalence of 42% in wild-caught fruit bats in Australia and Papua New Guinea (17). Virus was found in uterine fluids (10), and there is experimental evidence of vertical transmission to the fetus (37). This may be the mechanism of spread in the fruit bats, as they congregate in colonies of thousands of bats. The males are known to lick the body, wings, and genitals of the females throughout the breeding season (22). As for cross-species transfer, a possible source of the index field cases in horses could have been the (accidental) ingestion of HeV-infected fruit bat placentae or associated fetal fluids.

AUSTRALIAN BAT LYSSAVIRUS

After a human died of encephalitis caused by HeV infection (23), a bat with encephalitis was tested for HeV with negative results. Alternative immunohistochemical tests were considered as differential diagnoses on paraffin-embedded tissues. Rabies is a disease that is commonly seen in various families of bats overseas but thought not to occur in Australia. It is now history that the search for HeV ultimately led to the identification of Australian bat lyssavirus (ABL), a close relative of rabies virus (8). At the time it was named pteropid lyssavirus, as only fruit bats (*Pteropus* ssp.) were known to be infected and it was not clear that it was an Australian virus. It is now known to be widespread in the country in at least one species of nonpteropid, an insectivorous bat *(Taiphonous flaviventris),* and four species of fruit bats (15). It has also affected two humans with a rabies-like disease (17).

After the original indicative immunohistochemical test in which the brain tissue from the original bat was formalin fixed, there was a successful attempt at virus isolation by intracerebral inoculation from visceral tissue in a mouse (8). In subsequent examinations of numerous bats, virus has been readily isolated in mouse neuroblastoma (MNB) cells from the brains of bats reacting positively to the fluorescent antibody test (FAT) (15). Brain homogenate from intracerebrally inoculated mice and MNB cells with positive virus culture, when examined by negative-contrast electron microscopy, revealed the presence of classical bullet-shaped rhabdoviruses. The virus in MNB cells, when tested by indirect immunofluorescence and by laboratory rodent protection tests with a panel of monoclonal antibodies against various rabies virus serotypes and rabies-like viruses, has been shown to be the lyssavirus closest serologically to classical rabies virus. Its failure to react with a key monoclonal antibody, 62-15-2, delineated it as a rabies-related virus rather than rabies virus itself (9). The serological similarity means that the virus is readily detectable by FAT, immunohistochemistry, and immunoelectron microscopy with anti-rabies virus reagents.

Gene sequence comparisons done on the deduced N, phosphoprotein (M1), matrix protein (M2), glycoprotein (G), and L protein of ABL showed that it was more closely related to the serotype 1 classic rabies virus than to the other members of the *Lyssavirus* genus (3). Phylogenetic studies of nucleocapsid protein sequences clearly showed that ABL was a newly recognized genotype for the *Lyssavirus* genus, genotype 7, closely related to serotype 1 rabies virus (9). Within the genotype, there have been sufficient studies of isolates to show that at the molecular level, the viruses in the fruit bats differ from those in insectivorous bats; in fact, they form separate clades (A. R. Gould, J. A. Kattenbelt, S. Gumley, R. Lunt, P. T. Hooper, and A. D. Hyatt, XI Int. Cong. Virol., 1999, abstr. VET.08).

Pathogenesis and Epidemiology

Some virulence studies have been done in grey-headed flying foxes *(Pteropus poliocephalus)*, one of the Australian fruit bats in which the virus is enzootic, and in cats and dogs (K. A. McColl, T. Chamberlain, R. Lunt, K. Newberry, D. Middleton, and H. Westbury, XI Int. Cong. Virol., 1999, abstr. VP24.03). Three of 10 bats inoculated with 10^5 50% tissue culture infective doses ($TCID_{50}$) of ABL developed clinical signs and histological lesions consistent with lyssaviral infection at 15, 23, and 24 days postinoculation, which was confirmed by FAT, immunohistochemistry, and virus isolation. Three mature male cats that were inoculated intramuscularly with 10^5 $TCID_{50}$ of ABL did not develop abnormal clinical signs during a 3-month observation period, but the cats seroconverted at 29, 42, and 95 days postinoculation. Antilyssavirus antibody was found in the cerebrospinal fluid of one. There were no visible lesions histologically and no evidence of residual virus. More recently, a similar result was obtained in a susceptibility trial with dogs (McColl et al., abstr. VP24.03).

Clinical signs of ABL infection have been observed only in humans, bats, and the cats mentioned above. The first human case developed clinical signs consistent with diffuse encephalitis (1). There were a number of clinical features that were consistent with rabies virus or at least lyssavirus infection: initial numbness in the arm, swallowing difficulties, agitation, and depression. Clinical signs observed in bats have been weakness and paralysis (J. Barrett, P. Young, H. Field, B. Rodwell, G. Smith, and P. Hooper, 48th Annu. Wildl. Dis. Conf., 1999, abstr. 33). Aggression may be difficult to interpret, as it is often present in otherwise normal freshly trapped wild animals. The same group of researchers described one case that was thought to have been infected in the first 2 or 3 weeks of life (7). It exhibited a clinical course of 9 days after a presumed incubation period of 6 to 9 weeks. Serum samples obtained in the field have shown antibody in fruit bats, suggesting that some infections may be nonlethal or at least of prolonged duration. Lesions in humans and bats included an extremely variable nonsuppurative meningoencephalomyelitis and ganglioneuritis similar to that seen in rabies and rabies-like diseases (11, 28). Reactions to an immunohistochemical test for lyssavirus antigen also varied in intensity and distribution. They were also present in peripheral nervous tissues in bats and in one of eight salivary glands examined in a yellow-bellied sheath-tail bat.

ABL infections continue to be diagnosed in Australian bats. Thirty-seven positive cases were examined within a year of the first (15). Their origins covered the Australian mainland from near Darwin in the north to Melbourne in the southeast. Barrett et al. (abstr. 33) found 24 FAT positives in samples from 533 flying foxes trapped for epidemiological reasons. None of the 232 flying foxes that were assessed as clinically normal were FAT positive. Of the sick or injured, about 15 of 43 with nervous signs had ABL infection. Four of five clinically ill yellow-bellied sheath-tailed bats were positive.

Some of the previously mentioned observations suggest some features of the disease that may not be typical of rabies or rabies-like diseases. The disease can occur in very young animals, leading to quite long clinical courses. The presence of seropositive fruit bats suggests recovery in the field. Some animals can have virtually no lesions in their brains (11, 30) and yet be positive immunohistochemically, almost as though the viral infection was incidental to other disease. Anecdotally, while there have been two human cases, there have been many reports of people bitten frequently by various bats over the years without developing disease. One factor here could have been the limited attempts in the past to check cases of suspect encephalitis (31). Another factor is the possibility of only recent introduction of ABL to Australia. Although the cases in bats that have been retrospectively proved as infected only date back to 1995, the widespread nature of the disease across Australia in at least five different species of bats suggests that infection has been enzootic for a long time. While these are limited observations, they do indicate a virus adapted to its various hosts in the trees that seems unlikely to fully emerge by establishing in other species, especially ground-dwelling domestic animals.

MENANGLE VIRUS

The third of this sequence of viral discoveries occurred in 1997, when there was a disease characterized by stillborn piglets in a large (2,500 sows) piggery in New South Wales. An apparently new paramyxovirus was isolated as the causative agent and was called Menangle virus (MenV) (4, 26). Morphologically, it developed into pleomorphic particles 30 to 100 nm long, and so was consistent with the family *Paramyxoviridae*, but failed to react in a range of tests with reference antisera to other viruses in the family. There was some nucleotide sequence homology (A. R. Gould, personal communication) with La Piedad Michoacan virus (genus *Rubulavirus* of the *Paramyxoviridae*), the cause of blue-eye disease seen in pigs in Mexico (32).

Pathogenesis and Epidemiology

In the outbreak, there was a drop in the farrowing rate and an increase in the proportion of mummified and stillborn piglets (26). The latter could have arthrogryposis, brachygnathia, fibrinous straw-colored internal exudates, and pulmonary hypoplasia. There was severe degeneration of the central nervous system that could be seen histologically as encephalomyelitis with intranuclear and intracytoplasmic inclusion bodies, which contained 18-nm nucleocapsids. In an experimental study, eighth-passage MenV in BHK cells ($10^{4.5}$ $TCID_{50}$ of virus) was inoculated intravenously into sows at different stages of gestation (H. Westbury, P. Hooper, C. Morrissy, M. Williamson, and P. Daniels, unpublished data). No sows developed clinical signs of disease, but they seroconverted. Some piglets in all litters from infected sows were undersized, though they exhibited no gross lesions at necropsy. Although no virus was isolated from the brains, in nine piglets there was histological nonsuppurative meningoencephalitis, and nine showed positive intraneuronal immunohistochemical staining using the convalescent-phase serum from affected sows in the field and specific antiserum to MenV produced in rabbits. In the field outbreak, the virus was highly contagious, spreading from pen to pen and shed to shed. More than 90% of a sample of 88 pigs tested had seroconverted between May and September 1997 (26). The pattern of fetal deaths and abnormalities suggested that the virus could cross the placenta at various times. As there was a large colony of fruit bats near the outbreak herd, and as there was a greater awareness of the dangers of fruit bats after the experiences with HeV and ABL, stored sera from fruit bats were tested for neutralizing antibody for MenV, with positive results in two species (*Pteropus alecto* and *Pteropus conspicillatus*) (26; P. W. Selleck, personal communication). The experience of detecting two other viral diseases in bats provided an incentive to check the bats. Although immunogold-positive viral nucleocapsids have been seen in bat feces (A. W. Philbey, P. D. Kirkland, A. D. Ross, R. J. Love, R. J. Davis, A. B. Gleeson, and M. Srivastava, XI Int. Cong. Virol., 1999, abstr. VW31A), attempts to isolate the virus had been unsuccessful. More recently, a paramyxovirus reacting to anti-MenV antiserum has been isolated from the urine of fruit bats (*Pteropus hypomelanus*) in Malaysia (K. B. Chua, personal communication).

NIPAH VIRUS

The fourth of the sequence of viral discoveries with a potential bat host was made early in 1999. Many pig farmers were sickening and dying with encephalitis in Malaysia. An initial diagnosis of Japanese encephalitis was changed when a paramyxovirus was isolated from the brain of a person and quickly identified as a virus resembling HeV. It was subsequently confirmed in the farmers' pigs and also in dogs, cats, and horses. It was named Nipah virus (NV) after the village of origin of the person tested on that occasion.

The virus was first isolated by Chua et al. from the cerebrospinal fluid of two of three pig farmers when it caused syncytium formation in Vero cells after 5 days (5). It was then tested with anti-HeV antibodies by indirect immunofluorescence, with positive results. The cerebrospinal fluid from all three farmers reacted in an immunoglobulin M capture enzyme-linked immunosorbent assay for anti-HeV antibodies. The virus was typical of paramyxoviruses, pleomorphic and ranging in size from 160 to 300 nm (K. B. Chua, S. K. Lam, D. J. Gubler, and T. G. Ksiazek, XI Int. Cong. Virol., 1999, abstr. VW31B.01). HeV and NV cross neutralize but are sufficiently different that each virus requires approximately four times the concentration of heterologous antiserum for neutralization (Selleck, personal communication). They are also 21 to 25% different at the molecular level in the nucleoprotein and phosphoprotein genes (P. Rota, B. Harcourt, P. Rollin, W. Bellini, K. B. Chua, C. E. Goldsmith, J. J. Olson, M. Bunning, T. Ksiazek, and A. Tamin, XI Int. Cong. Virol., 1999, abstr. VW31B.04). The two viruses are considered sufficiently similar to one another and different from other viruses to form a separate genus. The close relationship meant that immunological tests developed for HeV could be used for NV immediately and prior to the preparation of anti-NV antibodies. At the time, this was important from a diagnostic point of view in the recognition of the disease in humans and domestic animals and from an epidemiological point of view in the search for a reservoir host. Subsequently, specific anti-NV antisera have been more useful because of their greater sensitivity.

Pathogenesis and Epidemiology

Humans with NV infection had fevers, headaches, and altered levels of consciousness (Chua et al., abstr. VW31B.01). Other signs were consistent with encephalitis and occasionally with atypical pneumonia (24). The histopathological features were those of systemic vasculitis with rarefactive necrosis, and although NV antigen was detected in neurons, there was minimal inflammatory reaction (P. T. Hooper, unpublished data; N.

Karim, Proc. Semin. Nipah Virus Natl. Cong. Anim. Health Prod., 1999; S. R. Zaki, XI Int. Cong. Virol., 1999, abstr. VW31B.05). Endothelial syncytia similar to those seen with HeV infections in animals were seen in many organs as well as the brain (Karim, abstr.) The predominant clinical syndrome in pigs was respiratory disease, marked by a severe cough (S. Shahirudin, M. Zamri-Saad, S. S Shamshad, H. Mahani, B. Norazian, P. Daniels, and A. J. Aziz, Proc. Semin. Nipah Virus Natl. Cong. Anim. Health Prod.). Nervous signs were seen in a small proportion of animals, resulting in death. The major pathological changes were bronchopneumonia, with syncytium formation in respiratory alveoli in some pigs, and some animals showed meningitis (S. Shahirudin and P. Hooper, unpublished data). Although there was evidence of some generalized vascular involvement in pigs, as occurred with humans, localization of NV infection in pigs was different from that in humans, in the respiratory epithelium rather than in neurons. The disease was confirmed by immunohistochemistry in two dogs in which generalized vascular disease resulting in severe pulmonary edema was found; in a cat with generalized vasculitis and meningitis; and in the brain of one horse with meningitis (Hooper, unpublished data).

NV isolated from a human was used in transmission trials. Two groups of three pigs, 5 to 6 weeks old, were challenged with 50,000 $TCID_{50}$, one group orally and the other by the subcutaneous route (D. Middleton, H. Westbury, C. Morrissy, A. Hyatt, B. King, G. Russell, M. Braun, J. Muscialli, and C. Carlson, XI Int. Cong. Virol., 1999, abstr. VW31B.06). Those challenged by the oral route showed no disease, while those challenged by the subcutaneous route developed neurological or respiratory signs. At the same time, two adult female cats challenged with the same dose oronasally developed severe respiratory disease. The successful transmissions were shown by histopathological, immunohistochemical, and electron microscopic examinations as being consistent with the field disease in Malaysia.

The occurrence of a respiratory syndrome in pigs marked by coughing together with evidence of production of virus in the lungs and the upper airways suggested that the respiratory route might be important epidemiologically in the transmission of NV. In humans, the disease was largely confined to pig farmers and to some abattoir workers, possibly from respiratory excretions from the pigs. There is also anecdotal evidence of a tendency for human infections in people assisting in farrowings, analogous with placental HeV infections. There was no evidence of person-to-person transmission in humans. Although most human infections occurred in 1998 and 1999, there is immunohistochemical evidence of infection in pigs extending back to 1995 (Hooper, unpublished data).

The answer to the question of the origin of NV has also been made simpler by the previous experience with HeV. The approach was to survey a range of animal species, with pteropid fruit bats as the highest priority (M. Y. Johara, H. Field, A. R. Sohayati, J. Maria, M. R. Azmin, and C. Morrissy, Proc. Semin. Nipah Virus Natl. Cong. Anim. Health Prod., 1999). Serum neutralizing antibodies were found in a significant proportion of pteropid bats sampled *(Pteropus vampyrus* and *Pteropus hypomelanus)* and in occasional individuals of two smaller species of fruit bat *(Eonycteris spelaea* and *Cynopterus brachuotis)*. A range of insectivorous bats and other wildlife were negative (Johara et al., abstr.). At this stage it is a question of speculation how the virus is transmitted within species and how the virus crossed the species barrier to pigs. Experiments with HeV indicate subclinical infection and placental transfer to the fetus, which is consistent with widespread field sampling of high levels of antibody and of placental isolation of the virus (37). A similar mechanism could occur with NV.

FRUIT BATS

While various genera of fruit bats occur around the world, the four viruses have only been associated with fruit bats that are distributed in a band from Madagascar in the western Indian Ocean, to south and southeast Asia, thence through Papua New Guinea to Australia (22). They are mostly *Pteropus* spp. but also *Dobsonia* spp. and the two small species mentioned above. Although migratory to a degree, individual species do not range over the entire area. This is important from the point of view of distribution of the viruses. The pteropid species in Malaysia are different from those in Australia, and there is evidence that the Malaysian species of *Pteropus* harbor NV and not HeV, while those in Australia harbor HeV and not NV. At this stage, the existence of the other viruses in the various species of fruit bat found outside Australia can only be a matter of speculation. It is difficult to believe that Australia would have ABL as a sole member of the lyssavirus group in isolation from the rest of the world without having either it or similar virus in bats extending across southeast Asia. A virus reacting to MenV antiserum has already been found in a fruit bat in Malaysia. Although fruit bats are migratory, their movements are sufficiently limited that while each geographic area may have similar viruses, the geographic separation has been long enough for substantial differentiation not only of the bats themselves but also of the viruses they contain.

ARE THESE FOUR IMPORTANT EMERGING VIRAL DISEASES?

Within 5 years, we have seen four newly identified viruses with the potential to be emerging disease agents, discovery of which seems to form a serendipitous sequence. How did the sequence arise? One answer was the technical ability and incentive to search for them. In the past, there may have been outbreaks of HeV infection in horses that did not spread because of the virus's low level of contagiousness in that species. These small outbreaks may not have justified massive laboratory examination. In the original outbreak reported, the trainer was known for the intimate care he took with his animals that could have inadvertently resulted in the spread of the virus to the large number of horses and to himself. He was known to put his hand deep into the mouths of horses as a routine when illness was suspected (Douglas, personal communication). The large numbers of horses and the coincidental human infection triggered an intensive investigation that was technically successful. Without the trainer's extraordinary and unlucky care, the virus probably would never have been discovered. Later, the attention directed at fruit bats laid the foundation for the work on the other three viruses. The discovery of ABL would probably not have been made without the prior discovery of HeV and the convenient technology available to make the original test for rabies. The immediate association of MenV with the fruit bats resulted from the heightened awareness of the danger of fruit bats after the two previous viral discoveries and the observation of the animals in the trees nearby. Finally, the prior discovery of HeV assisted with the recognition of an "HeV-like" virus, the paramyxovirus NV, in the brain of a human in Malaysia, especially as there had been a previous case of encephalitis in a human caused by HeV.

A second answer for the serendipitous sequence could basically lie in the environmental circumstances that could result in viral emergence from the fruit bats and therefore identification in new hosts. There is evidence of greater contact between humans and their domestic animals and the bats as we encroach on the rainforest. Biologists are noting increased urbanization of flying foxes. As forests are cleared, human home and commercial gardens become more attractive. The original site of the HeV incursion in Australia to a horse is now subject to commercial urban subdivision. Flying foxes, harboring MenV in Australia and NV in Malaysia, have been seen above intensive piggeries.

Two elements that could be significant in the emergence of a virus are its pathogenicity to the host, particularly if the host is a human, and its capacity to establish itself in the new host(s). The first of these viruses,

HeV, initially seemed to fulfil both elements during the disease outbreak first described at Hendra. However, subsequent experience with the field outbreaks and with transmission experiments indicate that although it was very lethal, it was not contagious in the new species and so did not become established in the species examined. Its presence in the fruit bats may pose a continued threat in that it will probably cause occasional small outbreaks in the future, but these should be limited. People handling horses with acute illnesses characterized by severe pneumonia, with HeV as a possible cause, should take precautions to avoid personal contamination. The situation with ABL is similar in that while it too is lethal, it is unlikely to establish itself in new hosts. It is widespread in various species of both fruit and insectivorous bats and has shown some potential emergence by causing fatal disease in two humans. However, it is likely to follow the pattern of nonhematophagous bats in North America, with transmission being primarily intraspecific only (2). We do not know if there were more cases apart from these prior to its discovery, because it seemed there were few attempts to diagnose rabies virus or lyssaviral infections in cases of clinical encephalitis in humans. Since then, there has been a wide public information program, and many people in contact with bats are protected by rabies vaccination or have received postexposure treatments. Experience with rabies and rabies-like viruses is that each virus is likely to remain in a sylvatic cycle within its favored host and not become established in a terrestrial species living close to humans, as with ABL. Reasonable care can then limit the number of new ABL infections in humans. MenV, once it was introduced into pigs, affected large numbers of sows and their offspring as well as two humans. It therefore seemed highly contagious in the pig and seems to have proven its ability to establish itself. Fortunately, the piggery was relatively isolated from others, and so the disease remained confined. Similar outbreaks could occur in the future, providing there is the initial viral transfer from fruit bats to pigs. However, it would be reasonable to believe that such outbreaks will be rare, as there has only been the one confirmed occurrence, and because of the awareness of the potential danger, there should be efforts to minimize contact of fruit bats with large commercial piggeries. The last of the four viruses, NV, nearly satisfied both elements by being lethal to humans and by becoming established, at least in pigs. As of July 1999, 265 humans had been affected, 105 fatally. Enormous numbers of pigs had contracted the disease, and the disease was so widespread that for public health protection, half of Malaysia's commercial pig population was destroyed. Fortunately, the human acted as an end host. After a phase of blood circulation, the virus localized in the brain, an inefficient method of spread. In pigs, there was localization in the epithelial cells of the airways and thence excretion, a violently efficient form of contagiousness. It is most fortunate that the tropism of the virus and so the disease's epidemiology in the human did not follow the same pattern, and so the second element of establishment was in pigs only. If it had been in humans, it does not take much imagination to believe that an even worse disaster could have occurred, well and truly consistent with the hypothetical prophecy of C. J. Peters.

SUMMARY AND CONCLUSIONS

Four new viruses have been isolated that have fruit bats in southeast Asia and Australia as natural hosts. These were HeV, ABLV, MenV, and NV. The first, HeV, was probably only isolated because it was from an outbreak of disease with numbers of animals and humans far greater than what would be expected from the virus's level of contagiousness. The remaining viruses were rapidly identified or their association with fruit bats was recognized because of the experience with HeV.

The potential for emergence of each virus was discussed. HeV, while being highly virulent in humans and horses, should not pose a major threat because it is not contagious in those species. ABL should be like rabies and rabies-like viruses, mostly confined to its own preferred host species. MenV has the potential to be contagious in pigs and so establish itself in that species, but it seems likely that its transfer across the species barrier will be a rare event. NV was the closest to being fully emergent. It became established as a contagious disease in pigs, causing disease in many in-contact humans. If its disease in the human were like that in the pigs, it would have been an even worse disaster, possibly worldwide.

Acknowledgments. There were many groups involved in the various investigations, coworkers at the CSIRO Australian Animal Health Laboratory, laboratory and field scientists in the Queensland State Department of Primary Industries and Queensland University (especially HeV and ABL), laboratory and field scientists in the NSW Agriculture Department and at Sydney University (especially MenV), the CDC and Malaysian scientists and doctors (NV).

REFERENCES

1. **Allworth, A., K. Murray, and J. Morgan.** 1996. A case of encephalitis due to a lyssavirus recently identified in fruit bats. *Commun. Dis. Bull.* **20:**504–505.
2. **Baer, G. M., and J. S. Smith.** 1991. Rabies in nonhematophagous bats, p. 341–366. *In* G. M. Baer (ed.), *The Natural History of Rabies.* CRC Press, Boca Raton, Fla.
3. **Bouhry, H., B. Kissi, and N. Tordo.** 1993. Molecular diversity of the *Lyssavirus* genus. *Virology* **194:**70–81.

4. **Chant, K., R. Chan, M. Smith, D. E. Dwyer, and P. Kirkland.** 1998. Probable human infection with a newly described virus in the family Paramyxoviridae. *Emerg. Infect. Dis.* **4:**273–275.

5. **Chua, K. B., K. J. Goh, K. T. Wong, A. Kamarulzaman, P. S. Tan, T. G. Ksiazek, S. R. Zaki, G. Paul, S. K. Lam, and C. T. Tan.** 1999. Fatal encephalitis due to Nipah virus among pig-farmers in Malaysia. *Lancet* **354:**1257–1259.

6. **Field, H. E., P. C. Barratt, R. J. Hughes, J. Shield, and N. D. Sullivan.** 2000. A fatal case of Hendra virus (equine morbillivirus) infection in a horse in North Queensland—an incident report with an epidemiological perspective. *Aust. Vet. J.* **78:**279–280.

7. **Field, H., B. McCall, and J. Barrett.** 1999. Australian bat lyssavirus infection in a captive juvenile black flying fox. *Emerg. Infect. Dis.* **5:**438–440.

8. **Fraser, G. C., P. T. Hooper, R. A. Lunt, A. R. Gould, L. J. Gleeson, A. D. Hyatt, G. M. Russell, and J. A. Kattenbelt.** 1996. Encephalitis caused by a lyssavirus in fruit bats in Australia. *Emerg. Infect. Dis.* **2:**327–331.

9. **Gould, A. R., A. D. Hyatt, R. Lunt, J. A. Kattenbelt, S. Hengstberger, and S. D. Blacksell.** 1998. Characterisation of a novel lyssavirus isolated from *Pteropid* bats in Australia. *Virus Res.* **54:**165–187.

10. **Halpin, K., P. L. Young, H. Field, and J. S. Mackenzie.** 1999. Newly discovered viruses of flying foxes. *Vet. Microbiol.* **68:**83–87.

11. **Hooper, P. T., G. C. Fraser, R. A. Foster, and G. J. Storie.** 1999. Histopathology and immunohistochemistry of bats infected by Australian bat lyssavirus. *Aust. Vet. J.* **77:**595–599.

12. **Hooper, P. T., A. R. Gould, A. D. Hyatt, M. A. Braun, J. A. Kattenbelt, S. G. Hengstberger, and H. A. Westbury.** 2000. The laboratory diagnosis and molecular characterisation of a new case of Hendra virus infection in a horse in Queensland. *Aust. Vet. J.* **78:**281–282.

13. **Hooper, P. T., A. R. Gould, G. M. Russell, J. A. Kattenbelt, and G. Mitchell.** 1996. The retrospective diagnosis of a second outbreak of equine morbillivirus infection. *Aust. Vet. J.* **74:**244–245.

14. **Hooper, P. T., P. J. Ketterer, A. D. Hyatt, and G. M. Russell.** 1997. Lesions of experimental equine morbillivirus pneumonia in horses. *Vet. Pathol.* **34:**312–322.

15. **Hooper, P. T., R. A. Lunt, A. R. Gould, H. Samaratunga, A. D. Hyatt, L. J. Gleeson, B. J. Rodwell, C. E. Rupprecht, J. S. Smith, and P. K. Murray.** 1997. A new lyssavirus—the first endemic rabies-related virus recognized in Australia. *Bull. Inst. Pasteur* **95:**209–218.

16. **Hyatt, A. D., and P. W. Selleck.** 1996. Ultrastructure of equine morbillivirus. *Virus Res.* **43:**1–15.

17. **Mackenzie, J. S.** 1999. Emerging viral diseases: an Australian perspective. *Emerg. Infect. Dis.* **5:**1–8.

18. **McCormack, J. G., A. M. Allworth, L. A. Selvey, and P. W. Selleck.** 1999. Transmissibility from horses to humans of a novel paramyxovirus, equine morbillivirus (EMV). *J. Infect. Dis.* **38:**22–23.

19. **Murphy, F. A., C. M. Fauquet, D. H. L. Bishop, S. A. Ghabrial, A. W. Jarvis, G. P. Martelli, M. A. Mayo, and M. D. Summers.** 1995. Classification and nomenclature of viruses, p. 268–274. *In Sixth Report of the International Committee on Taxonomy of Viruses.* Springer-Verlag, New York, N.Y.

20. **Murray, K., B. Eaton, P. Hooper, L. Wang, M. Williamson, and P. Young.** 1998. Flying foxes, horses, and humans: a zoonosis caused by a new member of the *Paramyxoviridae*, p. 43–58. *In* W. M. Scheld, D. Armstrong, and J. M. Hughes (ed.), *Emerging Infections 1*, ASM Press, Washington, D.C.

21. **Murray, K., P. Selleck, P. Hooper, A. Hyatt, A. Gould, L. Gleeson, H. Westbury, L. Hiley, L. Selvey, B. Rodwell, and P. Ketterer.** 1995. A morbillivirus that caused fatal disease in horses and humans. *Science* **268:**94–98.

22. **Nelson, J. E.** 1989. Pteropididae, p. 836–844. *In* D. W. Walton and B. J. Richardson (ed.), *Fauna of Australia,* vol. 1B. *Mammalia.* Australian Government Publishing Service, Canberra, Australia.
23. **O'Sullivan, J. D., A. M. Allworth, D. L. Paterson, T. M. Snow, R. Boots, L. J. Gleeson, A. R. Gould, A. D. Hyatt, and J. Bradfield.** 1997. Fatal encephalitis due to novel paramyxovirus transmitted from horses. *Lancet* **349:**93–95.
24. **Paton, N. I., Y. S. Leo, S. R. Zaki, A. P. Auchus, K. E., A. E. Ling, S. K. Chew, B. Ang, P. E. Rollin, T. Umapathi, I. Sng, C. C. Lee, E. Lim and T. G. Ksiazek.** 1999. Outbreak of Nipah-virus infection among abattoir workers in Singapore. *Lancet* **354:**1253–1256.
25. **Peters, C. J., and M. Olshaker.** 1997. Looking ahead, p. 302–323. *In* C. J. Peters and M. Olshaker (ed.), *Virus Hunter—Thirty Years of Battling Hot Viruses around the World.* Anchor Books, Garden City, N.Y.
26. **Philbey, A. W., P. D. Kirkland, A. D. Ross, R. J. Davis, A. B. Gleeson, R. J. Love, P. W. Daniels, A. R. Gould, and A. D. Hyatt.** 1998. An apparently new virus (family Paramyxoviridae) infectious for pigs, humans, and fruit bats. *Emerg. Infect. Dis.* **4:**269–271.
27. **Rogers, R. L., I. C. Douglas, F. C. Baldock, R. J. Glanville, K. T. Seppanen, L. J. Gleeson, P. W. Selleck, and K. J. Dunn.** 1996. Investigation of a second focus of equine morbillivirus infection in coastal Queensland. *Aust. Vet. J.* **74:**243–244.
28. **Samaratunga, H., J. W. Searle, and N. Hudson.** 1998. Non-rabies lyssavirus human encephalitis from fruit bats: Australian bat lyssavirus (pteropid lyssavirus) infection. *Neuropathol. Appl. Neurobiol.* **24:**331–335.
29. **Selvey, L. A., R. M. Wells, J. G. McCormack, A. J. Ansford, K. Murray, R. J. Rogers, P. S. Lavercombe, P. Selleck, and J. W. Sheridan.** 1995. Infections of humans and horses by a newly described morbillivirus. *Med. J. Aust.* **162:**642.
30. **Skerrat, L. F., R. Speare, L. Berger, and H. Winsor.** 1998. Lyssaviral infection and lead poisoning in black flying foxes from Queensland. *J. Wildl. Dis.* **34:**355–361.
31. **Skull, S. A., V. Krause, C. B. Dalton, and L. A. Roberts.** 1999. A retrospective search for lyssavirus in humans in the Northern Territory. *Aust. N. Z. J. Public Health* **23:**305–308.
32. **Stephan, H. A., G. M. Gay, and T. C. Ramirez.** 1988. Encephalomyelitis, reproductive failure and corneal opacity (blue eye) in pigs, associated with a paramyxovirus infection. *Vet. Rec.* **122:**6–10. (Erratum **122:**420.)
33. **Wang, L. F., W. P. Michalski, M. Yu, L. I. Pritchard, G. Crameri, B. Shiell, and B. T. Eaton.** 1998. A novel P/V/C gene in a new *Paramyxoviridae* virus which causes lethal infection in humans, horses, and other animals. *J. Virol.* **72:**1482–1490.
34. **Ward, M. P., P. F. Black, A. J. Childs, F. C. Baldock, W. R. Webster, B. J. Rodwell, and S. L. Brouwer.** 1996. Negative findings from serological studies of equine morbillivirus in the Queensland horse population. *Aust. Vet. J.* **74:**241–243.
35. **Westbury, H. A., P. T. Hooper, P. W. Selleck, and P. K. Murray.** 1995. Equine morbillivirus pneumonia: susceptibility of laboratory animals to the virus. *Aust. Vet. J.* **72:**278–279.
36. **Williamson, M. M., P. T. Hooper, P. W. Selleck, L. J. Gleeson, P. W. Daniels, H. A. Westbury, and P. K. Murray.** 1998. Transmission studies of Hendra virus (equine morbillivirus) in fruit bats, horses and cats. *Aust. Vet. J.* **76:**813–818.
37. **Williamson, M., P. Hooper, P. Selleck, H. Westbury, and R. Slocombe.** Experimental Hendra virus infection in pregnant fruit bats and guinea pigs. *J. Comp. Pathol.,* in press.
38. **Young, P. L., K. Halpin, P. W. Selleck, H. Field, J. L. Gravel, M. A. Kelly, and J. S. Mackenzie.** 1996. Serological evidence for the presence in *Pteropus* bats of a paramyxovirus related to equine morbillivirus. *Emerg. Infect. Dis.* **2:**239–240.
39. **Yu, M., E. Hansson, B. Shiell, W. Michalski, B. T. Eaton, and L. F. Wang.** 1998. Sequence analysis of the Hendra virus nucleoprotein gene: comparison with other members of the subfamily *Paramyxoviridae. J. Gen. Virol.* **79:**1775–1780.

Emerging Diseases of Animals
Edited by C. Brown and C. Bolin

Chapter 6

Understanding the Ecology and Epidemiology of Avian Influenza Viruses: Implications for Zoonotic Potential

David E. Swayne

Influenza originally referred to epidemics of acute, rapidly spreading catarrhal fevers in humans which have subsequently been shown to be caused by an orthomyxovirus (39). Today, orthomyxoviruses are known to cause significant numbers of natural infections and disease in humans (*Homo sapiens*), horses (*Equus caballus*), pigs (*Sus scrofa*), and various bird species and sporadic cases of naturally occurring disease in mink (*Mustela vision*) and various marine mammals (22, 50, 89). However, there is additional evidence to suggest the potential for other mammals to be infected based upon experimental studies and serologic surveys of domestic, laboratory, and wild mammalian populations (10, 62). However, rodents, carnivores, and ungulate ruminants do not appear to be natural reservoirs, nor are they involved as biological vectors of influenza virus transmission between animal species experiencing endemic influenza.

In general, influenza viruses replicate in epithelial cells of the upper respiratory tract or gastrointestinal tract, and the outcomes of such infections include no clinical signs, upper respiratory disease, pneumonia, and occasionally death. In birds, most infections are subclinical. However, in poultry, influenza infections have caused clinical respiratory disease or drops in egg production, and a few specific avian influenza virus strains

David E. Swayne • Southeast Poultry Research Laboratory, Agricultural Research Service, U.S. Department of Agriculture, Athens, GA 30605.

have caused severe disseminated infections with systemic disease and high death losses (19).

INFLUENZA VIRUS

Classification

Influenza viruses are negative-sense, single-stranded segmented RNA viruses in the family *Orthomyxoviridae* (40). There are four recognized genera: *Influenzavirus A* and *B,* with eight gene segments which code for 10 proteins; *Influenzavirus C,* with seven gene segments; and *Thogotovirus,* with six gene segments (40, 48). Serologically, the influenza viruses are typed as A, B, and C based on immunoprecipitating serological reaction to two internal viral proteins, the nucleoprotein and matrix protein (4). Type A influenza viruses are further classified into subtypes based on serologic reaction of the surface hemagglutinin (H) and neuraminidase (N) glycoproteins. There are 15 different H subtypes (H1 to H15) and nine different N subtypes (N1 to N9) (88). However, the distribution of H and N subtypes is not uniform among the major host species (Table 1). All 15 H and nine N subtypes have been reported in avian species and in most of the possible combinations (3).

Mechanisms for Genetic Variation

Influenza viruses have the propensity to change genetically, and this ability to change contributes to the interspecies transmission and zoonotic potential of influenza viruses. The concept of genetic change for influenza viruses was first reported in reference to the ability of human influenza viruses to change antigenically over time (39, 54), but the same processes occur with all influenza viruses. First, influenza viruses accumulate random RNA mutations that can translate into amino acid changes. These changes, termed antigenic drift, occur over long periods of time and are the most common type of antigenic change. All influenza virus proteins are subject to antigenic drift, but such changes accumulate most rapidly in the surface glycoproteins, especially the hemagglutinin. Second, on rare occasions, two different influenza viruses can infect the same cell, resulting in progeny with various combinations of the eight gene segments contributed by the two parent strains. Such hybrid influenza viruses or reassortants have resulted in the emergence of "new" dominant strains of influenza virus with a different surface hemagglutinin than that circulating in the endemic host-adapted strains. This type of antigenic change is

Table 1. Hemagglutinin subtype distribution among different birds (class Aves) and mammals (class Mammalia)[a]

Hemagglutinin subtype	Occurrence[b]					
	Mammalia			Aves		
	Humans	Swine	Equines	Wild dabbling ducks	Shorebirds	Domestic poultry (Galliformes)
H1	++	++		+	+	++
H2	+			+	+	+
H3	++	+	++	++		+
H4				++	+	+
H5	±			+	+	++
H6				++	+	+
H7	±		+	+	+	++
H8				+		+
H9	±	±		+	++	++
H10				+	+	+
H11				+	+	+
H12				+	+	
H13					++	+
H14				+		
H15				+	+	

[a] Adapted from reference 76.
[b] ±, rare; +, frequent; ++, common.

abrupt and is termed antigenic shift. Antigenic shift occurs with less frequency than antigenic drift.

Concepts of Virulence and Transmissibility

Confusion can arise when discussing two distinct issues in influenza virus biology, virulence and transmissibility (38). Virulence, or pathogenicity, refers to the disease-producing ability of the virus (12). This can be quantified as the ability to produce clinical signs or cause death. However, the virulence or pathotype classification of an influenza virus strain is only predictable for a single defined host. For example, some H5 and H7 avian influenza viruses that are highly pathogenic (i.e., highly virulent) for chickens (*Gallus gallus domesticus*) produce no infections or produce infections without disease in domestic ducks (*Anas platyrhynchos*), i.e., avirulent. Likewise, whether an avian influenza virus is mildly or highly pathogenic for chickens is not predictive of its potential to cause disease in humans or other mammals.

The other concept, transmissibility, implies that the virus is infectious and can be transferred to other animals of the same species under natural conditions. Although specific sequence changes in the hemagglutinin protein have caused the mildly pathogenic H5 and H7 influenza viruses to become highly pathogenic, this does not imply any increased transmissibility compared with the original mildly pathogenic parent avian influenza virus strain (1). For example, in the 1994–95 outbreak of H5N2 avian influenza virus in chickens in Mexico, no increase in virus transmissibility was reported when the mildly pathogenic parent strain mutated to become a highly pathogenic strain in Puebla during November 1994. By contrast, in some instances, a change in an avian influenza virus from mildly to highly pathogenic may decrease its transmission between chickens because birds may die before a significant quantity of the virus can be shed from respiratory secretions or digestive excretions.

In general, influenza viruses are adapted to individual host species, with intraspecies transmission occurring most frequently. Interspecies transmission does occur but most easily and frequently between two closely related host species, such as in chicken-to-turkey (*Meleagris gallopavo*) or chicken-to-northern bobwhite quail (*Colinus virginianus*) transmission. Interspecies transmission can occur across different orders within the same class, such as with free-flying duck (order Anseriformes)-to-turkey (order Galliformes) or pig-to-human transmission, but this is less frequent than occurs with closely related host species. Furthermore, interspecies transmission between different hosts within separate classes is even less frequent, as has occurred rarely with chicken-to-human or free-flying duck-to-pig transmission. One exception to the above rule has been the ease and frequency of transfer of swine H1N1 viruses to turkey breeder hens (52), but these are sporadic and isolated occurrences. Obviously, many factors such as geographic restriction, intermixing of species, age and density of birds, weather, and temperature also impact the ability of the avian influenza virus to move within and between host species and affect the overall incidence of infections.

PATHOGENESIS OF INFECTION AND DISEASE

Disease in Birds (Class Aves)

Influenza viruses can infect a variety of bird species (1, 76). Natural infections in most wild birds result in no clinical signs (1, 76). However, in experimental studies, infections in mallard ducks have resulted in mild depression in egg production for 1 week and transient decreases in T-lymphocyte function (45–47). One exception, an outbreak of influenza, resulted in significant mortality for common terns (*Sterna hirundo*) in South Africa during 1961 (5).

In poultry species (order Galliformes), infections by influenza viruses have been reported. For all influenza viruses of subtypes H1 to H4, H6, and H8 to H15 and most viruses of subtypes H5 and H7, virus replication and lesions have been limited to epithelial cells in the respiratory, intestinal, and renal systems (81–83). Mortality was usually low unless the illness was exacerbated by coinfection with bacteria or other viruses (55). Primary viral pneumonia or secondary bacterial pneumonia, tracheitis, sinusitis, and rhinitis are the most common lesions associated with influenza virus infections in poultry and suggest a predominant respiratory tract pathogenesis. In experimental studies with chickens, such viruses usually caused less than 75% mortality and have been termed mildly pathogenic, pathogenic, "apathogenic," or "not pathogenic" (67, 85).

By contrast, some H5 and H7 influenza virus strains in poultry replicated throughout many tissues and cell types, with widespread necrosis and severe inflammation (51). Mortality approached 100%. In pathotyping tests conducted in chickens, 75% or greater of the chickens died, and such influenza viruses were termed highly pathogenic (67, 85). In the field, such avian influenza viruses produced high mortality and severe disease and have historically been called fowl plague or fowl pest (78). By contrast, experimental infection of birds in the order Anseriformes, such as ducks, with avian influenza viruses highly pathogenic for chickens resulted in either no infection or limited local replication in the respiratory or alimentary system without disease (2, 94).

The virulence of influenza viruses in chickens is multigenic, but the hemagglutinin of the avian influenza virus is a major determinant of virulence or virulence potential (9). The ability of influenza viruses to spread and produce infectious progeny and lesions is related to the cleavage of the hemagglutinin in various cell types (1, 9). In chickens and turkeys, the presence of multiple basic amino acids and/or loss of carbohydrate chains at or near the proteolytic cleavage site of the hemagglutinin of highly pathogenic avian influenza viruses has been associated with the ability of ubiquitous proteases contained in cells throughout the body to cleave the viral hemagglutinin and support multiple cycles of avian influenza virus replication. By contrast, mildly pathogenic avian influenza viruses are cleaved only by trypsin-like enzymes which are restricted to specific cell types and a few locations such as epithelial cells within the respiratory, intestinal, and renal systems. Thus, the virulence of an individual influenza virus is dependent upon the virus and the host and can be accentuated by multiple environmental factors.

Disease in Mammals (Class Mammalia)

In mammals, influenza viruses primarily replicate in the respiratory tract and cause associated clinical signs and lesions (62). Virus replication occurs predominantly in the epithelium of the nasal cavity and results in clinical rhinitis, the most frequent clinical presentation in human patients, horses, and pigs. In addition, the virus may replicate in the tracheal and bronchial epithelium, resulting in tracheobronchitis. In some individuals, especially those most severely affected, the virus may replicate in the alveoli, causing histiocytic interstitial pneumonitis with alveolar edema. Widespread virus replication in multiple organs and associated systemic disease, as occurs with highly pathogenic avian influenza viruses in chickens and turkeys, is not a common pathogenic mechanism of virulent influenza viruses in mammals.

ECOLOGY AND EPIDEMIOLOGY OF INFLUENZA VIRUS INFECTIONS

Because influenza viruses are host adapted and maintained in multiple ecosystems, separate clades or lineages exist for influenza viruses circulating in different host species (Fig. 1) (60). Avian influenza viruses occupy two major lineages, Eurasian and North American, because of geographic limitations on the migration of birds; i.e., migrations are primarily north to south, with minimal crossover east to west between the Old and New Worlds. Individual influenza virus strains show various degrees of genetic relatedness to each other, and some influenza viruses isolated from different host species are clustered in the same clades, indicating a previous event with interspecies transmission. Interspecies exchange of influenza viruses and/or genes and influenza viral ecology follow two different principles: (i) specific animal species may be the primordial reservoirs, resulting in the influenza virus gene pools being maintained within a stable ecosystem, and (ii) various animal species (nonprimordial reservoirs) can have endemic infections with influenza viruses and serve as the immediate or transfer sources of influenza viruses and/or genes to other species and individuals.

Primordial Reservoirs of Influenza Virus

The restriction of hemagglutinin subtypes of influenza virus in humans, pigs, and horses to four subtypes (H1 to H3 and H7), and the presence of all 15 hemagglutinin subtypes, and the diversity of the

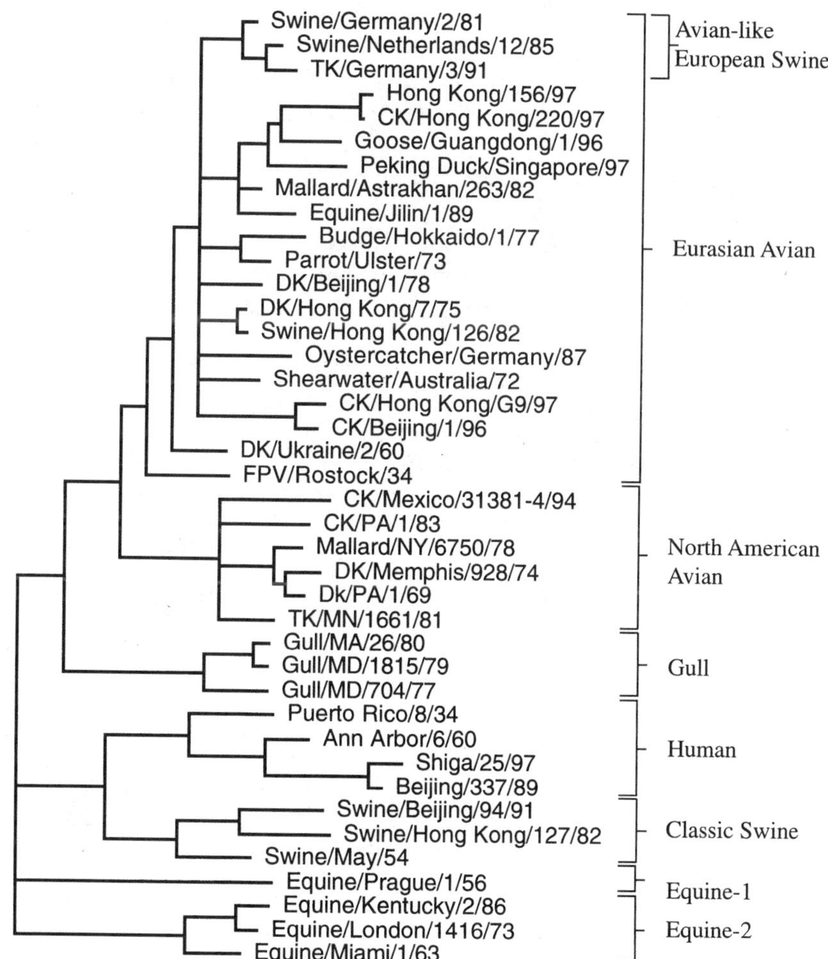

Figure 1. Phylogenetic tree of nucleoprotein genes, showing the major influenza virus lineages in birds and mammals. (Courtesy of David L. Suarez, U.S. Department of Agriculture [USDA].)

genetic information suggest that certain free-flying birds may be the primordial reservoirs for all influenza virus genes (89) (Table 1). Influenza viruses have been isolated from 90 species of free-flying birds representing 12 orders (76, 77), with most isolates obtained from birds occupying aquatic habitats (76). There may be some interspecies exchange of influenza viruses among these groups of birds if significant intermixing of species occurs and influenza virus-naive (susceptible) populations are available for infection by the specific influenza virus subtypes in question (Fig. 2).

There are no data to suggest that all bird species, especially upland game birds (order Galliformes), are primordial reservoirs of avian influenza viruses (60). However, the domestication of chickens and turkeys and the introduction of avian influenza viruses into the animal agricultural systems from the primordial free-flying bird reservoirs make poultry an immediate source and the major agricultural reservoir of influenza viruses. This promotes adaptation of influenza viruses to other gallinaceous bird species and provides a perpetual source of avian influenza viruses for different poultry ecosystems associated with animal agriculture and exhibition systems. Furthermore, poultry can be the source of avian influenza viruses for interspecies transmission to mammals.

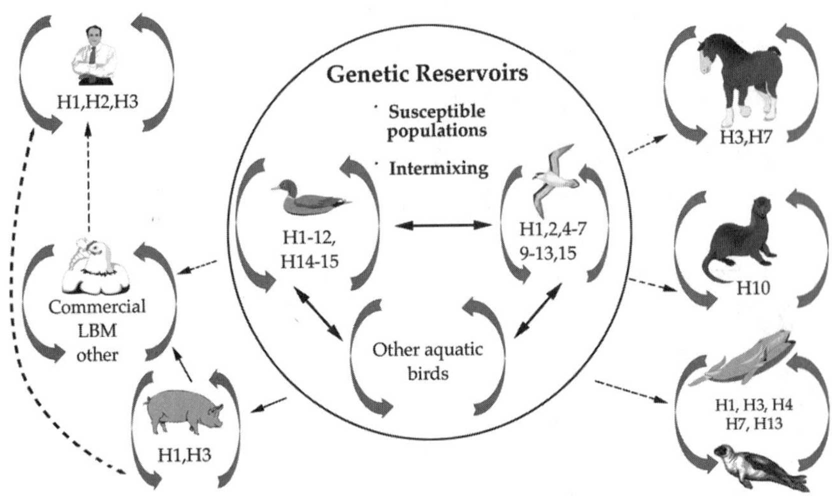

Figure 2. Ecology of influenza viruses in various avian and mammalian hosts. LBM, live-bird markets.

Avian Influenza Ecology—Nonprimordial Reservoirs

Diversity in the class Aves

Avian species should not be thought of as a single species but as a diverse group of animals with different degrees of relatedness. We clearly distinguish between viruses from humans, swine, and horses although they are "mammalian" species in the class Mammalia. Similarly, the class Aves contains 29 orders, 187 families, more than 2,000 genera, and more than 9,600 species based on differences in physiological, anatomical, behavioral, and ecological features (25). Birds should not be viewed as occupying a single ecosystem, and avian influenza viruses do not move freely within or across host species.

New ecological niches for birds have been created by humans, and such changes have resulted in transmission, adaptation, and perpetuation of avian influenza viruses outside the primordial reservoir in other bird species, including poultry. Biologically, species-specific adaptation has been identified for some free-flying waterfowl-origin and poultry-origin avian influenza viruses because of differences in the ability of individual avian influenza viruses to infect and cause disease in chickens, turkeys, quail, and ducks (2, 45, 47, 74, 75, 82, 83, 94). Furthermore, the frequency and ease of interspecies transmission of individual avian influenza viruses may be determined by the closeness of the phylogenetic relationship of the individual host animals. For example, within the order Galliformes, turkeys and chickens are the two most common poultry species raised in animal agriculture. These two species are most frequently infected during outbreaks of highly pathogenic avian influenza and have severe disease, while birds in the next closest related order, Anseriformes (15), can occasionally be infected by these highly pathogenic avian influenza viruses, but they have not been associated with experimental or natural mortality (76, 94). Furthermore, species adaptation has been demonstrated for avian influenza viruses, such that crossing from wild mallard ducks (order Anseriformes) to domestic turkeys (order Galliformes) is a multistep adaptive process (28, 60). The index case in a turkey flock may be detected in a few turkeys at slaughter, as evidenced by seroconversion to type A influenza virus. Over the next several months, the waterfowl-origin virus adapts to the turkey, resulting in increased transmissibility as evidenced by increased numbers of infected flocks and the appearance of clinical signs of respiratory disease, drops in egg production, and mortality (28, 60; D. Halvorson, personal communication). In addition, many influenza virus isolates from shorebirds and gulls (order Charadriiformes) do not replicate in experimentally exposed ducks (order Anseriformes) (30, 76, 89). These data suggest that host adaptation of avian influenza viruses occurs between different bird species and that avian influenza virus is a misnomer with the same inaccuracy as the term mammalian influenza viruses.

Bird domestication and agriculture in avian influenza virus ecology

Humans have intervened in the natural ecosystems of birds through captivity, domestication, industrial agriculture, national and international commerce, and nontraditional raising practices. Each situation has created "new" and unique ecosystems in which organisms, both pathogens and nonpathogens, can adapt and pass within and between bird species. There are six primary types of ecosystems for birds which impact the maintenance of avian influenza virus: (i) wild bird populations, (ii) bird collection and trade systems, (iii) live-bird market systems, (iv) backyard and hobby flocks, (v) range-raised commercial poultry systems, and (vi) integrated indoor commercial poultry systems.

First, the wild-bird populations have been dealt with previously in the primordial reservoirs of free-flying birds, but other bird species that are not free-flying may be exposed to and infected with avian influenza viruses, including Antarctic penguins, which have not been exposed to animal agricultural sources and are not "free-flying" (53). Other groups of birds can be infected and include the orders Passeriformes (perching birds), Columbiformes (doves), and Piciformes (woodpeckers), but these are involved only in local transmission of the virus and not as long-term reservoirs (76).

Second, with the development of rapid transportation and international commerce, domestic poultry species and wild-captive birds come into close if not direct contact with the exchange of pathogens, including avian influenza and Newcastle disease viruses. In these situations, some wild birds may not be reservoirs of avian influenza virus but become infected from the domesticated poultry or free-flying wild birds that are being held in close contact at the collection and trading points. This scenario is repeated around the world, primarily in developing countries. Such collection and trading circles would include wild psittacine and passerine birds caught in Central and South America and Asia and other wild birds caught around the world for use in private exhibition and zoological collections. Importation restrictions, quarantine, and testing procedures prevent entry of avian influenza- or Newcastle disease-infected birds through legitimate import channels, but illegal smuggling of such birds continues to be a source of avian influenza and Newcastle disease virus isolations and introduction into some countries (58, 68, 86).

Third, in some cultures, birds produced for human food may be sold through a live-bird market system in which the consumer can select a live bird, which is killed and dressed to the customer's cultural expectations and culinary standards (69, 84). In such systems, poultry are raised in small groups on scattered farms, collected in central locations by middlemen, and traded or exchanged at one or more distribution centers or auctions. In the final disposition, the birds are distributed through a retail market and sold live to the consumer to take home or killed and dressed on site (Fig. 3). Residual live birds in the retail outlet may move in multiple directions, returning to the distribution center, the middleman, or one or more farms. This system provides minimal control of bird movement and no uniform health inspections or standards. Birds of multiple species are mixed, and movement of birds is bidirectional, with continual introduction of naive birds for pathogen exposure, infection, and transmission. In general, cleanup and disinfection are inconsistently practiced, so any disruption in the infectious organism's transmission cycle is minimal. The

Figure 3. Poultry species from multiple sources are mixed in the live-bird markets, where poultry pathogens such as avian influenza virus can be exchanged and maintained. (Photo courtesy of Martin Smeltzer, USDA.)

use of open ponds for raising ducks raises the possibility of transmission of avian influenza viruses from commingled free-flying ducks to domestic ducks, entry into the live-bird market system, interspecies transmission, adaptation to new bird species, and the creation of endemic avian influenza (Fig. 4). During 1997, avian influenza viruses were demonstrated to have been transmitted from poultry in the live-bird markets to commercially raised meat turkeys and egg-laying chickens in Pennsylvania (96) and humans in Hong Kong (16, 79, 80) (Fig. 4). Many developed nations have instituted avian influenza and Newcastle disease control programs for the live-bird market systems.

Fourth, many individuals own small flocks of birds for food and fiber production, exhibition, and recreation (Fig. 5). In general, if the birds are held to a closed one-way direction of movement, their role in avian influenza virus transmission to other birds and farms is minimal. However, many of these flocks may be housed outdoors and have direct contact with free-flying birds or other poultry. Any infections with avian influenza viruses would dead-end unless these birds were traded or exhibited. If birds are brought from multiple sources to central social gatherings or trading areas, avian influenza viruses and other pathogens can move from farm to farm. Cockfighting provides the means for birds to enter the

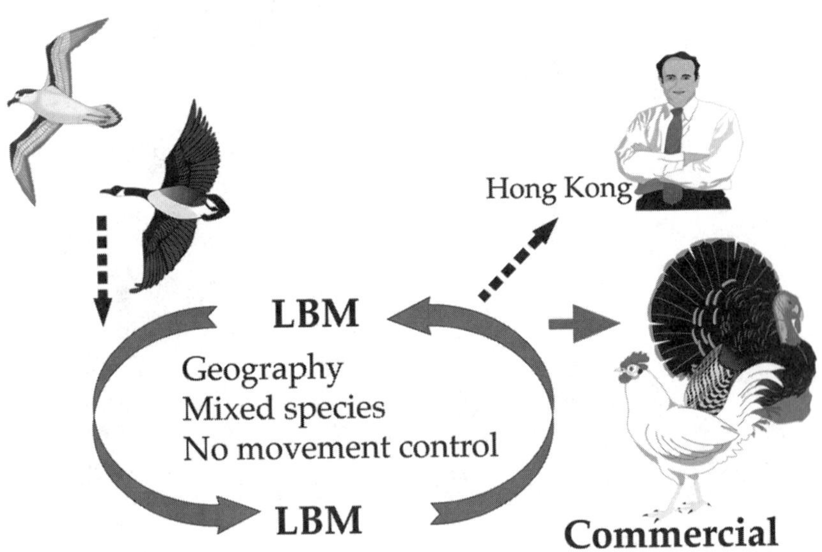

Figure 4. Ecology of avian influenza viruses in the live-bird markets (LBM).

Figure 5. Gathering of poultry during exhibition or recreational events such as illegal cockfights in the United States provides the opportunity to exchange pathogens, including avian influenza virus. (Photo from an anonymous source.)

collection and international trading ecosystem. All-in-all-out is the best method for controlling diseases among cockfighting birds.

Fifth, some geographic areas produce poultry commercially or industrially in outdoor ranges. In the United States, some poultry are range-reared commercially, usually turkeys raised for seasonal markets at Thanksgiving and Christmas (Fig. 6). However, other commercial endeavors have reared poultry in outdoor aquatic environments, such as large groups of waterfowl (ducks and geese) in Asia and some chickens in developed countries where legislative or consumer pressures for animal welfare standards have resulted in poultry's having access to outdoor environments for enrichment purposes. Such scenarios have increased the likelihood of the introduction of avian influenza virus from free-flying birds into domestic poultry. In Minnesota, traditional and molecular epidemiology has demonstrated the direct transfer of mildly pathogenic avian influenza viruses from free-flying birds (primarily mallard ducks) to range turkeys reared outdoors during annual wild-bird migrations (29). However, during 1998, turkeys were not raised on the range in Minnesota, and no cases of avian influenza were reported, strengthening the need

Figure 6. Commercial turkeys raised outdoors for seasonal markets on occasion have had direct contact with migratory waterfowl and have been infected with mildly pathogenic avian influenza viruses. (Photo courtesy of Alex Bermudez, University of Missouri.)

to limit outdoor rearing of domestic poultry to prevent the introduction of avian influenza from the primordial reservoir (86; D. Halvorson, personal communication).

Sixth, poultry raised in controlled indoor environments have the least probability of introduction of avian influenza virus from free-flying birds or other sources (Fig. 7). However, the probability of introduction of avian influenza virus into commercial indoor poultry operations increases significantly if direct or indirect interactions occur with the previous five agriculture systems. Risk factors especially important in the introduction of avian influenza to commercial poultry from the other agriculture systems include contact with chicken shipping crates or other feces-contaminated equipment, contact with contaminated clothing and shoes of poultry workers, use of unchlorinated pond water frequented by migratory ducks, and close geographic proximity to infected farms for airborne transmission. Once avian influenza virus enters this commercial poultry system, it moves with ease throughout the integrated company because of the practice of poor interfarm biosecurity by service personnel and visitors and the use of common equipment and labor forces between farms within the same company.

Figure 7. A commercial broiler house sufficient to raise 20,000 chickens.

Mammalian Influenza Ecology

In mammals, as in avian species, influenza viruses most frequently are passed intraspecies with influenza infections being endemic in humans, pigs, and horses (Fig. 2). On rare or uncommon occasions, interspecies transfer of influenza viruses or virus genes has been reported in mammals, originating from either the primordial avian influenza virus reservoir or other mammal species (89) (Fig. 2). Most frequently, interspecies exchange of these "new" influenza viruses has resulted in limited infections in the new host species because most "new" strains are not sufficiently host adapted to be freely transmissible, and the result has been a small-scale incident or limited number of infections within the new host population. For example, a swine H1N1 influenza virus infected and caused respiratory illness in several people at a state fair in Wisconsin during 1988, including a lethal infection in a pregnant woman (92). Similar isolated incidences have occurred worldwide, and most have gone undetected. By contrast, some avian influenza viruses have infected and developed some adaptation to mink and marine mammals, causing sporadic disease outbreaks with significant mortality (Fig. 2) (22, 24, 31, 32, 44, 63). However, avian influenza has not become established in these species as an endemic infection or disease.

In a few rare instances, the interspecies-transmitted influenza viruses have undergone antigenic drift or shift, producing novel virus lineages which have become established in the new host species. These adapted or reassortant influenza viruses have resulted in epidemics or pandemics of disease and, when sufficient protection has been achieved in the host population, became endemic viruses. For example, several avian-like European swine influenza virus lineages have arisen by interspecies transmission of avian influenza viruses, reassortment with various swine-adapted influenza viruses, and readaptation into domestic swine (Fig. 1, avian-like European swine lineage) (11, 64).

The importance of understanding influenza virus ecology in all animal species cannot be overemphasized in developing strategies to control avian and mammalian influenza.

> The ultimate control of the disease (influenza) depends on definitive answers to questions raised herein concerning . . . the ecologic relationship among human and animal strains.
>
> *Edwin Kilbourne, 1970* (37)

> Perhaps it is necessary to consider influenza as a zoonotic disease that affects several species rather than considering that the influenzas among the several (host) species are separate and unique entities.
>
> *Bernard Easterday and Bela Tumova, 1972* (21)

INFLUENZA IN THE HUMAN POPULATION

Historical Accounts of Animal-to-Human Influenza Virus Transmission: Emphasis on Avian Influenza Viruses

Sporadic cases of transmission of animal influenza viruses to humans have been documented (13). However, such cases are rare compared to the common human-adapted influenza virus cases that occur each year and during epidemics or pandemics. These sporadic cases of animal influenza in humans are most often associated with swine influenza viruses and usually with occupational exposure, such as farmers and animal handlers in research laboratories (13). For example, in an experimental study of H1N1 swine influenza viruses in pigs, two animal caretakers developed influenza and H1N1 influenza viruses were isolated (93). Molecular analysis determined that the experimental H1N1 swine virus and the viruses isolated from the caretakers were the same and that the latter viruses were not the result of recombination between swine and human influenza viruses (93). Cases of swine influenza in humans are clinically indistinguishable from human-adapted influenza virus infections, usually resolve without complications, and have limited human-to-human transmission (13). However, several deaths have occurred with swine influenza virus infections in immunosuppressed or stressed humans (13).

Direct transmission of avian influenza viruses to humans occurs even less frequently than swine-to-human transmission. One explanation may be the greater ease of transmissibility of swine influenza viruses and of infection of human respiratory epithelial cells based on cell receptors. Avian influenza viruses demonstrate preferential binding to respiratory epithelium with the cell surface receptor *N*-acetylneuraminic acid-α2,3-galactose linkage (α2,3 linkage) on sialoligosaccharides, while swine and human influenza viruses preferentially bind to *N*-acetylneuraminic acid-α2,6-galactose linkage (α2,6 linkage) on sialoligosaccharides (34). Avian respiratory epithelium has predominantly α2,3 linkage, and human respiratory epithelium has predominantly α2,6 linkage, while swine respiratory epithelium has a mixture of α2,3 linkage and α2,6 linkage (34). This suggests that swine influenza viruses have a higher probability than avian influenza viruses of binding and infecting human respiratory epithelium and that avian influenza viruses have a higher probability of infecting swine than human respiratory epithelium.

Despite the lower preference of avian influenza viruses for human respiratory epithelial cells, five incidences of limited human infection with avian influenza viruses have been reported and are highlighted below.

Steele virus in 1959

The first reported infection and isolation of an avian influenza virus from a human was in 1959 (17). A 46-year-old man developed infectious hepatitis after his return from a 2-month trip overseas through Asia, the Middle East, Africa, and Europe. Upon return to the United States, he experienced lassitude, headaches, and occasional chills during the first 2 weeks and bloating and indigestion during the next 2 weeks (17). A fowl plague-like virus (H7N7) was isolated from his blood and shown to be highly pathogenic for chickens. His recovery was full, but no neutralizing antibodies against avian influenza viruses were detected.

Seal influenza viruses in 1978 and 1979

During 1978 and 1979, an H7N7 avian influenza virus caused an outbreak of respiratory illness and death in harbor seals in the New England region of the United States (90). Self-limiting conjunctivitis was reported in workers handling the seals during the disease outbreak (90, 91). The isolate was mildly pathogenic for chickens.

United Kingdom H7N7 avian influenza in 1996

During 1996, an H7N7 influenza virus was isolated from a 43-year-old woman with conjunctivitis in England (3, 43). The woman tended a collection of domesticated ducks of various breeds that mixed freely with wild ducks on a small lake. The hemagglutinin gene of the influenza virus isolate was most closely related to that of an H7N7 avian influenza virus isolated from turkeys in Ireland during 1995 and was not closely related to any H7N7 equine influenza viruses (3, 87). The other seven gene segments were closely related to influenza viruses isolated from birds ($n = 4$) or to mammalian viruses that were recently introduced from the avian primordial reservoir (7, 36, 57). The isolate was mildly pathogenic for chickens.

Hong Kong H5N1 avian influenza virus in 1997

During March to May 1997, chickens on three farms were diagnosed with highly pathogenic avian influenza of the H5N1 subtype (60, 70). On 15 May 1997, a 3-year-old boy was admitted to a hospital in Hong Kong for respiratory distress and persistent fever (80). An H5N1 influenza virus was isolated from a tracheal aspirate taken on 19 May. Unfortunately, the child died on 21 May with Reyes' syndrome and acute respiratory distress secondary to viral pneumonia. No other cases of H5N1 influenza were diagnosed in birds or people during the spring or summer of 1997. During August 1997, investigators from the Agriculture and Fisheries Department surveyed 8- to 14-week-old chickens grown locally in Hong Kong and found no serologic evidence of H5N1 infections (L. Sims, personal communication).

On 6 November 1997, a 2-year-old boy was admitted to a hospital in Hong Kong with high fever and a cough. An H5N1 influenza virus was isolated from a nasopharyngeal swab. The boy recovered and was discharged on 9 November. By January 1998, a total of 18 human cases of H5N1 influenza were diagnosed, with six associated deaths. Molecular analysis of the chicken and human H5N1 influenza viruses indicated that all eight gene segments were of avian influenza virus lineage (6, 16, 79, 80). Patients with the Hong Kong H5N1 influenza virus had fever, symptoms of upper respiratory disease, and gastrointestinal symptoms such as vomiting, diarrhea, and pain (95). Patients who died had severe bilateral pneumonia, and other complications and lesions included hemophagocytosis, liver dysfunction, renal failure, septic shock, and pancytopenia (95). The human H5N1 isolates were highly pathogenic for chickens and mice (*Mus musculus*) in experimental studies (18, 79).

In December 1997, additional cases of H5N1 avian influenza were identified in chickens in the live-bird wholesale markets, live-bird retail markets, and farms of Hong Kong (Fig. 8) (62). All chickens on Hong Kong farms and in the live-bird markets as well as other bird species in

Figure 8. Wholesale poultry market at Cheung Sha Wan (left) and one of 1,000 retail markets in Hong Kong that provide consumers with fresh poultry meat (right).

contact with chickens were depopulated between 29 December 1997 and 2 January 1998. The farms and markets were disinfected. The elimination of infected poultry was a major factor in ending the incident, and no new cases of H5N1 influenza were diagnosed in humans after the depopulation of birds and disinfection of poultry facilities (70).

Hong Kong and China H9N2 influenza in 1998 and 1999

In March 1999, two children (a 13-month-old and a 4-year-old) in Hong Kong were hospitalized for fever and respiratory disease (59; Centers for Disease Control and Prevention, http://www.cdc.gov/ncidod/diseases/flu/H9N2info.htm). Since birth, these children had multiple health problems, a condition often termed failure to thrive. An H9N2 virus was isolated from nasopharyngeal aspirates from each patient. All eight gene segments were of the avian influenza viral lineage. The human H9N2 influenza viruses were mildly pathogenic for chickens (D. E. Swayne, unpublished data). Both patients recovered and were discharged. During December 1998, five cases of H9N2 influenza were reported from China in patients ranging in age from 1 to 70 years. These patients had uneventful recoveries (N. Cox, personal communication).

Origins of New Human Influenza Viruses

Concepts for origin of human influenza viruses

Limitations or restrictions on human movement between countries based on health criteria are minimal. With the advent of rapid international travel by airplane, automobile, train, and ship, human influenza viruses that have undergone antigenic drift or shift to create new influenza virus strains can spread rapidly between countries, causing significant human illness. Such antigenic drift and the freedom of human travel account for seasonal variations in human influenza cases and the need for annual reevaluation of influenza virus strains for inclusion in human influenza vaccines. Similar international travel along with antigenic shift can explain the appearance of pandemic human influenza.

By contrast, movement of domestic animals such as pigs, poultry, and horses between countries is restricted by agreements on health and sanitary standards established by the Office International des Epizooties in Paris (56). In general, poultry importation requires veterinary examination and certification that the animal is healthy and free of infectious diseases before leaving the nation of origin and that the country of origin is free of specific diseases, including highly pathogenic avian influenza.

Although the interchange or spread of influenza viruses between humans and other species occurs rarely in nature (79), wild waterfowl and other aquatic birds are the primordial reservoir of all influenza virus genes, based on molecular evidence, and ultimately they serve as the source of the influenza virus genes for the mammalian populations (89). The entry of such avian influenza virus genes into the human population is a rare event, covering large time spans. Although in toto transmission of animal influenza virus to humans has occurred, most occurrences have been isolated cases, without new viruses becoming established in the human population.

The most plausible theory of how new pandemic viruses arise is through recombination of gene segments from more than one influenza virus infecting a single cell (14, 39). The pig has been proposed as the mixing vessel or intermediary host in which reassortment between human, swine, and avian influenza viruses can take place with the emergence of new virus strains with the right constellation of genes to allow transmission and replication in the human host (3, 65). Theoretically, this would require a "new" hemagglutinin subtype, so immunity to established human hemagglutinin subtypes would not inhibit the adaptation process (3, 65). Circumstantial evidence in support of the pig mixing-vessel theory has emerged from the 1957 and 1968 human pandemics in Asia, where humans, pigs, and waterfowl were kept together or located in close proximity to each other (71).

Although this pig mixing-vessel theory is plausible, the detection of human infections with avian influenza viruses raises the possibility of a direct recombination event within the human host without the need for a swine intermediary. Furthermore, if pigs have been so closely linked as an intermediate host or mixing vessel for introduction of new H subtypes from avian to human influenza viruses, why do so few influenza virus subtypes naturally infect swine (3)? Furthermore, no H2 swine viruses have been identified since 1957 to account for the human H2 pandemic of 1957 (Table 1). Although the probability of an avian influenza virus's entering the human population, reassorting, and establishing a new lineage of human influenza virus, causing a pandemic, is extremely rare, this is consistent with the long time span between emergences of human pandemic influenza viruses (3). The probability of an avian or nonhuman mammalian virus in toto infecting and adapting to the new human host is less likely than a recombination event with an existing human-adapted influenza virus.

The 1918, 1957, and 1968 pandemic human influenza viruses

Molecular epidemiology has emerged as an important tool in studying viral evolution and determining the source of outbreak viruses. For example, analysis of sequence data has demonstrated that the 1957 (H2N2) and 1968 (H3N2) human influenza viruses resulted from reassortment of three and two avian influenza virus genes with five and six human influenza virus genes, respectively (36, 61, 66). However, the source of the 1918 pandemic virus is unknown. The lack of swine or human influenza viruses before the 1930s and the sparsity of avian influenza viruses before the 1930s make sequence analysis and phylogenetics of little help in answering the question of the origin of the 1918 pandemic influenza virus. Speculation suggests that the 1918 influenza virus could have been a previously circulating human influenza virus, an avian influenza virus, a swine influenza virus, or a recombinant of two or three of the above sources (26, 38, 61). Based on sequence analysis of RNA extracted from 1918 paraffin tissue blocks, Reid and Taubenberger have hypothesized that the 1918 virus was not of immediate avian origin but either was established in humans prior to 1918 as a whole avian virus and gradually adapted to more efficient replication and transmission in the human or was a recombinant of human and avian virus lineages (61).

LABORATORY ANIMAL MODELS OF INFLUENZA

Various animal models have been used to study the pathobiology of influenza, including ferrets (*Mustela putorius furo*), various nonhuman primates, mice, guinea pigs (*Cavia porcellus*), Syrian hamsters (*Mesocricetus auratus*), chinchillas (*Chinchilla langera*), chickens, and rats (*Rattus norvegicus*) (62). Most models are selected to provide information with application to human influenza, especially understanding of the pathogenesis of infection and disease and development and evaluation of vaccines and chemotherapeutics. Although influenza viruses do not cause natural infections in rodents and they are not reservoirs for influenza virus, the ability to infect various rodent species raises a concern about the potential for natural infections to occur in the future (13).

The mouse has been the most widely used model for studying human influenza (62). Mice make good laboratory models for influenza because they are small, easy and safe to handle, available as inbred lines of various genotypes and phenotypes, and able to reproduce the respiratory component of human influenza, including lesions and cell types for virus replication (62). However, mouse models for studying human influenza have some disadvantages. First, most laboratory mouse strains have far more genetic homogeneity than the human population, making extrapolation of findings to the human population tenuous. Second, mice are not natural hosts of influenza viruses and require passage and adaptation of the influenza virus to the mouse model before consistent viral replication, clinical signs, and death are produced. Third, humans have the *MxA* and *MxB* gene loci, which confer resistance to influenza (33), while most strains of laboratory mice lack a functional gene analogous to the influenza resistance gene (35). Such laboratory mice are susceptible to influenza infection and have been designated *Mx1*$^{-}$ (35). Wild mice and a few novel laboratory strains of mice have the *Mx1*$^{+}$ gene locus, which imparts resistance to influenza (35). Despite these disadvantages, mouse models can provide information about the potential of some animal influenza virus strains to cross species barriers and infect and cause disease in multiple mammalian host species, with some extrapolation to humans (18).

Various researchers have shown the ability of the human- and chicken-origin Hong Kong H5N1 avian influenza viruses of 1997 to infect, cause disease in, and kill laboratory mice without prior passage of the virus strains in mice (23, 27, 41, 49, 72). These viruses were also highly pathogenic for chickens under natural and experimental conditions and were associated with several human deaths in nature (79, 80, 95). Is there a potential for other highly pathogenic avian influenza viruses to exhibit similar biological properties for mice and chickens and thus possess a potential to infect and cause disease in humans? In examining seven highly pathogenic avian influenza viruses, Dybing et al. showed that three highly pathogenic avian influenza viruses isolated in Hong Kong during 1997 were unique in being highly lethal for mice without requiring passage for adaptation to mice, while highly pathogenic H5 avian influenza viruses isolated from poultry in Scotland in 1959, England in 1991, Mexico in 1995, Italy in 1997, and Hong Kong in 1999 were not highly lethal for mice (18; A. N. Cauthen, D. E. Swayne, S. Schultz-Cherry, M. L. Perdue, and D. L. Suarez, submitted for publication). It is currently unknown if the mouse model will be predictive for the ability of mildly pathogenic avian influenza viruses to cross species and replicate in a mammalian host and therefore support the chances of a recombination event with a human or other mammalian-adapted virus. High pathogenicity of an avian influenza virus for chickens does not correlate with infectivity or virulence potential for mice or humans.

CAN WE PREDICT AND PREVENT HUMAN INFLUENZA PANDEMICS?

Pandemics of human influenza have been a continual source of human suffering and death, from the first incident recorded by Hippocrates in 412 B.C. through the 21 well-documented pandemics between 1700 and 1968 (8). In the 20th century alone, the 1918–1919 pandemic has been estimated to have killed between 20 million and 100 million people worldwide, while in the United States the pandemic killed more Americans in 1 year than were killed during World War I, World War II, the Korean war, and the Vietnam war combined (42). The 1918–1919 pandemic decreased the U.S. population's average life expectancy from 51 years in 1917 to 39 years in 1918. History has shown that influenza pandemics have been major causes of human suffering and death. Can we predict impending pandemics of influenza and prevent their occurrence in the future? This is a difficult and complex question to answer, but recent scientific findings have placed a few pieces into the puzzle.

First, wild birds have been shown to be important in influenza virus ecology, as originally shown by Slemons et al. (73) and Easterday et al. (20). From this finding has come the concept of a primordial wild-bird reservoir of all influenza virus genes and the need for surveillance of animals and humans for the appearance of new influenza virus strains (89). Second, the direct infection of humans by animal influenza viruses, such as the H5N1 avian influenza viruses in Hong Kong during 1997 and H1N1 swine influenza viruses over multiple years, occurs around the globe, with most infections going undetected and being self-limited. However, the potential exists for rare infections by animal influenza viruses in a small subset of the human population sufficient to maintain the viruses until they have adapted to the human host and can be transmitted from human to human. Third, the recombination of avian and human influenza viral genes to create new human influenza viruses (i.e., hybrid viruses through antigenic shift) was important in the emergence of the 1957 and 1968 human influenza pandemics. Although pandemics of human influenza occur with rarity, can vigilance in surveillance of animal and human populations for influenza viruses identify points of entry for animal viruses or genes into the human population such that steps can be taken to prevent a catastrophe? Intervention into the H5N1 avian influenza incidence in Hong Kong during 1997 may have averted the development of a human influenza pandemic.

SUMMARY AND CONCLUSIONS

Influenza is a respiratory disease of humans, swine, horses, mink, sea mammals, and some birds caused by infection with orthomyxoviruses. Influenza is endemic in humans, horses, swine, and some birds, but sporadically epidemics do occur in influenza-naive populations or when new subtypes of influenza viruses are introduced. Generally, influenza viruses are adapted to a single host species and are transmitted freely between individuals within the same species, but on occasion influenza viruses have crossed the species barrier and infected a new host. The genes for all influenza virus are contained within influenza viruses maintained in wild aquatic birds, which serve as a primordial genetic reservoir. Influenza virus infections in the wild aquatic birds do not cause disease. The advent of animal agriculture has altered natural ecosystems and created new ecosystems with opportunities for contact between avian influenza viruses, and susceptible populations of wild birds and domestic poultry resulting in infection, adaptation, and continual circulation of influenza in new populations of avian hosts.

Avian influenza viruses have been shown to cross the species barrier and infect mammalian species, most commonly swine and humans. Infections by avian influenza viruses have been documented in humans and include a case of hepatitis in the United States during 1959, several cases of conjunctivitis in the United States with avian-like seal viruses during 1978 and 1979, a case of conjunctivitis in the United Kingdom during 1996, 18 cases of H5N1 influenza in Hong Kong during 1997, and seven cases of H9N2 influenza in China and Hong Kong during 1998 and 1999. These occurrences emphasize the potential for interspecies transmission of avian influenza viruses from birds to humans, although most such occurrences have been sporadic and self-limiting. Occasionally, avian influenza viruses have adapted over time to another host species and become a new endemic virus. However, the main occurrence and threat of avian influenza viruses to humans are not in toto infection, spread, and adaptation of the avian influenza viruses to the host, but the recombination of one or more avian influenza viral genes, especially the genes that encode the surface hemagglutinin and neuraminidase glycoproteins, with genes from human-adapted influenza viruses to create a hybrid influenza virus (antigenic shift). These hybrid viruses have the capacity to replicate in and be transmitted between humans and have resulted in the emergence of the 1957 and 1968 pandemic outbreaks of influenza in humans. Laboratory animal models, especially the mouse, may assist in predicting the potential of avian influenza viruses to infect and cause disease and death in mammals. Such data may be extrapolated to identify viruses with the potential to infect humans.

Acknowledgments. I thank David Suarez, Michael Perdue, Jody Dybing, and Stacey Schultz-Cherry for providing contributing data. Joan Beck, John Latimer, Liz Turpin, Patsy Decker, and Laura Perkins are thanked for providing technical support. Harsh K. Jain is thanked for the computer drawings.

REFERENCES

1. **Alexander, D. J.** 1993. Orthomyxovirus infections, p. 287–316. *In* J. B. McFerran and M. S. McNulty (ed.), *Virus Infections of Birds.* Elsevier Science, London, United Kingdom.
2. **Alexander, D. J., W. H. Allan, D. G. Parsons, and G. Parsons.** 1978. The pathogenicity of four avian influenza viruses for fowls, turkeys and ducks. *Res. Vet. Sci.* **24:**242–247.
3. **Banks, J., E. Speidel, and D. J. Alexander.** 1998. Characterisation of an avian influenza A virus isolated from a human—is an intermediate host necessary for the emergence of pandemic influenza viruses? *Arch. Virol.* **143:**781–787.
4. **Beard, C. W.** 1970. Demonstration of type-specific influenza antibody in mammalian and avian sera by immunodiffusion. *Bull. W. H. O.* **42:**779–785.
5. **Becker, W. B.** 1966. The isolation and classification of Tern virus: influenza A-Tern South Africa—1961. *J. Hyg. (Lond.)* **64:**309–320.
6. **Bender, C., H. Hall, J. Huang, A. Klimov, N. Cox, A. Hay, V. Gregory, K. Cameron, W. Lim, and K. Subbarao.** 1999. Characterization of the surface proteins of influenza A (H5N1) viruses isolated from humans in 1997–1998. *Virology* **254:**115–123.

7. **Berg, M., L. Englund, I. A. Abusugra, B. Klingeborn, and T. Linne.** 1990. Close relationship between mink influenza (H10N4) and concomitantly circulating avian influenza viruses. *Arch. Virol.* **113:**61–71.
8. **Beveridge, W. I. B.** 1978. *Influenza: the Last Great Plague,* p. 25. Prodist, New York, N. Y.
9. **Bosch, F. X., M. Orlich, H. D. Klenk, and R. Rott.** 1979. The structure of the hemagglutinin, a determinant for the pathogenicity of influenza viruses. *Virology* **95:**197–207.
10. **Brown, I. H., T. R. Crawshaw, P. A. Harris, and D. J. Alexander.** 1998. Detection of antibodies to influenza A virus in cattle in association with respiratory disease and reduced milk yield. *Vet. Rec.* **143:**637–638.
11. **Brown, I. H., P. A. Harris, J. W. McCauley, and D. J. Alexander.** 1998. Multiple genetic reassortment of avian and human influenza A viruses in European pigs, resulting in the emergence of an H1N2 virus of novel genotype. *J. Gen. Virol.* **79:**2947–2955.
12. **Brugh, M.** 1998. Re-examination of pathogenicity, virulence and lethality, p. 129–132. *In* D. E. Swayne and R. D. Slemons (ed.), *Proceedings of the Fourth International Symposium on Avian Influenza.* U.S. Animal Health Association, Richmond, Va.
13. **Brugh, M., and R. D. Slemons.** 1994. Influenza, p. 385–395. *In* G. W. Beran and J. H. Steele (ed.), *Handbook of Zoonoses: Viral,* 2nd ed. CRC Press, Boca Raton, Fla.
14. **Burnet, M.** 1959. *Natural History of Infectious Disease,* p. 276–290. Cambridge University Press, London, United Kingdom.
15. **Chatterjee, S.** 1997. *The Rise of Birds,* p. 262–272. Johns Hopkins University Press, Baltimore, Md.
16. **Claas, E., A. Osterhaus, R. van Beek, J. C. de Jong, G. F. Rimmelzwaan, D. A. Senne, S. Krauss, K. F. Shortridge, and R. G. Webster.** 1998. Human influenza A H5N1 related to a highly pathogenic avian influenza virus. *Lancet* **351:**472–477.
17. **Delay, P. D., H. Casey, and H. S. Tubiash.** 1967. Comparative study of fowl plague virus and a virus isolated from man. *Public Health Rep.* **82:**615–620.
18. **Dybing, J. K., S. Schultz-Cherry, D. E. Swayne, D. L. Suarez, and M. L. Perdue.** 2000. Distinct pathogenesis of Hong Kong-origin H5N1 viruses in mice as compared to other highly pathogenic H5 avian influenza viruses. *J. Virol.* **74:**1443–1450.
19. **Easterday, B. C., V. S. Hinshaw, and D. A. Halvorson.** 1997. Influenza, p. 583–605. *In* B. W. Calnek, H. J. Barnes, C. W. Beard, L. R. McDougald, and Y. M. Saif (ed.), *Diseases of Poultry,* 10th ed. Iowa State University Press, Ames.
20. **Easterday, B. C., D. O. Trainer, B. Tumova, and H. G. Pereira.** 1968. Evidence of infection with influenza viruses in migratory waterfowl. *Nature* **219:**523–524.
21. **Easterday, B. C., and B. Tumova.** 1972. Avian influenza viruses: in avian species and the natural history of influenza. *Adv. Vet. Sci. Comp. Med.* **16:**201–222.
22. **Englund, L., B. Klingeborn, and T. Mejerland.** 1986. Avian influenza A virus causing an outbreak of contagious interstitial pneumonia in mink. *Acta Vet. Scand.* **27:**497–504.
23. **Gao, P., S. Watanabe, T. Ito, H. Goto, K. Wells, M. McGregor, A. J. Cooley, and Y. Kawaoka.** 1999. Biological heterogeneity, including systemic replication in mice, of H5N1 influenza A virus isolates from humans in Hong Kong. *J. Virol.* **73:**3184–3189.
24. **Geraci, J. R., D. J. St. Aubin, I. K. Barker, R. G. Webster, V. S. Hinshaw, W. J. Bean, H. L. Ruhnke, J. H. Prescott, G. Early, A. S. Baker, S. Madoff, and R. T. Schooley.** 1982. Mass mortality of harbor seals: pneumonia associated with influenza A virus. *Science* **215:**1129–1131.
25. **Gill, F. B.** 1995. *Ornithology,* p. 3–20. W. H. Freeman, New York, N.Y.
26. **Gorman, O. T., W. J. Bean, Y. Kawaoka, I. Donatelli, Y. J. Guo, and R. G. Webster.** 1991. Evolution of influenza A virus nucleoprotein genes: implications for the origins of H1N1 human and classical swine viruses. *J. Virol.* **65:**3704–3714.

27. **Gubareva, L. V., J. A. McCullers, R. C. Bethell, and R. G. Webster.** 1998. Characterization of influenza A/Hong Kong/156/97 (H5N1) virus in a mouse model and protective effect of zanamivir on H5N1 infection in mice. *J. Infect. Dis.* **178:**1592–1596.
28. **Halvorson, D. A.** 1995. Avian influenza control in Minnesota. *Poult. Dig.* **54:**12–19.
29. **Halvorson, D. A., C. J. Kelleher, and D. A. Senne.** 1985. Epizootiology of avian influenza: effect of season on incidence in sentinel ducks and domestic turkeys in Minnesota. *Appl. Environ. Microbiol.* **49:**914–919.
30. **Hinshaw, V. S., G. M. Air, A. J. Gibbs, L. Graves, B. Prescott, and D. Karunakaran.** 1982. Antigenic and genetic characterization of a novel hemagglutinin subtype of influenza A viruses from gulls. *J. Virol.* **42:**865–872.
31. **Hinshaw, V. S., W. J. Bean, J. Geraci, P. Fiorelli, G. Early, and R. G. Webster.** 1986. Characterization of two influenza A viruses from a pilot whale. *J. Virol.* **58:**655–656.
32. **Hinshaw, V. S., W. J. Bean, R. G. Webster, J. E. Rehg, P. Fiorelli, G. Early, J. R. Geraci, and D. J. St. Aubin.** 1984. Are seals frequently infected with avian influenza viruses? *J. Virol.* **51:**863–865.
33. **Horisberger, M. A.** 1995. Interferons, Mx genes, and resistance to influenza virus. *Am. J. Respir. Crit. Care Med.* **152:**S67–S71.
34. **Ito, T., J. N. S. S. Couceiro, S. Kelm, L. G. Baum, S. Krauss, M. R. Castrucci, I. Donatelli, H. Kida, J. C. Paulson, R. G. Webster, and Y. Kawaoka.** 1998. Molecular basis for the generation in pigs of influenza A viruses with pandemic potential. *J. Virol.* **72:**7367–7373.
35. **Jin, H. K., T. Yamasihita, K. Ochiai, O. Haller, and T. Watanabe.** 1998. Characterization and expression of the Mx1 gene in wild mouse species. *Biochem. Genet.* **36:**311–322.
36. **Kawaoka, Y., S. Krauss, and R. G. Webster.** 1989. Avian-to-human transmission of the PB1 gene of influenza A viruses in the 1957 and 1968 pandemics. *J. Virol.* **63:**4603–4608.
37. **Kilbourne, E. D.** 1970. Influenza 1970: unquestioned answers and unanswered questions. *Arch. Environ. Health* **21:**286–292.
38. **Kilbourne, E. D.** 1973. The molecular epidemiology of influenza. *J. Infect. Dis.* **127:**478–487.
39. **Kilbourne, E. D.** 1987. *Influenza.* Plenum, New York, N.Y.
40. **Klenk, H. D., N. J. Cox, R. A. Lamb, B. W. J. Mahy, K. Nakamura, P. A. Nuttall, P. Palese, and R. Rott.** 1995. Orthomyxoviridae, p. 293–299. *In* F. A. Murphy, C. M. Fauquet, D. H. Bishop, S. A. Ghabrial, A. W. Jarvis, G. P. Martelli, M. A. Mayo, and M. D. Summers (ed.), *Virus Taxonomy.* Springer-Verlag, New York, N.Y.
41. **Kodihalli, S., H. Goto, D. L. Kobasa, S. Krauss, Y. Kawaoka, and R. G. Webster.** 1999. DNA vaccine encoding hemagglutinin provides protective immunity against H5N1 influenza virus infection in mice. *J. Virol.* **73:**2094–2098.
42. **Kolata, G.** 1999. *Flu: the Story of the Great Influenza Pandemic of 1918 and the Search for the Virus That Caused It,* p. ix-xi, 1-32. Farrar, Straus and Giroux, New York, N.Y.
43. **Kurtz, J., R. J. Manvell, and J. Banks.** 1996. Avian influenza virus isolated from a woman with conjunctivitis. *Lancet* **348:**901–902.
44. **Lang, G., A. Gagnon, and J. R. Geraci.** 1981. Isolation of an influenza A virus from seals. *Arch. Virol.* **68:**189–195.
45. **Laudert, E., D. Halvorson, V. Sivanandan, and D. Shaw.** 1993. Comparative evaluation of tissue tropism characteristics in turkeys and mallard ducks after intravenous inoculation of type A influenza viruses. *Avian Dis.* **37:**773–780.
46. **Laudert, E., V. Sivanandan, and D. Halvorson.** 1993. Effect of an H5N1 avian influenza virus infection on the immune system of mallard ducks. *Avian Dis.* **37:**845–853.
47. **Laudert, E. A., V. Sivanandan, and D. A. Halvorson.** 1993. Effect of intravenous inoculation of avian influenza virus on reproduction and growth in mallard ducks. *J. Wildl. Dis.* **29:**523–526.

48. **Leahy, M. B., J. T. Dessens, F. Weber, G. Kochs, and P. A. Nuttall.** 1997. The fourth genus in the Orthomyxoviridae: sequence analysis of two Thogoto virus polymerase proteins and comparison with influenza viruses. *Virus Res.* **50:**215–224.
49. **Lu, X., T. M. Tumpey, T. Morken, S. R. Zaki, N. J. Cox, and J. M. Katz.** 1999. A mouse model for the evaluation of pathogenesis and immunity to influenza A (H5N1) viruses isolated from humans. *J. Virol.* **73:**5903–5911.
50. **Lvov, D. K., V. M. Zdanov, A. A. Sazonov, N. A. Braude, E. A. Vladimirtceva, L. V. Agafonova, E. I. Skljanskaja, N. V. Kaverin, V. I. Reznik, T. V. Pysina, A. M. Oserovic, A. A. Berzin, I. A. Mjasnikova, R. Y. Podcernjaeva, S. M. Klimenko, V. P. Andrejev, and M. A. Yakhno.** 1978. Comparison of influenza viruses isolated from man and from whales. *Bull. W. H. O.* **56:**923–930.
51. **Mo, I. P., M. Brugh, O. J. Fletcher, G. N. Rowland, and D. E. Swayne.** 1997. Comparative pathology of chickens experimentally inoculated with avian influenza viruses of low and high pathogenicity. *Avian Dis.* **41:**125–136.
52. **Mohan, R., Y. M. Saif, G. A. Erickson, G. A. Gustafson, and B. C. Easterday.** 1981. Serologic and epidemiologic evidence of infection in turkeys with an agent related to the swine influenza virus. *Avian Dis.* **25:**11–16.
53. **Morgan, I. R., and H. A. Westbury.** 1981. Virological studies of Adelie penguins (*Pygoscelis adeliae*) in Antarctica. *Avian Dis.* **25:**1019–1026.
54. **Murphy, B. R., and R. G. Webster.** 1996. Orthomyxoviruses, p. 1397–1445. *In* B. N. Fields, D. M. Knipe, and P. M. Howley (ed.), *Fields Virology,* 3rd ed. Lippincott-Raven, Philadelphia, Pa.
55. **Newman, J., D. Halvorson, and D. Karunakaran.** 1981. Complications associated with avian influenza infections, p. 8-12. *In Proceedings of the First International Symposium on Avian Influenza.* U.S. Animal Health Association, Richmond, Va.
56. **OIE Code Commission.** 1992. *International Animal Health Code.* Office International des Epizooties, Paris, France.
57. **Okazaki, K., Y. Kawaoka, and R. G. Webster.** 1989. Evolutionary pathways of the PA genes of influenza A viruses. *Virology* **172:**601–608.
58. **Panigrahy, B., D. A. Senne, and J. E. Pearson.** 1992. Subtypes of avian influenza virus (AIV) isolated from exotic passerine and psittacine birds, 1982–1991, p. 136–143. *In* B. Easterday (ed.), *Proceedings of the Third International Symposium on Avian Influenza.* U.S. Animal Health Association, Richmond, Va.
59. **Peiris, M., K. Y. Yuen, C. W. Leung, K. H. Chan, P. L. Ip, R. W. Lai, W. K. Orr, and K. F. Shortridge.** 1999. Human infection with influenza H9N2. *Lancet* **354:**916–917.
60. **Perdue, M. L., D. L. Suarez, and D. E. Swayne.** Avian influenza in the 1990's. *Poult. Avian Biol. Rev.,* in press.
61. **Reid, A. H., and J. K. Taubenberger.** 1999. The 1918 flu and other influenza pandemics: "over there" and back again. *Lab. Investig.* **79:**95–101.
62. **Renegar, K. B.** 1992. Influenza virus infections and immunity: a review of human and animal models. *Lab. Anim. Sci.* ig. **42:**222–232.
63. **Ridgway, S. H.** 1979. Reported causes of death of captive killer whales (Orcinus orca). *J. Wildl. Dis.* **15:**99–104.
64. **Scholtissek, C., H. Burger, P. A. Bachmann, and C. Hannoun.** 1983. Genetic relatedness of hemagglutinins of the H1 subtype of influenza A viruses isolated from swine and birds. *Virology* **129:**521–523.
65. **Scholtissek, C., H. Burger, O. Kistner, and K. F. Shortridge.** 1985. The nucleoprotein as a possible major factor in determining host specificity of influenza H3N2 viruses. *Virology* **147:**287–294.
66. **Scholtissek, C., I. Koennecke, and R. Rott.** 1978. Host range recombinants of fowl plague (influenza A) virus. *Virology* **91:**79–85.

67. **Senne, D. A., J. E. Pearson, Y. Kawaoka, E. A. Carbrey, and R. G. Webster.** 1986. Alternative methods for evaluation of pathogenicity of chicken Pennsylvania H5N2 viruses, p. 246–257. *In* B. C. Easterday (ed.), *Proceedings of the Second International Symposium on Avian Influenza.* U.S. Animal Health Association, Richmond, Va.
68. **Senne, D. A., J. E. Pearson, L. D. Miller, and G. A. Gustafson.** 1983. Virus isolations from pet birds submitted for importation into the United States. *Avian Dis.* **27:**731–744.
69. **Senne, D. A., J. E. Pearson, and B. Panigrahy.** 1992. Live poultry markets: a missing link in the epidemiology of avian influenza, p. 50–58. *In* B. C. Easterday (ed.), *Proceedings of the Third International Symposium on Avian Influenza.* University of Wisconsin, Madison.
70. **Shortridge, K. F.** 1999. Poultry and the influenza H5N1 outbreak in Hong Kong, 1997: abridged chronology and virus isolation. *Vaccine* **17:**S26–S29.
71. **Shortridge, K. F., and C. H. Stuart Harris.** 1982. An influenza epicentre? *Lancet* **ii:**812–813.
72. **Shortridge, K. F., N. N. Zhou, Y. Guan, P. Gao, T. Ito, Y. Kawaoka, S. Kodihalli, S. Krauss, D. Markhill, G. Murti, M. Norwood, D. Senne, L. Sims, A. Takada, and R. G. Webster.** 1998. Characterization of avian H5N1 influenza viruses from poultry in Hong Kong. *Virology* **252:**331–342.
73. **Slemons, R. D., D. C. Johnson, J. S. Osborn, and F. Hayes.** 1974. Type-A influenza viruses isolated from wild free-flying ducks in California. *Avian Dis.* **18:**119–124.
74. **Slemons, R. D., and D. E. Swayne.** 1992. Nephrotropic properties demonstrated by A/chicken/Alabama/75 (H4N8) following intravenous challenge of chickens. *Avian Dis.* **36:**926–931.
75. **Slemons, R. D., and D. E. Swayne.** 1995. Tissue tropism and replicative properties of waterfowl-origin influenza viruses in chickens. *Avian Dis.* **39:**521–527.
76. **Stallknecht, D. E.** 1998. Ecology and epidemiology of avian influenza viruses in wild bird populations: waterfowl, shorebirds, pelicans, cormorants, etc., p. 61–69. *In* D. E. Swayne and R. D. Slemons (ed.), *Proceedings of the Fourth International Symposium on Avian Influenza.* U.S. Animal Health Association, Richmond, Va.
77. **Stallknecht, D. E., and S. M. Shane.** 1988. Host range of avian influenza virus in free-living birds. *Vet. Res. Commun.* **12:**125–141.
78. **Stubbs, E. L.** 1948. Fowl pest, p. 603–614. *In* H. E. Biester and L. H. Schwarte (ed.), *Diseases of Poultry,* 2nd ed. Iowa State University Press, Ames.
79. **Suarez, D. L., M. L. Perdue, N. Cox, T. Rowe, C. Bender, J. Huang, and D. E. Swayne.** 1998. Comparison of highly virulent H5N1 influenza A viruses isolated from humans and chickens from Hong Kong. *J. Virol.* **72:**6678–6688.
80. **Subbarao, K., A. Klimov, J. Katz, H. Regnery, W. Lim, H. Hall, M. Perdue, D. Swayne, C. Bender, J. Huang, M. Hemphill, T. Rowe, M. Shaw, X. Xu, K. Fukuda, and N. Cox.** 1998. Characterization of an avian influenza A (H5N1) virus isolated from a child with a fatal respiratory illness. *Science* **276:**393–396.
81. **Swayne, D. E.** 1997. Pathobiology of H5N2 Mexican avian influenza viruses for chickens. *Vet. Pathol.* **34:**557–567.
82. **Swayne, D. E., and R. D. Slemons.** 1994. Comparative pathology of a chicken-origin and two duck-origin influenza virus isolates in chickens: the effect of route of inoculation. *Vet. Pathol.* **31:**237–245.
83. **Swayne, D. E., and R. D. Slemons.** 1995. Comparative pathology of intravenously inoculated wild duck- and turkey-origin type A influenza virus in chickens. *Avian Dis.* **39:**74–84.
84. **Trock, S. C.** 1998. Epidemiology of influenza in live bird markets and ratite farms, p. 76–78. *In* D. E. Swayne and R. D. Slemons (ed.), *Proceedings of the Fourth International Symposium on Avian Influenza.* U.S. Animal Health Association, Richmond, Va.

85. **U. S. Animal Health Association.** 1994. Report of the Committee on Transmissible Diseases of Poultry and Other Avian Species: criteria for determining that an AI virus isolation causing an outbreak must be considered for eradication, p. 522. *In Proceedings of the 98th Annual Meeting of the U.S. Animal Health Association.* U.S. Animal Health Association, Richmond, Va.
86. **U. S. Animal Health Association.** 1998. Report of the Committee on Transmissible Diseases of Poultry and Other Avian Species, p. 614–659. *In Proceedings of the 102nd Annual Meeting of the U.S. Animal Health Association.* U.S. Animal Health Association, Richmond, Va.
87. **Webster, R. G.** 1993. Are equine 1 influenza viruses still present in horses? *Equine Vet. J.* **25:**537–538.
88. **Webster, R. G.** 1997. Predictions for future human influenza pandemics. *J. Infect. Dis.* **176** (Suppl. 1):S14–S19.
89. **Webster, R. G., W. J. Bean, O. T. Gorman, T. M. Chambers, and Y. Kawaoka.** 1992. Evolution and ecology of influenza A viruses. *Microbiol.Rev.* **56:**152–179.
90. **Webster, R. G., J. Geraci, G. Petursson, and K. Skirnisson.** 1981. Conjunctivitis in human beings caused by influenza A virus of seals. *N. Engl. J. Med.* **304:**911.
91. **Webster, R. G., V. S. Hinshaw, W. J. Bean, K. L. van Wyke, J. R. Geraci, D. J. St. Aubin, and G. Petursson.** 1981. Characterization of an influenza A virus from seals. *Virology* **113:**712–724.
92. **Wells, D. L., D. J. Hopfensperger, N. H. Arden, M. W. Harmon, J. P. Davis, M. A. Tipple, and L. B. Schonberger.** 1991. Swine influenza virus infections: transmission from ill pigs to humans at a Wisconsin agricultural fair and subsequent probable person-to-person transmission. *JAMA* **265:**478–481.
93. **Wentworth, D. E., M. W. McGregor, M. D. Macklin, V. Neumann, and V. S. Hinshaw.** 1997. Transmission of swine influenza virus to humans after exposure to experimentally infected pigs. *J. Infect. Dis.* **175:**7–15.
94. **Wood, G. W., G. Parsons, and D. J. Alexander.** 1995. Replication of influenza A viruses of high and low pathogenicity for chickens at different sites in chickens and ducks following intranasal inoculation. *Avian Pathol.* **24:**545–551.
95. **Yuen, K. Y., P. K. S. Chan, M. Peiris, D. N. C. Tsang, T. L. Que, K. F. Shortridge, P. T. T. W. K. Cheung, E. T. F. Ho, R. Sung, and A. F. B. Cheng.** 1998. Clinical features and rapid viral diagnosis of human disease associated with avian influenza A H5N1 virus. *Lancet* **351:**467–471.
96. **Ziegler, A. F., S. Davison, H. Acland, and R. J. Eckroade.** 1999. Characteristics of H7N2 (nonpathogenic) avian influenza virus infections in commercial layers, in Pennsylvania, 1997–98. *Avian Dis.* **43:**142–149.

Emerging Diseases of Animals
Edited by C. Brown and C. Bolin

Chapter 7

Transmissible Spongiform Encephalopathies

Linda A. Detwiler, Richard Rubenstein, and Elizabeth S. Williams

INTRODUCTION

The transmissible spongiform encephalopathies (TSEs) are a family of diseases caused by an agent yet to be fully characterized. The diseases have also been referred to as the prion diseases as the acceptance of one of the etiological theories has grown. They share a number of characteristics:

1. a prolonged incubation period of months or years,
2. a progressive debilitating neurological illness which is always fatal,
3. the presence of scrapie-associated fibrils in detergent-treated extracts of brain tissue from animals or humans affected by these diseases detected by negative-stain electron microscopy,
4. pathological changes apparently confined to the central nervous system (CNS),
5. transmissibility of the agent by either natural or experimental means or both, and
6. failure of the transmissible agent to elicit a detectable specific immune response in the host.

Linda A. Detwiler • Veterinary Services, Animal and Plant Health Inspection Service, U.S. Department of Agriculture, 320 Corporate Blvd., Robbinsville, NJ 08691. **Richard Rubenstein** • Department of Virology, Institute for Basic Research in Developmental Disabilities, New York State Office of Mental Retardation and Developmental Disabilities, 1050 Forest Hill Rd., Staten Island, NY 10314. **Elizabeth S. Williams** • Department of Veterinary Sciences, University of Wyoming, Laramie, 1174 Snowy Range Rd., Laramie, WY 82070.

Although a number of characteristics are common among the TSEs, they also differ from species to species. Differences occur in pathogenesis, routes of natural transmission, and distribution of infectivity in tissues. These differences are important and need to be taken into account for diagnosis, prevention, and control.

There are specific TSEs which affect humans and others which affect animals. Evidence suggests that bovine spongiform encephalopathy (BSE) has crossed the species barrier to cause variant Creutzfeldt-Jakob disease (CJD) in humans. TSEs include the following.

Creutzfeldt-Jakob Disease

CJD is a sporadic presenile dementia which affects humans. It occurs at an annual incidence of about 1 per 1 million population. Approximately 85% of the cases are sporadic, with no known source of exposure. Another percentage of the cases are familial in nature and are associated with certain inherited gene mutations. An extremely small number of cases are the result of iatrogenic transmission, such as cases associated with corneal transplants, contaminated pituitary growth hormone injections, and poorly disinfected brain electrodes (14, 15).

Kuru

Kuru is a disease of the Fore tribe in the Eastern Highlands of Papua New Guinea. The disease is characterized by a loss of coordination followed by dementia. The infection was thought to spread through the practice of cannibalism and exposure to high-risk tissues such as brain. Since the practice was discontinued, the disease has essentially disappeared.

Gerstmann-Sträussler-Scheinker Syndrome

Gerstmann-Sträussler-Scheinker syndrome is a rare familial disease of humans caused by an autosomal dominant inherited mutation in the PrP gene. It occurs at an estimated annual incidence of 5 cases per 100 million population. The disease is characterized by loss of coordination and dementia (91).

Fatal Familial Insomnia

Fatal familial insomnia is an extremely rare human disorder characterized by trouble sleeping and disturbances of the autonomic nervous system. An inherited mutation in the PrP gene has been identified as the precipitating factor of this disease (64, 107).

Variant Creutzfeldt-Jakob Disease

Sixty-eight definite and probable cases of vCJD have been identified in the United Kingdom, one in Ireland, and two in France between 1994 and April 2000. Unlike sporadic CJD, the cases were unique in that the patients were younger than usual, the clinical manifestation was different, and the cases displayed a new neuropathological profile. The Spongiform Encephalopathy Advisory Committee (SEAC) in the United Kingdom concluded that although there was no direct scientific evidence of a link between BSE and CJD, based on current data and in the absence of any credible alternative, the most likely explanation at present is that these cases are linked to exposure to BSE before the introduction of the specified bovine offal ban in 1989 (119).

Transmissible Mink Encephalopathy

Transmissible mink encephalopathy is an uncommon disease of ranch-raised mink first described by Hartsough and Burger in 1965 (43). The disease has been reported in Canada, Finland, Germany, Russia, and the United States. The last case reported in the United States was in Wisconsin in 1985 (70). It has been suggested that transmissible mink encephalopathy is a result of feeding either scrapie-infected sheep or infected cattle to mink.

Chronic Wasting Disease

Chronic wasting disease (CWD) was first reported by Williams and Young in 1980 (120) and originally was confined to mule deer and Rocky Mountain elk held in captivity in Colorado and Wyoming (121). The disease has been confirmed in free-ranging cervids (mule deer, white-tailed deer, and elk) in Colorado and Wyoming. CWD has also been diagnosed in ranch-raised elk in several U.S. states and one herd in Canada. The disease is characterized by emaciation, changes in behavior, and excessive salivation.

Feline Spongiform Encephalopathy

Feline spongiform encephalopathy (FSE) is a naturally occurring TSE first reported in domestic cats in 1990 (124). Over 80 cases of the disease have been diagnosed in the United Kingdom since 1990. In addition, one case of FSE has been identified in Norway (10), and one has been identified in Liechtenstein. Cats with the disease display locomotor disturbances, behavioral changes, and a hypersensitivity to sudden movements or noises (123). Current evidence suggests that the disease in cats is most probably linked to BSE.

Scrapie

Scrapie is an insidious, degenerative disease affecting the CNS of sheep and goats. The disease is also called *la tremblante* (French, trembling), *Traberkrankheit* (German, trotting disease), and *rida* (Icelandic, ataxia or tremor). It was first recognized as a disease of sheep in Great Britain and other countries of western Europe over 250 years ago (83). Scrapie has been reported worldwide and affects most sheep-producing regions with few notable exceptions. Australia and New Zealand are commonly accepted to be scrapie-free. Scrapie is endemic in the United States.

Other cases of spongiform encephalopathy have been reported in kudu, eland, nyala, gemsbok, and a few exotic cats. These too are thought to be linked to BSE-contaminated feed.

The primary focus of this chapter will be on BSE and its assumed relationship with vCJD. The chapter will also provide more detailed summaries on scrapie and CWD.

ETIOLOGY

The clinical, pathological, and molecular genetic features of the transmissible spongiform encephalopathies have led to speculation on the nature of the etiologic agent and the pathogenic mechanisms of the disease. There are three main theories on the nature of the scrapie agent. (i) The first is the virus theory, in which the virus would have to have unusual biochemical and biophysical characteristics that would help explain the remarkable physicochemical properties (28, 68, 95, 96). (ii) The second is the prion theory, in which the agent is composed exclusively of a host-coded normal cellular protein (PrP^{C}) that becomes partially protease resistant (PrP^{res}), most likely through a posttranslational conformation change after infection. In this theory, there are no nonhost components of the agent. That is, a specific informational molecule (nucleic acid, e.g., RNA or DNA) is not present (7, 90). (iii) The third is the virino theory, which states that the agent consists of a host-derived protein coat, with PrP being one of the candidates for this protective protein, and a small noncoding regulatory nucleic acid (30, 55).

All of the proposed theories have some degree of validity. Proponents of the virus and virino theories conclude that the existence of different scrapie strains unequivocally proves the existence of a nucleic acid component of the infectious agent which, as in conventional viruses, may undergo mutations responsible for phenotypic variations. The problem with these theories is that no agent-specific nucleic acid has been convincingly identified to copurify with infectivity (32, 69, 73, 78, 102). However, absence of evidence is not evidence of absence. It is difficult to hypothesize a protein-only theory to explain the genetics associated with TSEs other than by altered protein conformations (see below). Moreover, chemical, enzymatic, and physical treatments which usually inactivate or degrade nucleic acids have no effect on the transmissible properties of the infectious agent (3, 4, 72, 77). Possible reasons for this are that the amount of nucleic acid of the putative agent is too small to be detected with available techniques and that its tight bond to the protein protects it from chemical or physical inactivation. Weakening the virus and virino theories also is the inability to identify any virus particles under the electron microscope (9, 24) and the failure of an infected host to generate an immune response, although small particles resembling virus structures have been observed by electron microscopy (82).

The prion model involves propagation of a protein-only agent (PrP^{Sc}) whereby PrP^{C} can assume various tertiary structures caused by a combination of host genetics and the introduction of altered (infectious) PrP (PrP^{Sc}). Hence, the structure of the infecting PrP^{res} imprints on the normal cellular precursor (PrP^{C}), which results in a change of shape to the protease-resistant form. It is therefore apparent that the interactions between PrP^{res} and PrP^{C} are important for the pathogenesis of the disease and the conversion of PrP^{C} to PrP^{Sc}. Several studies have shown that sequence compatibility and the direct interaction between PrP^{C} and PrP^{res} are requirements for the efficient conversion of PrP^{C} to the protease-resistant form (6, 57, 87, 88, 100). It has further been demonstrated that this PrP sequence compatibility for the conversion correlates with the inter- and intraspecies disease transmissibility in vivo (8, 58, 93). Furthermore, it has recently been shown that the interaction between the two PrP isoforms is selective and highly localized when examined under near physiological conditions (47). It is suspected that mutations in the PrP gene may render resulting proteins susceptible to "flipping," and the shape changes account for what are commonly referred to as "strain" differences. Several explanations for scrapie strain genetics in the context of the prion theory have been suggested, but none have been proven (89, 110). A recent publication suggests that passaging the agent through different hosts causes the conformational change, possibly limiting prion diversity (99).

However, this study could not eliminate the possibility of the existence of a mixture of scrapie strains. In addition, studies by Telling et al. (109) on the transmission of the human infectious agent from the brains of CJD patients to transgenic mice containing the human PrP gene postulated the involvement of a macromolecule termed protein X. This hypothetical protein is thought to participate in the conversion of PrP^{C} to PrP^{Sc}. It should be pointed out that the prion theory fails to explain how the PrP of the infecting agent originally assumed the aberrant structure associated with infectivity and how the different structures originated as a function of the different strains. Although numerous scrapie strains can be differentiated in a single host (i.e., sheep), the PrPs associated with these strains have not shown any biochemical and molecular differences. The prion theory also does not explain cross-species transmission experiments in which recipient mice develop disease without the formation of detectable PrP^{Sc} (63, 67). Studies on this final point by Hedge and colleagues (44) have suggested that accumulation of a PrP mutation-induced transmembrane form of PrP^{C}, termed ${}^{Ctm}PrP$, can cause neurodegeneration in the absence of PrP^{Sc}. They further suggest a model for an interrelationship of ${}^{Ctm}PrP$ and PrP^{res}, in which Ctm causes the neurodegeneration seen in both genetic and transmissible prion diseases.

BSE seems to be caused by a single strain type. This BSE strain is different from historical and contemporary isolates from sheep and goats with natural scrapie, as determined by study of incubation periods and brain "lesion profiles" in mice.

Regardless of whether the prion (PrP^{Sc}) is or is not the etiologic agent, the partially protease-resistant form of the prion protein is a marker of infection. There are currently a number of tests which may be used to detect the presence of the PrP^{res}, including immunohistochemistry and Western blot analysis.

BOVINE SPONGIFORM ENCEPHALOPATHY

BSE, widely known as mad cow disease, is a chronic, degenerative disease affecting the CNS of cattle. Worldwide there have been more than 180,000 cases since the disease was first diagnosed in 1986 in Great Britain. BSE has had a substantial impact on the livestock industry in the United Kingdom. The disease has also been confirmed in native-born cattle in Belgium, Denmark, France, Ireland, Luxembourg, Liechtenstein, The Netherlands, Northern Ireland, Portugal, and Switzerland. However, over 95% of all BSE cases have occurred in the United Kingdom. BSE is not known to exist in the United States.

Epizootiology

There are different scientific hypotheses about the origins of BSE. The epidemiologic data suggest that BSE in Great Britain is an extended common-source epidemic involving feed containing TSE-contaminated meat and bone meal as a protein source. The causative agent is suspected to be from either scrapie-affected sheep or cattle with a previously unidentified TSE. Changes in rendering operations in the early 1980s—particularly the removal of a solvent extraction process that included a steam heat treatment—may have played a part in the appearance of the disease and the subsequent amplification of the agent in the food chain. A ban on feeding animal protein of ruminant origin to ruminants was enacted in Great Britain in July 1988 (116). Recent studies have questioned the role of the solvents in reducing levels of infectivity. These studies did not look at the process as a whole (108).

In Great Britain, the epidemic peaked in 1992 and 1993, with approximately 1,000 cases being reported per week (Fig. 1). In 2000, it remains on the decline, with approximately 50 cases reported per week. Cases which have been detected in other countries appear be a result of importations of live cattle or, more significantly, contaminated feed from Great Britain.

There is no evidence that BSE spreads horizontally, i.e., by contact between unrelated adult cattle or from cattle to other species. Some evidence suggests that maternal transmission may occur at an extremely low level. Results of British research show low levels of transmission of BSE from affected cows to their offspring. These results show that there is an approximately 9% increase in the occurrence of BSE in the offspring of BSE-affected dams compared to calves born to dams in which BSE was not detected. The study did not ascertain if this was the result of genetic factors or true transmission. The research did, however, point out that at this level, if maternal transmission does occur, it alone will not sustain the epidemic (117).

A transmissible spongiform encephalopathy has been diagnosed in eight species of captive wild ruminants as well as exotic (cheetahs, pumas, a tiger, and an ocelot) and domestic cats. There have been about 81 domestic cat cases of FSE in Great Britain, one in Norway, one in Northern Ireland, and one in Liechtenstein. The agent isolated from several of these cases is indistinguishable from BSE in cattle using strain typing in mice, suggesting that FSE is actually BSE in exotic and domestic cats. This also appears to be true for the other ruminants. Epidemiological evidence suggests that BSE-contaminated feed is the primary source of infection in these species (76).

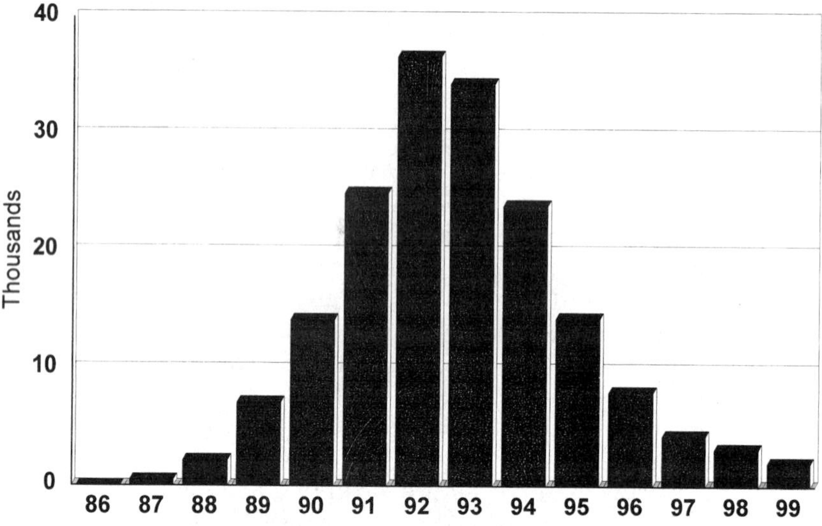

Figure 1. Number of confirmed cases of BSE in Great Britain by year of restriction, 1986 through 1999. The data include 46 cases that were not subject to official restrictions and were identified by proactive surveillance or at autopsy. Source: United Kingdom Ministry of Agriculture, Fisheries and Food.

It has also been suggested that 68 definite and probable cases (as of April 2000) of a variant form of CJD (vCJD) in Great Britain (United Kingdom Department of Health monthly Creutzfeldt-Jakob figures, April 2000), one case in Ireland, and two cases in France may be linked to exposure to BSE before the introduction of a specified bovine offal ban at slaughter in 1989. The ban excluded from human consumption brain, spinal cord, and other tissues with potential BSE infectivity.

BSE has been experimentally transmitted to the following species via intracerebral inoculation: cattle (111, 112), sheep, goats (34, 35), mink (94), pigs (29), marmosets (1), macaques (62), and mice (38). Intracerebral transmission was attempted in hamsters but was not successful. BSE has successfully been transmitted via the oral route to cattle, sheep, goats (35), mice (2), and mink (94). Oral transmission has been attempted in swine. The inoculated swine were euthanized at 84 months of age and had not exhibited any signs of a TSE. Parenteral and oral transmission has also been attempted in chickens with no evidence of disease thus far. Tissues from the BSE-challenged pigs and chickens are being inoculated into susceptible mice to look for residual infectivity. The subpassages from chickens are over 42 months postinoculation with no evidence of disease

(United Kingdom Ministry of Agriculture, Fisheries, and Food home page, www.maff.gov.uk/animalh/bse).

BSE and CJD: human health concerns

On 20 March 1996, the United Kingdom's SEAC announced the identification of 10 cases of a new variant form of CJD. All of the patients developed onset of illness in 1994 or 1995. The following features describe how these 10 cases differed from the sporadic form of CJD:

- The affected individuals were much younger than the classical CJD patient. Typically, CJD patients are over 63 years old. The average patient age for the variant form of CJD was 27.5 (range, 16 to 42) years.
- The course of the disease in vCJD averaged 13 months. Classical CJD cases average a 6-month duration.
- In the variant cases, electroencephalographic electrical activity in the brain was not typical of sporadic CJD.
- Although brain pathology was recognizable as CJD, the pattern was different from that of sporadic CJD, with large aggregates of prion protein plaques.

Another difference between CJD and vCJD is the finding of PrP^{res} in peripheral tissue such as tonsil (46). This has led to concern about potential contamination of the blood supply. Measures such as leukodepletion of the blood supply in the United Kingdom and blood deferrals in the United States have been taken. In the United States, donors who have resided in the United Kingdom for 6 months or more during the high-risk period are deferred from donating blood.

Epidemiologic and case studies have not revealed a common risk factor among the cases of vCJD. According to the SEAC, all victims were reported to have eaten beef or beef products in the last 10 years, but none had knowingly eaten brain material. One of the affected individuals had been a vegetarian since 1991 (119). Although the patients have not been reported to have eaten brain material, it is possible that exposure was through dietary contamination or perhaps another peripheral route (118).

The SEAC concluded that although there was no direct scientific evidence of a link between BSE and vCJD, based on current data and in the absence of any credible alternative, the most likely explanation was that the cases were linked to exposure to BSE before the introduction of control measures, in particular, the specified bovine offal ban in 1989. Research reported in later 1996 and 1997 has found evidence to further support a causal association between vCJD and BSE. Two significant studies published in the 2 October 1997 edition of *Nature* led the SEAC to conclude that the BSE agent is highly likely to be the cause of vCJD. Moira Bruce and colleagues at the Institute for Animal Health in Edinburgh, Scotland, inoculated three panels of inbred mice and one panel of crossbred mice with isolates obtained from cases of BSE, vCJD, and sporadic CJD. Interim results indicate that mice inoculated with BSE material showed the same pattern of incubation time, clinical signs, and brain lesions as mice inoculated with tissues from patients with vCJD. This provides evidence that BSE and vCJD have the same signature or are the same "strain." In addition, sporadic CJD and known scrapie strains were not similar to vCJD or BSE (16).

Results from another study published by Hill et al. (Imperial College School of Medicine, London, United Kingdom) strongly support Bruce's results. The paper by Hill et al. reports findings of BSE transmission to transgenic mice expressing only human PrP (45).

Another paper, by Collinge et al., in the 24 October 1996 edition of *Nature* discusses a molecular analysis of prion strain variation in relationship to the etiology of vCJD. In the paper, Collinge and colleagues demonstrate that two of the three forms of CJD (sporadic and iatrogenic) can be distinguished, after treatment by proteolytic cleavage, by different band sizes on Western blot analysis. Types 1 and 2 are associated with different clinicopathological phenotypes of sporadic CJD, and type 3 is seen in iatrogenic CJD cases where there is a peripheral route of inoculation. Iatrogenic cases of CJD with a direct CNS exposure resemble the pattern manifested by sporadic CJD. vCJD, while resembling type 3 to some degree, can be differentiated from each of the three types by a specific pattern of band intensities. Furthermore, glycoform patterns of vCJD were "closely similar to" patterns from wild-type mice inoculated with BSE and "closely resembled" FSE as well as patterns from macaques inoculated with BSE. However, natural BSE was not detected on Western blot using the antibodies in the above-mentioned studies (3F4 monoclonal or R073 polyclonal). The use of rabbit antibody to synthetic human PrP peptide (95 to 108) with BSE from its natural host, the cow, did generate a signal and glycoforms closely similar to those from transmitted BSE and vCJD (26).

More recently, studies using transgenic animals expressing the bovine PrP have supported the view that BSE-infected cattle are responsible for vCJD (101). These mice not only propagated the BSE infectious agent in the absence of a species barrier but also were highly susceptible to vCJD and natural sheep scrapie. Furthermore, the transgenic mice inoculated with either vCJD or BSE had indistinguishable disease characteristics.

The Health and Safety Executive (HSE) in the United Kingdom now advises that BSE must be considered a biological agent (human pathogen) within the meaning of the Control of Substances Hazardous to Health Regulations 1994 (HSE press release, October 1997).

Infectivity of tissues, products, and body fluids

In naturally infected animals, the agent has been identified in the brain, spinal cord, and retina. The agent was identified by mouse bioassay. The route of inoculation into the mice was intracranial. The naturally infected animals were adult cattle exhibiting clinical signs of disease (38).

Experiments with mice that were fed milk, mammary gland, placenta, lymph nodes, or spleen from sick animals have failed to transmit the disease within the natural life span of the mice or to establish subclinical infection of the lymphoreticular system (74). Studies were also carried out in mice to determine the propagation of the TSE agents associated with oral contamination of mouse-adapted sheep scrapie strains and mouse-adapted BSE strains (66). Using PrP^{Sc} as a marker for infectivity, these studies indicated that the pathway of infection for both strains began with the lymphoreticular system associated with the digestive tract, i.e., Peyer's patches and the mesenteric lymph nodes. This was followed by entrance into the bloodstream and distribution to organs of the lymphoreticular system that are not related to the digestive tract, such as the spleen and the axillary lymph nodes.

Another study was conducted to examine the pathogenesis of BSE in cattle, i.e., the replication (tissue distribution) of the agent during the incubation period. This study, which has not yet been completed, has identified the agent via mouse bioassay in the distal ileum of experimentally infected calves. It is thought that the agent may be associated with the lymphoid tissue of the intestines. The calves were 4 months old at the time of oral dosing (111). First isolation of the agent in the distal ileum was made at 6 months after challenge. Subsequent isolations from the distal ileum were made at 10, 14, and 18 months after dosing (111). This study has also identified infectivity in bone marrow, trigeminal ganglion, dorsal root ganglion, brain, and spinal cord (112).

No infectivity has been found by parenteral or oral challenge in over 40 other tissues from clinically ill cattle using the mouse bioassay. It appears as if the distribution of the BSE agent is not as diverse as that of the scrapie agent in sheep. Another consideration which must be noted is the species barrier of the mouse bioassay. There is a possibility that the agent is present but at such low levels that the bioassay is not sensitive enough to detect it (www.maff.gov.uk/animalh/bse).

Clinical Signs

Cattle affected by BSE develop a progressive degeneration of the nervous system. Affected animals may display changes in temperament, abnormalities of posture and movement, and changes in sensation. More specifically, the signs include apprehension, nervousness, or aggression; incoordination, especially hind limb ataxia; tremor; difficulty in rising; and hyperaesthesia to sound and touch. In addition, many animals have decreased milk production and/or loss of body condition despite continued appetite. There is no treatment, and affected cattle die.

The incubation period ranges from 2 to 8 years. Following the onset of clinical signs, the animal's condition gradually deteriorates until the animal becomes recumbent and dies or is destroyed. This usually takes from 2 weeks to 6 months. Most cases in Great Britain have occurred in dairy cows (Friesians) between 3 and 6 years of age (116). The youngest confirmed bovine with BSE has been 20 months old, and the oldest has been 18 years and 10 months old.

Diagnosis

The diagnosis of BSE is based on the occurrence of clinical signs of the disease and currently is confirmed by postmortem histopathological examination of brain tissue. Bilaterally symmetrical degenerative changes are usually seen in the gray matter of the brain stem. These changes are characterized by vacuolation or microcavitation of nerve cells in the brain stem nuclei. The neural perikarya and axons of certain brain stem nuclei contain intracytoplasmic vacuoles of various sizes, giving the impression of a spongy brain. Hypertrophy of astrocytes often accompanies the vacuolation (113). A diagnosis may also be made by the detection of scrapie-associated fibrils by electron microscopy.

Supplemental tests are available to enhance the ability to diagnose BSE. Research has shown that the partially protease-resistant form of the prion protein (PrP^{BSE}) is found in the brain of BSE-infected cattle. Two tests may currently be used to detect PrP^{BSE}, immunohistochemistry and the Western blot technique. In the past, if the brain tissue was not harvested shortly after the animal's death, autolysis could make it very difficult to confirm a diagnosis. The Western blot allows a diagnosis of BSE even if the brain has been frozen or autolyzed.

Other tests currently being evaluated:

1. DELFIA technology, which is a two-site noncompetitive immunometric procedure using two different monoclonal antibodies. This test takes less than 24 h to complete.

2. High-throughput chemiluminescent enzyme-linked immunosorbent assay using a polyclonal antibody. This test can be completed in under 4 h.
3. Sandwich immunoassay using two monoclonal antibodies, which can be completed in less than 24 h.

The potential live-animal tests, including some in development, are: tests specific for the partially protease-resistant form of the prion protein, a capillary electrophoresis test and a Western blot test with increased sensitivity, and tests which identify unique substances of infected animals or humans, a cyclic voltametric method which describes unique substances in urine and an immunoblot test describing unique substances in cerebrospinal fluid. Tests for PrP^{BSE} in the preclinical cow may not be as successful as they are appearing to be in scrapie diagnosis, as PrP^{BSE} has not been detected in peripheral tissues to date.

Differential diagnoses for BSE include rabies, listeriosis, nervous ketosis, milk fever, grass tetany, lead poisoning, and other toxicities or etiological agents which affect the nervous system or musculoskeletal system of adult cattle.

Treatment, Prevention, and Control

There is no known treatment for BSE or any of the TSEs, and there is no preventative vaccine. However, various compounds, including amphotericins, sulfated polyanions, Congo red dye, anthracycline antibiotics, synthetic peptides, and branched polyamines, have been tested for their ability to inhibit TSE agent replication (18–20, 103, 105, 106). Although they have been shown to be somewhat effective in vitro, none have demonstrated promising results in in vivo model systems.

BSE from foreign sources may be prevented by the implementation of regulations prohibiting the import of live ruminants and ruminant products (especially meat and bone meal and offal) from countries where BSE may exist. Since the origin of BSE remains unknown, preventing an epidemic of BSE would involve, at a minimum, a prohibition of feeding ruminant proteins to ruminants. The prevention program of any country should also include an active surveillance program focused on high-risk cattle for the early detection of BSE. Most countries of the world have prohibited the importation of cattle and bovine products from countries known to have BSE. In addition, many countries have taken steps to enact regulations prohibiting the feeding of ruminant proteins to ruminants. This is true even in countries such as Australia and New Zealand with no known animal TSEs.

Agricultural officials in countries known to have BSE have taken a series of actions to control and hopefully eliminate BSE. These include making BSE a notifiable disease, prohibiting the inclusion of certain animal proteins in ruminant rations (the feed bans vary depending on the amount of BSE detected), and the depopulation of certain populations of cattle thought to be of higher risk due to epidemiological findings.

To prevent human exposure to the BSE agent, numerous countries have established prohibitions on the inclusion of high-risk material in foods, pharmaceuticals, cosmetics, etc. Numerous studies have been carried out to identify the target cells for TSE agent replication following peripheral infection and prior to neuroinvasion (56, 65, 71). By identifying and characterizing the cells supporting replication, development of potential targets for therapy may be possible. The immune system is central in the pathogenesis of TSEs, as evidenced by replication in the spleen and lymph nodes after peripheral infection (56). However, the actual cells involved in the replication process are uncertain. Studies by Brown et al. (12) provide evidence that PrP^{C}-expressing follicular dendritic cells, which reside in the spleen and depend on B-cell signals for maturation, are required for the replication of a mouse scrapie strain. One must be cautious in relating this to other TSEs, since the tissue distribution of the BSE agent can vary considerably, even for the same route of inoculation, depending on the genetic background of the host (34, 66, 112).

U.S. Actions

An active surveillance program that has been in place for 10 years has not detected BSE in the United States. The U.S. Department of Agriculture (USDA), Food and Drug Administration, and industry groups actively work to maintain this status. The measures that the USDA Animal and Plant Health Inspection Service (APHIS) has taken in this regard include prohibitions and restrictions on certain animal and product imports; ongoing surveillance for the disease in the United States; preparation of an emergency response plan in the unlikely event that an introduction were to occur; and ongoing educational efforts. APHIS actively shares information and coordinates closely with other federal agencies as well as the states, livestock and affiliated industries, veterinary and research communities, and consumer groups in order to ensure that the United States has a uniform approach to TSEs which is based on sound scientific information.

APHIS has a comprehensive surveillance program in place in the United States to ensure timely detection and swift response in the unlikely event that an introduction of BSE were to occur. This surveillance program incorporates both the location of imports from countries known to have BSE and targeted active and passive surveillance for BSE or other TSE in cattle.

APHIS has conducted a traceback effort to locate each of the 496 British cattle that were imported into the United States between 1 January 1981 and July 1989. In July 1989, the United States prohibited the importation of ruminants from countries affected with BSE. As of January 2000, only four of these animals are known to be alive in the United States, and these animals are being carefully monitored by APHIS personnel on an ongoing basis. In addition, 2 head of cattle imported from Belgium in 1996 and 34 imported from other European countries in 1996 and 1997 are now under quarantine. APHIS, in cooperation with the states and industry, continues to purchase these animals for diagnostic purposes. No evidence of BSE has been found in any of the imported animals.

The United States has had an active surveillance program for BSE since May 1990. BSE is a notifiable disease, and more than 250 federal and state regulatory veterinarians are specially trained to diagnose foreign animal diseases, including BSE. APHIS leads an interagency surveillance program, which includes the Food Safety Inspection Service and the Centers for Disease Control and Prevention. The surveillance samples include field cases of cattle exhibiting signs of neurological disease, cattle condemned at slaughter for neurological reasons, rabies-negative cattle submitted to public health laboratories, neurological cases submitted to veterinary diagnostic laboratories and teaching hospitals, and random samplings of cattle which are nonambulatory at slaughter. Figure 2 shows that as of March 2000, 10,285 brains had been examined for BSE or another form of TSE in cattle. No evidence of either condition has been detected by histopathology or immunohistochemistry.

As of 12 December 1997, APHIS has prohibited the importation of live ruminants and most ruminant products from all of Europe until a thorough assessment of the risks can be made. The new restrictions apply to Albania, Austria, Bosnia-Herzegovina, Bulgaria, Croatia, the Czech Republic, Denmark, Yugoslavia, Finland, Germany, Greece, Hungary, Italy, Macedonia, Norway, Poland, Romania, Slovakia, Slovenia, Spain, and Sweden as well as European countries which have reported cases of BSE in native cattle. This action was taken because in 1997, The Netherlands, Belgium, and Luxembourg reported their first cases of BSE in native-born cattle. There is evidence that European countries may have had high BSE risk factors for several years and less than adequate surveillance. Additionally, Belgium reported that a cow diagnosed with BSE was processed into the animal food chain.

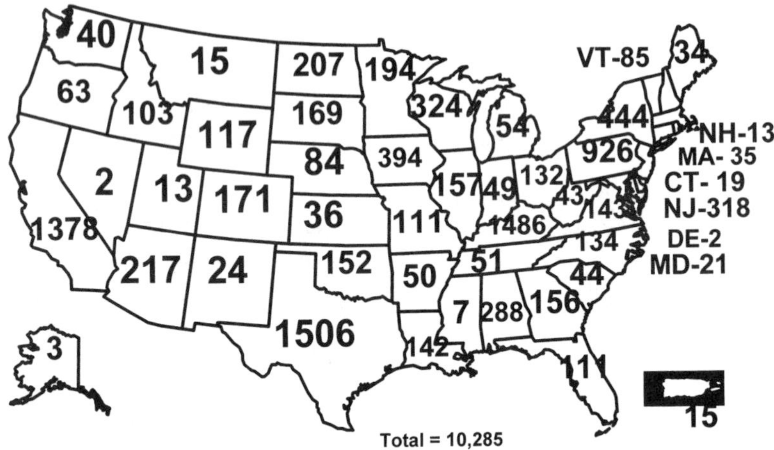

Figure 2. Number of bovine brains submitted for BSE testing, by state, from 10 May 1990 through 31 March 2000. None tested positive for BSE. Source: National Veterinary Services Laboratory, APHIS, USDA.

The U.S. Food and Drug Administration has established regulations which prohibit the feeding of most mammalian proteins to ruminants. The effective date of this regulation was 4 August 1997.

CHRONIC WASTING DISEASE

Epidemiology

CWD was recognized as a syndrome by biologists working with captive deer in the late 1960s. It occurs in deer and elk on a few wildlife research facilities and in free-ranging cervids in limited areas of nine contiguous counties in southeastern Wyoming and north-central Colorado. Recently, CWD was diagnosed among privately owned elk on game farms in Saskatchewan, Canada, and in South Dakota, Nebraska, Colorado, Montana, and Oklahoma. Fewer than 250 cases of CWD have been diagnosed over the last 20 years, mostly in captive cervids from research facilities and in free-ranging mule deer.

Surveillance of free-ranging deer and elk for evidence of CWD has taken two forms. Wild cervids showing clinical signs compatible with the case definition are examined in veterinary diagnostic laboratories for spongiform encephalopathy. No cases of CWD in free-ranging cervids has been diagnosed outside the known CWD-endemic areas of Wyoming and Colorado. The second surveillance technique involves voluntary or mandatory submission of heads of hunter-harvested deer and elk so that brains can be examined for evidence of preclinical or subclinical CWD. Hunter-harvested cervid surveillance began in 1983, and thousands of animals have been tested. In specific management units in the CWD-endemic area, estimated prevalence is 1 to 8% in mule deer and 1% in elk. In surrounding management units, estimated prevalence in deer and elk is 1% (104). The epidemiology and prevalence of CWD in privately owned elk is under study.

The mode of transmission of CWD is not known. Epidemiologic evidence strongly suggests that lateral transmission occurs among deer and elk and probably from mule deer to elk and white-tailed deer. Maternal transmission may occur but does not explain many cases of CWD. Concentration of animals in captivity may facilitate transmission; however, CWD is maintained in populations of deer even at moderate to low population densities. There is no evidence that CWD is a food-borne disease associated with consumption of animal protein. The origin of CWD is not known, and the source(s) of CWD in captivity and in the wild is uncertain. As determined from strain typing in mice, it is not the same as BSE or known scrapie strains (16).

The host range of CWD is currently only known to be mule and white-tailed deer and Rocky Mountain elk. Subspecies of *Cervus elaphus* are probably susceptible, but it is not known if other cervids can develop CWD. There is currently no evidence that other wild species, domestic animals, or humans are susceptible to CWD, though research to better characterize the host range is currently under way.

Clinical Signs

CWD naturally affects mule deer (*Odocoileus hemionus*), white-tailed deer (*Odocoileus virginianus*), and Rocky Mountain elk (*Cervus elaphus nelsoni*). Affected cervids (members of the deer family) are older than 17 months of age; the majority are 3 to 5 years of age. Sex does not appear to affect susceptibility to CWD. The earliest clinical signs are behavioral changes, which may include alterations in interaction with humans and members of the herd. These subtle changes are often only appreciated by caretakers familiar with the individual animal. With disease progression, behavioral alterations may include periods of stupor and depression. As the name suggests, progressive weight loss is characteristic of CWD and may occur over a long period of time. The duration of clinical signs varies from a few days in unusual cases to as long as a year, but is most often 2 to 3 months. At the terminal stages of disease, animals are emaciated. However, intercurrent disease, especially aspiration pneumonia, may cause an affected animal to die while still in good to fair body condition. In the later stages of disease, clinical signs may include increased drinking and urinating, excessive salivation, and incoordination and trembling. These clinical signs are nonspecific and could be caused by many other diseases affecting wild and captive deer and elk; thus, laboratory examination is required for diagnosis of CWD.

Prevention and Control

Free-ranging deer and elk are not transplanted or moved from the CWD-endemic areas of Wyoming and Colorado to prevent extension of CWD. Surveillance to monitor distribution and prevalence of CWD in free-ranging deer and elk is being conducted so that changes over time can be detected. Because transmission of the disease may be facilitated by high densities of deer and elk, the concentration of these species by artificial feeding is either not done or prohibited in areas where CWD is endemic.

Guidelines are being developed by the elk-farming industry in conjunction with state and federal animal health and wildlife management agencies to address CWD surveillance, quarantine, inventory, and diagnosis in privately owned elk. Under state animal or provincial health agency authority, CWD-affected privately owned elk herds have been quarantined or depopulated.

Education of the public, hunters, meat processors and taxidermists, wildlife biologists and game wardens, animal health officials, and veterinary diagnosticians and pathologists is being conducted through brochures, press releases, journal articles, public presentations, workshops, and a video. Because of the many unknowns surrounding CWD, much research is under way to better characterize the disease, determine host range, develop and validate diagnostic tests, and understand the epidemiology. The results of this work will assist in developing science-based methods for prevention and control.

SCRAPIE

Epidemiology

Over the years there has been much debate over the transmissibility of scrapie. Initially, arguments centered around a genetic versus infectious origin. H. B. Parry felt that scrapie was an autosomal recessive genetic disease which was not naturally infectious. He did concede that affected animals harbored a transmissible agent which was infectious by artificial routes (84). Evidence of transmissibility went on record when Cuille and Chelle (27) were successful in transmitting the disease from affected sheep to healthy sheep via intraocular injection. Chandler (21, 22, 23) added to this discovery by transmitting scrapie to mice. Later, there was information to suggest that scrapie was a naturally occurring contagious disease which was caused by an infectious agent (11, 31, 48). At present it is generally accepted that scrapie is an infectious, contagious disease, with genetics playing an influential role which is not completely understood. The means of natural transmission have not been fully defined.

The scrapie agent is thought to be spread most commonly from ewe to offspring and to other lambs in contemporary lambing groups through contact with the placenta and placental fluids (79, 85, 86, 92). Signs or effects of the disease usually do not appear until 2 to 5 years after the animal is infected. Scrapie has not been shown to be transmitted via semen. With regard to transmission via embryo transfer, different research has obtained different results. Two studies done in the United Kingdom found that offspring of embryos from infected donors did develop scrapie (36, 37). Another study involving embryo transfer was done in the United States by Foote and colleagues (33). This study, which only involved washed embryos, did not result in the transmission of scrapie to progeny via the embryo or uterus.

Sheep may live 1 to 6 months or longer after the onset of clinical signs, but death is inevitable.

Our understanding of the natural transmission of scrapie is like an unfinished puzzle: there are a number of unconnected pieces, but no full picture is in sight. In addition to the questions posed above, there are other unknowns regarding the transmission of scrapie.

1. What is the infective dose for sheep and goats?
2. At what point in the incubation period does the host shed the agent and by what route(s)?
3. Is this shedding continuous or intermittent?
4. Is there a carrier state?

In naturally infected sheep, scrapie infectivity has been found primarily in lymphoreticular and CNS tissue. However, smaller amounts have been detected in liver, pancreas, and nasal mucosa (41).

Genetic variations among different breeds of sheep may play a role in whether sheep will become infected and how quickly clinical signs may appear. Researchers in Edinburgh, Scotland, identified a gene called *Sip* (for scrapie incubation period) that controls the incubation period of scrapie in Cheviot and Swaledale sheep. Individuals with "short-incubation" alleles usually develop signs between 2 and 5 years of age. Sheep with "long-incubation" alleles often die from what appear to be natural causes before the incubation period is complete. Because the incubation period can be longer than 5 years, it is not known to what extent or under what conditions infected sheep with the long-incubation alleles might be able to transmit the disease to healthy sheep. It is likely that the prion protein (*PrP*) gene and the gene controlling scrapie incubation periods (*Sip*) are the same (17, 49, 52, 114). Further research involving additional breeds has suggested that genetic influence may extend beyond incubation length to conferring some degree of disease resistance (5, 25, 50, 51, 53, 54, 60, 61, 80, 115).

The strain of the scrapie agent also appears to affect the development of clinical signs and the length of the incubation period (39, 40).

There is no scientific evidence to indicate that scrapie poses a risk to human health (42, 59). There is no epidemiologic evidence that scrapie of sheep and goats is transmitted to humans, such as through contact on the farm, at slaughter plants, or butcher shops (13).

Clinical Signs

Signs of scrapie may vary widely among individual animals and breeds of sheep and are slowly progressive in their development. Early signs include subtle changes in behavior or temperament; these changes are followed by other signs, which may include tremor, especially of the head and neck, loss of coordination, scratching and rubbing against fixed objects, weight loss despite retention of appetite, biting of feet and limbs, lip smacking, and gait abnormalities, including high-stepping of the forelegs, hopping like a rabbit, and swaying of the back end. An infected animal may appear normal if left undisturbed at rest. However, when stimulated by a sudden noise, excessive movement, or the stress of handling, the animal may tremble or fall down in a convulsion-like state. Several other problems can cause clinical signs similar to scrapie in sheep, including ovine progressive pneumonia, listeriosis, and rabies; the presence of external parasites (lice and mites); pregnancy toxemia; and toxins.

Diagnosis, Prevention, and Control

Routine methods of preventing a disease which is laterally transmitted are vaccination, quarantine, testing, removal, and prohibition of animal and animal product movements. Since the scrapie agent elicits no detectable immune response in the host, vaccines and serological tests have not been possible to date. Current diagnostic tests being used for sheep and goats are not effective for the preclinical live animal. This has made it difficult to ascertain which animals are incubating the disease and may be shedding the agent until clinical signs appear. Because we are unable to identify apparently normal sheep which may be shedding the agent, the only absolute way to prevent an introduction of scrapie has been to prohibit all movements of sheep and goats and their products into an area.

The primary method of postmortem confirmation was histopathology. Currently, other confirmatory methods include animal bioassay, the detection of scrapie-associated fibrils by electron microscopy, immunohistochemical detection of the partially protease-resistant form of the prion protein (PrP^{res}), and the use of immunoblotting techniques to detect PrP^{res}. Tests for the detection of preclinical scrapie are under development or in the process of being validated. Studies conducted in The Netherlands and United States indicate that immunohistochemistry appears to be useful in detecting scrapie in preclinical sheep. This research has revealed the presence of PrP^{res} in the tonsils of preclinical sheep (98) and lymphoid tissue of the third eyelid (81). Capillary electrophoresis, with laser-induced fluorescence detection and immune complex formation, has been shown to detect scrapie. The method involves a competition assay for PrP^{res} using fluorescently tagged peptides from the prion protein and a specific antibody to the peptides. Results thus far have shown this method of detection to be diagnostic for scrapie in brain tissue (97). Other tissues which have been tested are lymph node, spleen, tonsil, and blood. Until some of the new preclinical tests are validated, the ideal means for preventing the introduction of scrapie is to maintain a closed flock, especially with regard to ewes. Any replacement ewes or breeding rams should originate from flocks not known to be affected with scrapie and managed so as to preclude the introduction of scrapie. However, in reality, this is difficult to do since there is no definitive prepurchase test to ensure freedom from the disease. A buyer must rely on the knowledge, integrity, and honesty of the seller. Despite the lack of a definitive live-animal diagnostic test, there are some preventive steps a potential purchaser or new owner of purchased animals may take.

1. Know the flock health status for the flocks in which an animal has been housed.
2. Test for scrapie susceptibility? If research finds that there is definitive evidence that certain genotypes are resistant to scrapie infection or reduce the transmission of scrapie, genetic testing would prove very useful to a flock owner. Currently there is evidence that breeding for certain PrP

genotypes will reduce if not eliminate clinical disease. However, research has not been completed to eliminate the possibility of a "carrier" animal.

3. If a ewe of unknown or questionable disease status is present in the flock, spread to other animals can be minimized by keeping her separate from the rest of the flock at lambing time and for a period of 3 months following lambing. In addition, her offspring should also be segregated from other lambs. If these animals show no abnormalities by the time they are 4.5 years old, they can safely be included with the rest of the flock at lambing time, provided that their opportunity for exposure occurred within the first 6 months of their life. Ideally, all offspring of these animals of questionable disease status should be kept separate at lambing time until the dam is believed to be very low risk (i.e., has attained 4.5 years of age and remains clinically normal).
4. If the flock develops scrapie, the risk of further spread and/or reintroduction of the disease can be minimized through removal of high-risk animals, careful cleaning and disinfection of facilities, and improved management of animals at lambing time, especially by segregating animals into small groups and keeping the risk classification of animals in each group at the same level. In certain countries, the method of control is total depopulation of the infected flock and in some cases total depopulation of the source flock.
5. Although meat and bone meal contaminated with the scrapie agent have not been definitively proven to play a role in the transmission of scrapie, it is prudent to prohibit the feeding of ruminant meat and bone meal to sheep and goats in countries where scrapie or other TSEs are endemic.

The reporting of clinical disease can be influenced by many outside factors. One of the most important factors which influence reporting is education of producers and practicing veterinarians. Producers and practitioners must be aware that scrapie is a reportable disease and know where to report it. They must also be familiar with the clinical signs of scrapie and how to differentiate these from other diseases. Hence, education is an essential component of scrapie surveillance.

Other pressures that negatively impact the willingness to report scrapie relate to the consequences and impact on individual flocks and may in some cases impact the status of the entire country or region. Depending on control measures in place, all movements from a flock may be restricted or the flock owner may be forced to depopulate the flock. At a minimum, a report of scrapie usually deflates the value of the animals in the flock.

REFERENCES

1. **Baker, H. F., R. M. Ridley, and G. A. H. Wells.** 1993. Experimental transmission of BSE and scrapie to the common marmoset. *Vet. Rec.* **132:**403–406.
2. **Barlow, R. M., and D. J. Middleton.** 1990. Dietary transmission of bovine spongiform encephalopathy to mice. *Vet. Rec.* **126:**111–112.
3. **Bellinger Kawahara, C. G., J. E. Cleaver, T. O. Diener, and S. B. Prusiner.** 1987. Purified scrapie prions resist inactivation by UV irradiation. *J. Virol.* **61:**159–166.
4. **Bellinger Kawahara, C. G., T. O. Diener, M. P. McKinley, D. F. Groth, D. R. Smith, and S. B. Prusiner.** 1987. Purified scrapie prions resist inactivation by procedures that hydrolyze, modify, or shear nucleic acids. *Virology* **160:**271–274.
5. **Belt, P. B. G. M., I. H. Muileman, B. E. C. Schreuder, J. B. Ruijter, A. L. J. Gielkens, and M. A. Smits.** 1995. Identification of five allelic variants of the sheep PrP gene and their association with natural scrapie. *J. Gen. Virol.* **76:**509–517.
6. **Bessen, R. A., D. A. Kocisko, G. J. Raymond, S. Nandan, P. T. Lansbury, Jr., and B. Caughey.** 1995. Non-genetic propagation of strain-specific properties of scrapie prion protein. *Nature* **375:**698–700.
7. **Bolton, D. C., and P. E. Bendheim.** 1988. A modified host protein model of scrapie. *Ciba Found. Symp.* **135:**164–181.
8. **Bossers, A., P. B. G. M. Belt, G. J. Raymond, B. Caughey, R. de Vries, and M. A. Smits.** 1997. Scrapie susceptibility-linked polymorphisms modulate the in vitro conversion of sheep prion protein to protease-resistant forms. *Proc. Natl. Acad. Sci. USA* **94:**4931–4936.
9. **Bots, G. T., J. C. Man, and A. Verjaal.** 1971. Virus-like particles in brain tissue from two patients with Creutzfeldt-Jakob disease. *Acta Neuropathol.* (Berlin) **18:**267–270.
10. **Bratberg, B., K. Ueland, and G. A. H. Wells.** 1995. Feline spongiform encephalopathy in a cat in Norway. *Vet. Rec.* **136:**444.
11. **Brotherston, J. G., C. C. Renwick, J. T. Stamp, I. Zlotnik, and I. H. Pattison.** 1968. Spread of scrapie by contact to goats and sheep. *J. Comp. Pathol.* **78:**9–17.
12. **Brown, K. L., K. Stewart, D. L. Ritchie, N. A. Mabbott, A. Williams, H. Fraser, W. I. Morrison, and M. E. Bruce.** 1999. Scrapie replication in lymphoid tissues depends on prion protein-expressing follicular dendritic cells. *Nat. Med.* **5:**1308–1312.
13. **Brown, P., F. Cathala, R. F. Raubertas, D. C. Cajdusek, and P. Castaigne.** 1987. The epidemiology of Creutzfeldt-Jakob disease: conclusion of a 15-year investigation in France and review of the world literature. *Neurology* **37:**895–904.
14. **Brown, P.** 1988. The clinical neurology and epidemiology of Creutzfeldt-Jakob disease, with special reference to iatrogenic cases. *Ciba Found. Symp.* **135:**3–23.
15. **Brown, P.** 1988. Human growth hormone therapy and Creutzfeldt-Jakob disease: a drama in three acts. *Pediatrics* **81:**85–92.
16. **Bruce, M. E., R. G. Will, J. W. Ironside, I. McConnell, D. Drummond, A. Suttie, L. McCardle, A. Chree, J. Hope, C. Birkett, S. Cousens, H. Fraser, and C. J. Bostock.**

1997. Transmissions to mice indicate that a new variant CJD is caused by the BSE agent. *Nature* **389:**498–501.

17. **Carlson, G. A., D. T. Kingsbury, P. A. Goodman, S. Coleman, S. T. Marshall, S. J. DeArmond, D. Westaway, and S. B. Prusiner.** 1986. Linkage of prion protein and scrapie incubation time genes. *Cell* **46:**503–511.
18. **Caughey, B., K. Brown, G. J. Raymond, G. E. Katzenstein, and W. Thresher.** 1994. Binding of the protease-sensitive form of prion protein PrP to sulfated glycosaminoglycan and Congo red. *J. Virol.* **68:**2135–2141.
19. **Caughey, B., D. Ernst, and R. Race.** 1993. Congo red inhibition of scrapie agent replication. *J. Virol.* **67:**6270–6272.
20. **Chabry, J., B. Caughey, and B. Chesebro.** 1998. Specific inhibition of in vitro formation of protease resistant prion protein by synthetic peptides. *J. Biol. Chem.* **273:**13203–13207.
21. **Chandler, R. L.** 1961. Encephalopathy in mice produced by inoculation with scrapie brain material. *Lancet* **i:**1378–1379.
22. **Chandler, R. L.** 1962. Encephalopathy in mice. *Lancet* **i:**107–108.
23. **Chandler, R. L.** 1963. Experimental scrapie in the mouse. *Res. Vet. Sci.* **4:**276.
24. **Cho, H. J., and A. S. Greig.** 1975. Isolation of 14-nm virus-like particles from mouse brain infected with scrapie agent. *Nature* **257:**685–686.
25. **Clouscard, C., P. Beaudry, J. M. Elsen, D. Milan, M. Dussaucy, C. Bounneau, F. Schelcher, J. Chatelain, J. J. Launay, and J. L. Laplanche.** 1995. Different allelic effects of the codon 136 and 171 of the prion protein gene in sheep with natural scrapie. *J. Gen. Virol.* **76:**2097–2101.
26. **Collinge, J., K. C. L. Sidle, J. Meads, J. Ironside, and A. F. Hill.** 1996. Molecular analysis of prion strain variation and the aetiology of 'new variant' CJD. *Nature* **383:**685–690.
27. **Cuille, J., and P. L. Chelle.** 1936. La maladie dite tremblante du mouton: est-elle inoculable? *Acad. Sci. Paris* **203:**1552–1554.
28. **Czub, M., H. R. Braig, and H. Diringer.** 1988. Replication of scrapie agent in hamsters infected intracerebrally confirms the pathogenesis of an amyloid-inducing virosis. *J. Gen. Virol.* **69:**1753–1756.
29. **Dawson, M., G. A. H. Wells, B. N. J. Parker, and A. C. Scott.** 1990. Primary parenteral transmission of bovine spongiform encephalopathy to the pig. *Vet. Rec.* **127:**338.
30. **Dickinson, A. G., and G. W. Outram.** 1979. The scrapie replication-site hypothesis and its implication for pathogenesis, p. 13–32. *In* S. B. Prusiner and W. J. Hadlow (ed.), *Slow Transmissible Diseases of the Nervous System,* vol. 2. Academic Press, New York, N.Y.
31. **Dickinson, A. G., J. T. Stamp, and C. C. Renwick.** 1974. Maternal and lateral transmission of scrapie in sheep. *J. Comp. Pathol.* **84:**19–25.
32. **Duguid, J. R., R. G. Rohwer, and B. Seed.** 1988. Isolation of cDNAs of scrapie-modulated RNAs by subtractive hybridization of a cDNA library. *Proc. Natl. Acad. Sci. USA* **85:**5738–5742.
33. **Foote, W. C., W. Clark, A. Maciulis, J. W. Call, J. Hourrigan, C. Evans, M. Marshall, and M. de Camp.** 1993. Prevention of scrapie transmission in sheep using embryo transfer. *Am J. Vet. Res.* **54:**1863–1868.
34. **Foster, J. D., M. Bruce, I. McConnell, A. Chee, and H. Fraser.** 1996. Detection of BSE infectivity in brain and spleen of experimentally infected sheep. *Vet. Rec.* **133:**546–548.
35. **Foster, J. D., J. Hope, and H. Fraser.** 1993. Transmission of bovine spongiform encephalopathy to sheep and goats. *Vet. Rec.* **133:**339–341.
36. **Foster, J. D., W. A. C. McKelvey, J. A. Mylne, A. Williams, N. Hunter, J. Hope, and H. Fraser.** 1992. Studies on maternal transmission of scrapie in sheep by embryo transfer. *Vet. Rec.* **130:**341–343.
37. **Foster, J. D., N. Hunter, A. Williams, J. A. Mylne, W. A. C. McKelvey, J. Hope, H. Fraser,**

and C. Bostock. 1996. Observations on the transmission of scrapie in experiments using embryo transfer. *Vet. Rec.* **138:**559–562.

38. **Fraser, H., I. McConnell, G. A. H. Wells, and M. Dawson.** 1988. Transmission of bovine spongiform encephalopathy to mice. *Vet. Rec.* **123:**472.
39. **Goldmann, W., N. Hunter, G. Smith, J. Foster, and J. Hope.** 1994a. PrP genotypes and Sip gene in Cheviot sheep form the basis for scrapie strain typing in sheep. *Ann. N.Y. Acad. Sci.* **724:**296–299.
40. **Goldmann, W., N. Hunter, G. Smith, J. Foster, and J. Hope.** 1994b. PrP genotype and agent effects in scrapie: change in allelic interaction with different isolates of agent in sheep, a natural host of scrapie. *J. Gen. Virol.* **75:**989–995.
41. **Hadlow, W. J., R. C. Kennedy, and R. E. Race.** 1982. Natural infection of Suffolk sheep with scrapie virus. *J. Infect. Dis.* **146:**657–664.
42. **Harries Jones, R., R. Knight, R. G. Will, S. N. Cousens, P. G. Smith, and W. B. Mathews.** 1988. Creutzfeldt-Jakob disease in England and Wales, 1980–1984: a case-control study of potential risk factors. *J. Neurol. Neurosurg. Psychiatry* **51:**1113–1119.
43. **Hartsough, G. R., and D. Burger.** 1965. Encephalopathy of mink. I. Epizootologic and clinical observations. *J. Infect. Dis.* **115:**387–392.
44. **Hedge, R. S., P. Tremblay, D. Groth, S. J. DeArmond, S. B. Prusiner, and V. R. Lingappa.** 1999. Transmissible and genetic prion diseases share a common pathway of neurodegeneration. *Nature* **402:**822–826.
45. **Hill, A. F., M. Desbruslais, S. Joiner, K. C. L. Sidle, I. Gowland, and J. Collinge.** 1997. The same prion strain causes vCJD and BSE. *Nature* **389:**448–450.
46. **Hill, A. F., M. Zeidler, J. W. Ironside, and J. Collinge.** 1997. Diagnosis of new variant Creutzfeldt-Jakob disease by tonsil biopsy. *Lancet* **349:**99–100.
47. **Horiuchi, M., and B. Caughey.** 1999. Specific binding of normal prion protein to the scrapie form via a localized domain initiates its conversion to the protease-resistant state. *EMBO J.* **18:**3193–3203.
48. **Hourrigan, J., A. Klingsporn, W. W. Clark, and M. deCamp.** 1979. Epidemiology of scrapie in the United States, p. 331–356. *In* S. B. Prusiner and W. J. Hadlow (ed.), vol. 1. *Slow Transmissible Diseases of the Nervous System.* Academic Press, New York, N.Y.
49. **Hunter, N., J. D. Foster, A. G. Dickinson, and J. Hope.** 1989. Linkage of the gene for the scrapie-associated fibril protein (PrP) to the Sip gene in Cheviot sheep. *Vet. Rec.* **124:**363–366.
50. **Hunter, N., W. Goldmann, G. Benson, J. D. Foster, and J. Hope.** 1993. Swaledale sheep affected by natural scrapie differ significantly in PrP genotype frequencies from healthy sheep and those selected for reduced incidence of scrapie. *J. Gen. Virol.* **74:**1025–1031.
51. **Hunter, N., W. Goldmann, J. Foster, D. Cairns, and G. Smith.** 1997. Natural scrapie and PrP genotype: case-control studies in British sheep. *Vet. Rec.* **141:**137–140.
52. **Hunter, N., J. Hope, I. McConnell, and A. G. Dickinson.** 1987. Linkage of the scrapie-associated fibril protein (PrP) gene and Sinc using congenic mice and restriction fragment length polymorphism analysis. *J. Gen. Virol.* **68:**2711–2716.
53. **Hunter, N., W. Goldmann, G. Smith, and J. Hope.** 1994. The association of a codon 136 PrP gene variant with the occurrence of natural scrapie. *Arch. Virol.* **137:**171–177.
54. **Ikeda, T., M. Horiuchi, N. Ishiguro, Y. Muramatsu, G. D. Kai-Uwe, and M. Shinagawa.** 1995. Amino acid polymorphisms of PrP with reference to the onset of scrapie in Suffolk and Corriedale sheep in Japan. *J. Gen. Virol.* **76:**2577–2581.
55. **Kimberlin, R. H.** 1982. Scrapie agent: prions or virinos? *Nature* **297:**107–108.
56. **Kimberlin, R. H., and C. A. Walker.** 1979. Pathogenesis of mouse scrapie: dynamics of agent replication in spleen, spinal cord, and brain after infection by different routes. *J. Comp. Pathol.* **89:**551–562.

57. **Kocisko, D. A., J. H. Come, S. A. Priola, B. Chesebro, G. J. Raymond, P. T. Lansbury, Jr., and B. Caughey.** 1994. Cell-free formation of protease-resistant prion protein. *Nature* **370:**471–474.
58. **Kocisko, D. A., S. A. Priola, G. J. Raymond, B. Chesebro, P. T. Lansbury, Jr., and B. Caughey.** 1995. Species specificity in the cell-free conversion of prion protein to protease-resistant forms: a model for the scrapie species barrier. *Proc. Natl. Acad. Sci. USA* **92:**3923–3927.
59. **Kondo, K., and Y. Kuroiiwa.** 1982. A case control study of Creutzfeldt-Jakob disease: association with physical injuries. *Ann. Neurol.* **11:**377–381.
60. **Laplanche, J. L., J. Chatelain, D. Westaway, S. Thomas, M. Dussaucy, J. Brugere-Picoux, and J. M. Launay.** 1993. PrP polymorphisms associated with natural scrapie discovered by denaturing gradient gel electrophoresis. *Genomics* **15:**30–37.
61. **Laplanche, J. L., J. Chatelain, P. Beaudry, M. Dussaucy, C. Bounneau, and J. Launay.** 1993. French autochthonous scrapied sheep without the 136Val PrP polymorphism. *Mamm. Genome* **4:**463–464.
62. **Lasmézas, C. I., J. P. Deslys, R. Demalmay, K. T. Adjou, F. Lamoury, and D. Dormont.** 1996. BSE transmission to macaques. *Nature* **381:**743–744.
63. **Lasmézas, C. I., J. P. Deslys, O. Robain, A. Jaegly, V. Beringue, J. M. Peyrin, J. G. Fournier, J. J. Hauw, J. Rossier, and D. Dormont.** 1997. Transmission of the BSE agent to mice in the absence of detectable abnormal prion protein. *Science* **275:**402–405.
64. **Lugaresi, E., R. Medori, P. Montagna, A. Baruzzi, P. Cortelli, A. Lugaresi, P. Tinuper, M. Zucconi, and P. Gambetti.** 1986. Fatal familial insomnia and dysautonomia with selective degeneration of thalamic nuclei. *N. Engl. J. Med.* **315:**997–1003.
65. **Mabbott, N. A., K. L. Brown, J. Manson, and N. E. Bruce.** 1997. T-lymphocyte activation and the cellular form of the prion protein. *Immunology.* **92:**161–165.
66. **Maignien, T., C. I. Lasmézas, V. Beringue, D. Dormont, and J. P. Deslys.** 1999. Pathogenesis of the oral route of infection of mice with scrapie and bovine spongiform encephalopathy agents. *J. Gen. Virol.* **80:**3035–3042.
67. **Manson, J. C., E. Jamieson, H. Baybutt, N. L. Tuzi, R. Barron, I. McConnell, R. Somerville, J. Ironside, R. Will, M.S. Sy, D. W. Melton, J. Hope, and C. Bostock.** 1999. A single amino acid alteration (101L) introduced into murine PrP dramatically alters incubation time of transmissible spongiform encephalopathy. *EMBO J.* **18:**6855–6864.
68. **Manuelidis, L., G. Murdoch, and E. E. Manuelidis.** 1988. Potential involvement of retroviral elements in human dementias. *Ciba Found. Symp.* **135:**117–129.
69. **Manuelidis, L., and E. E. Manuelidis.** 1981. Search for specific DNAs in Creutzfeldt-Jakob infectious brain fractions using A nick translation. *Virology* **109:**435–443.
70. **Marsh, R. F., and W. J. Hadlow.** 1992. Transmissible mink encephalopathy. *Rev. Sci. Tech. Off. Int. Epizoot.* **11:**539–550.
71. **McBride, P. A., P. Eikelenboom, G. Kraal, H. Fraser, and M. E. Bruce.** 1992. PrP protein is associated with follicular dendritic cells of spleens and lymph nodes in uninfected and scrapie-infected mice. *J. Pathol.* **168:**413–418.
72. **McKinley, M. P., F. R. Masiarz, S. T. Isaacs, J. E. Hearst, and S. B. Prusiner.** 1983. Resistance of the scrapie agent to inactivation by psoralens. *Photochem. Photobiol.* **37:**539–545.
73. **Meyer, N., V. Rosenbaum, B. Schmidt, K. Gilles, C. A. Mirenda, D. Groth, S. B. Prusiner, and D. Riesner.** 1991. Search for a putative scrapie genome in purified prion fractions reveals a paucity of nucleic acids. *J. Gen. Virol.* **72:**37–49.
74. **Middleton, D. J., and R. M. Barlow.** 1993. Failure to transmit bovine spongiform encephalopathy to mice by feeding them with extraneural tissues of affected cattle. *Vet. Rec.* **132:**545–547.

75. **Miller, M. W., M. A. Wild, and E. S. Williams.** 1998. Epidemiology of chronic wasting disease in captive Rocky Mountain elk. *J. Wildl. Dis.* **34:**532–536.
76. **Ministry of Agriculture, Foods and Fisheries.** 1997. *Bovine Spongiform Encephalopathy: an Update, June 1997.* Ministry of Agriculture, Food and Fisheries, London, England.
77. **Neary, K., B. Caughey, D. Ernst, R. E. Race, and B. Chesebro.** 1991. Protease sensitivity and nuclease resistance of the scrapie agent propagated in vitro in neuroblastoma cells. *J. Virol.* **65:**1031–1034.
78. **Oesch, B., D. F. Groth, S. B. Prusiner, and C. Weissman.** 1988. Search for a scrapie-specific nucleic acid: a progress report. *Ciba Found. Symp.* **135:**209–223.
79. **Onodera, T., T. Ikeda, Y. Muramatsu, and M. Shinagawa.** 1993. Isolation of the scrapie agent from the placenta of sheep with natural scrapie in Japan. *Microbiol. Immunol.* **37:**311–316.
80. **O'Rourke, K. I., R. P. Melco, and J. R. Mickelson.** 1996. Allelic frequencies of an ovine scrapie susceptibility gene. *Anim. Biotechnol.* **7:**155–162.
81. **O'Rourke, K. I., T. V. Baszler, S. M. Parish, and D. P. Knowles.** 1998. Preclinical detection of PrP^{sc} in nictitating membrane lymphoid tissue of sheep. *Vet. Rec.* **142:**489–491.
82. **Ozel, M., and H. Diringer.** 1994. An extraordinarily small, suspicious, virus-like structure in fractions from scrapie hamster brain. *Lancet* **343:**894–895.
83. **Parry, H. B.** 1983. Recorded occurrences of scrapie from 1750, p. 31–59. *In* D. R. Oppenheimer (ed.), *Scrapie Disease in Sheep*. Academic Press, New York, N.Y.
84. **Parry, H. B.** 1964. Natural scrapie in sheep. I. Clinical manifestation and general incidence, treatment, and related syndromes, p. 95–97. *In Report of the Scrapie Seminar,* ARS 91–53. U.S. Department of Agriculture, Washington, D.C.
85. **Pattison, I. H., M. N. Hoare, J. N. Jebbett, and W. A. Watson.** 1972. Spread of scrapie to sheep and goats by oral dosing with foetal membranes from scrapie-affected sheep. *Vet. Rec.* **90:**465–468.
86. **Pattison, I. H., M. N. Hoare, J. N. Jebbett, and W. A. Watson.** 1974. Further observations on the production of scrapie in sheep by oral dosing with foetal membranes from scrapie-affected sheep. *Br. Vet. J.* **130:**65–67.
87. **Priola, S. A., B. Caughey, R. E. Race, and B. Chesebro.** 1994. Heterologous PrP molecules interfere with accumulation of protease-resistant PrP in scrapie-infected murine neuroblastoma cells. *J. Virol.* **68:**4873–4878.
88. **Prusiner, S. B.** 1990. Transgenic studies implicate interactions between homologous PrP isoforms in scrapie prion replication. *Cell* **63:**673–686.
89. **Prusiner, S. B.** 1991. Molecular biology of prion disease. *Science* **252:**1515–1522.
90. **Prusiner, S. B.** 1982. Novel proteinaceous infectious particles cause scrapie. *Science* **216:**135–144.
91. **Prusiner, S. B.** 1995. The prion diseases. *Sci. Am.* **272:**48–57.
92. **Race, R., A. Jenny, and D. Sutton.** 1998. Scrapie infectivity and proteinase K-resistant prion protein in sheep placenta, brain, spleen and lymph node: implications for transmission and antemortem diagnosis. *J. Infect. Dis.* **178:**949–953.
93. **Raymond, G. J., J. Hope, D. A. Kocisko, S. A. Priola, L. D. Raymond, A. Bossers, J. Ironside, R. G. Will, S. G. Chen, R. B. Petersen, P. Gambetti, R. Rubenstein, M. A. Smits, P. T. Lansbury, Jr., and B. Caughey.** 1997. Molecular assessment of the potential transmissibilities of BSE and scrapie to humans. *Nature* **388:**285–288.
94. **Robinson, M. M., W. J. Hadlow, T. P. Huff, G. A. H. Wells, M. Dawson, R. F. Marsh, and J. R. Gorham.** 1994. Experimental infection of mink with bovine spongiform encephalopathy. *J. Gen. Virol.* **75:**2151–2155.
95. **Rohwer, R. G.** 1984. Scrapie infectious agent is virus-like in size and susceptibility to inactivation. *Nature* **308:**658–662.

96. **Rohwer, R. G.** 1984. Virus-like sensitivity of the scrapie agent to heat inactivation. *Science* **223:**600–602.
97. **Schmerr, M. J., A. L. Jenny, M. S. Bulgin, J. M. Miller, A. N. Hamir, R. C. Cutlip, and K. R. Goodwin.** 1999. Use of capillary electrophoresis and fluorescent peptides to detect the abnormal prion protein in the blood of animals that are infected with a transmissible spongiform encephalopathy. *J. Chromatogr. A* **853:**207–214.
98. **Schreuder, B. E. C., L. J. M. van Keulen, M. E. W. Viromans, L. P. M. Langeveld, and M. A. Smits.** 1996. Preclinical test for prion diseases. *Nature* **381:**536.
99. **Scott, M. R., D. Groth, J. Tatzelt, M. Torchia, P. Tremblay, S. J. DeArmond, and S. B. Prusiner.** 1997. Propagation of prion strains through specific conformers of the prion protein. *J. Virol.* **71:**9032–9044.
100. **Scott, M. R., R. Kohler, D. Foster, and S. B. Prusiner.** 1992. Chimeric prion protein expression in cultured cells and transgenic mice. *Protein Sci.* **1:**986–997.
101. **Scott, M. R., R. Will, J. Ironside, H.-O. B. Nguyen, P. Tremblay, S. J. DeArmond, and S. B. Prusiner.** 1999. Compelling transgenetic evidence for transmission of bovine spongiform encephalopathy prions to humans. *Proc. Natl. Acad. Sci. USA* **96:**15137–15142.
102. **Sklaviadis, T., A. Akowitz, E. E. Manuelidis, and L. Manuelidis.** 1993. Nucleic acid binding proteins in highly purified Creutzfeldt-Jakob disease preparations. *Proc. Natl. Acad. Sci. USA* **90:**5713–5717.
103. **Snow, A. D., R. Kisilevsky, J. Willmer, S. B. Prusiner, and S. J. DeArmond.** 1989. Sulfated glycosaminoglycans in amyloid plaques of prion diseases. *Acta Neuropathol.* **77:**337–342.
104. **Spraker, T. R., M. W. Miller, E. S. Williams, D. M. Getzy, W. J. Adrian, G. G. Schoonveld, R. A. Spowart, K. I. O'Rourke, J. M. Miller, and P. A. Merz.** 1997. Spongiform encephalopathy in free-ranging mule deer (*Odocoileus hemionus*), white-tailed deer (*Odocoileus virginianus*), and Rocky Mountain elk (*Cervus elaphus nelsoni*) in northcentral Colorado. *J. Wildl. Dis.* **33:**1–6.
105. **Supattapone, S., H.-O. B. Nguyen, F. E. Cohen, S. B. Prusiner, and M. R. Scott.** 1999. Elimination of prions by branched polyamines and implications for therapeutics. *Proc. Natl. Acad. Sci. USA* **96:**14529–14534.
106. **Tagliavini, F., R. A. McArthur, B. Canciani, G. Giaccone, M. Porro, M. Bugiani, P. M.-J. Lievens, O. Bugiani, E. Peri, P. Dall'Ara, M. Rocchi, G. Poli, G. Forloni, T. Bandiera, M. Vasari, A. Suarato, P. Cassutti, M. A. Cervini, J. Lansen, M. Salmona, and C. Post.** 1997. Effectiveness of anthracycline against experimental prion disease in Syrian hamsters. *Science* **276:**1119–1122.
107. **Tateishi, J., P. Brown, T. Kitamoto, Z. Hoque, R. Roos, R. Wollman, L. Cervenakova, and D. C. Gajdusek.** 1995. First experimental transmission of fatal familial insomnia. *Nature* **376:**434–435.
108. **Taylor, D. M., K. Fernie, I. McConnell, C. E. Ferguson, and P. J. Steele.** 1998. Solvent extraction as an adjunct to rendering: the effect on BSE and scrapie agents of hot solvents followed by dry heat and steam. *Vet. Rec.* **143:**6–9.
109. **Telling, G. C., M. Scott, J. Mastrianni, R. Gabizon, M. Torchia, F. E. Cohen, S. J. DeArmond, and S. B. Prusiner.** 1995. Prion propagation in mice expressing human and chimeric PrP transgenes implicates the interaction of cellular PrP with another protein. *Cell* **83:**79–90.
110. **Weissman, C.** 1991. A unified theory of prion propagation. *Nature* **352:**679–683.
111. **Wells, G. A. H., M. Dawson, S. A. C. Hawkins, R. B. Green, I. Dexter, M. F. Francis, M. M. Simmons, A. R. Austin, and M. W. Horigan.** 1994. Infectivity in the ileum of cattle challenged orally with bovine spongiform encephalopathy. *Vet. Rec.* **135:**40–41.
112. **Wells, G. A. H., S. A. C. Hawkins, R. B. Green, A. R. Austin, I. Dexter, Y. I. Spencer, M. J. Chaplin, M. J. Stack, and M. Dawson.** 1998. Preliminary observations on the

pathogenesis of experimental bovine spongiform encephalopathy (BSE): an update. *Vet. Rec.* **142:**103–106.

113. **Wells, G. A. H., A. C. Scott, C. T. Johnson, R. F. Gunning, R. D. Hancock, M. Jeffrey, M. Dawson, and R. Bradley.** 1987. A novel progressive spongiform encephalopathy in cattle. *Vet. Rec.* **121:**419–420.
114. **Westaway, D., P. A. Goodman, C. A. Mirenda, M. P. McKinley, G. A. Carlson, and S. B. Prusiner.** 1987. Distinct prion proteins in short and long scrapie incubation period mice. *Cell* **51:**651–662.
115. **Westaway, D., V. Zuliani, C. M. Cooper, M. DaCosta, S. Neuman, A. L. Jenny, L. Detwiler, and S. B. Prusiner.** 1994. Homozygosity for prion protein alleles encoding glutamine-171 renders sheep susceptible to natural scrapie. *Genes Dev.* **8:**959–969.
116. **Wilesmith, J. W., J. B. M. Ryan, W. D. Hueston, and L. J. Hoinville.** 1992. Bovine spongiform encephalopathy: epidemiological features 1985 to 1990. *Vet. Rec.* **130:**90–94.
117. **Wilesmith, J. W., G. A. H. Wells, J. B. M. Ryan, D. Gavier-Widen, and M. M. Simmons.** 1997. A cohort study to examine maternally associated risk factors for bovine spongiform encephalopathy. *Vet. Rec.* **141:**239–243.
118. **Will, R.** 1999. New variant Creutzfeldt-Jakob disease. *Biomed. Pharmacother.* **53:**9–13.
119. **Will, R. G., J. W. Ironside, M. Zeidler, S. N. Cousens, K. Estibeiro, A. Alperovitch, S. Poser, M. Pocchiari, A. Hofman, and P. G. Smith.** 1996. A new variant of Creutzfeldt-Jakob disease in the UK. *Lancet* **347:**921–925.
120. **Williams, E. S., and S. Young.** 1980. Chronic wasting disease of captive mule deer: a spongiform encephalopathy. *J. Wildl. Dis.* **16:**89–98.
121. **Williams, E. S., and S. Young.** 1982. Spongiform encephalopathy of Rocky Mountain elk. *J. Wildl. Dis.* **18:**465–471.
122. **Williams, E. S., and S. Young.** 1992. Spongiform encephalopathies in Cervidae. *Rev. Sci. Tech. Off. Int. Epizoot.* **11:**551–567.
123. **Wyatt, J. M., G. R. Pearson, T. N. Smerdon, T. J. Gruffydd-Jones, G. A. H. Wells, and J. W. Wilesmith.** 1991. Naturally occurring scrapie-like spongiform encephalopathy in five domestic cats. *Vet. Rec.* **129:**233–236.
124. **Wyatt, J. M., G. R. Pearson, T. Smerdon, T. J. Gruffydd-Jones, and G. A. H. Wells.** 1990. Spongiform encephalopathy in a cat. *Vet. Rec.* **126:**513.

Emerging Diseases of Animals
Edited by C. Brown and C. Bolin

Chapter 8

Brucellosis in Terrestrial Wildlife and Marine Mammals

Jack C. Rhyan

The first *Brucella* organism described was *Brucella melitensis*, identified in 1887 by David Bruce as the causative agent of Malta fever, a disease infecting humans in the Mediterranean region and usually contracted by the consumption of unpasteurized goat's milk. Ten years later, Bernard Bang isolated and described the organism, now known as *Brucella abortus*, as the cause of infectious abortion in cattle. Serologic reactions to *B. abortus* were first detected in bison from Yellowstone National Park (YNP) in 1917 (36), and in elk on the National Elk Refuge (NER) in 1930 (38). *Brucella suis* was first isolated from aborted swine fetuses in 1914 (58) and positive serology for brucellosis was demonstrated in wild swine in Hawaii in 1962 (39). In 1963, Huntley and coworkers (27) isolated a *Brucella* sp. from caribou, and in 1966, Meyer (33) established *B. suis* biovar 4 as the causative agent of rangiferine brucellosis in Eskimos and reindeer.

In recent years, the presence of brucellosis in bison and elk in YNP and surrounding portions of Wyoming, Montana, and Idaho has stimulated numerous meetings, several lawsuits, protests by activist groups, a congressionally mandated investigation by the National Academy of Sciences (8), and considerable national public debate. Similarly, though with less national public attention, brucellosis in wild swine and arctic caribou and reindeer has emerged as an important disease issue.

How is it that these diseases, recognized in wildlife for over 35 years, have recently become such "hot button" issues? Undoubtedly, a combination of lesser factors are in part responsible, including increased public

---

Jack C. Rhyan • National Animal Health Programs, Veterinary Services, National Wildlife Research Center, Animal and Plant Health Inspection Service, U.S. Department of Agriculture, Fort Collins, CO 80521.

awareness and popularity of wildlife; urban sprawl into prime wildlife habitat, resulting in increased contact of wildlife with humans and their domestic animals; global trade issues; the recent development of the game farming industry; and, in some cases such as feral swine, recent expansion of the animals' population and range in combination with its increased popularity as a game animal. However, the single most important reason these diseases have emerged as such hot issues is what can be described as erosive or Ozark Mountain emergence. That is, similar to the geologic development of the Ozark Mountains by continued erosion of the softer soil and rock of the Ozark Plateau, leaving only the harder rock as remaining peaks, the control and eradication of brucellosis from domestic animal populations have left the harder-to-control *Brucella* infections in wildlife species as "emerged" disease peaks. Indeed, the livestock industry's attention is shifting from elimination of the disease in livestock to prevention of reinfection of livestock. The combination of these remaining peaks with the several above-mentioned factors have resulted in the unprecedented emergence of wildlife brucellosis as an issue of national and international importance. Not only are regulatory veterinarians and animal industry groups concerned; when prime-time television transmits footage of Yellowstone bison being killed to our living rooms, every viewer becomes a concerned citizen desiring immediate resolution of the problem.

BRUCELLOSIS IN ANIMALS AND HUMANS

There are currently six named species of *Brucella* plus the newly discovered group of brucellae infecting marine mammals. *Brucella melitensis* is primarily a disease of goats and sheep that is present in most areas of the world where goats are raised. It also causes disease in cattle herds located in brucellosis-endemic areas. In goats and sheep, the infection usually results in late-term abortion and may cause mastitis, lameness, hygroma, and orchitis. It is transmitted through milk and contact with products of parturition. In humans, *B. melitensis* is considered the principal cause of brucellosis and is clinically the most severe. In the United States, disease caused by *B. melitensis* in animals is considered exotic, and nearly all cases of human infection are due to the consumption of imported unpasteurized goat cheese. Acute brucellosis in humans is characterized by recurrent fever, chills, night sweats, headache, arthralgia, arthritis, anorexia, nausea, weight loss, weakness, backache, and dementia. Chronic brucellosis may cause progressive endocarditis and spondylitis. The heart lesions may result in death.

B. abortus causes a disease of cattle, elk, and bison that, until recent years, was common in humans. Before widespread pasteurization of milk, brucellosis was prevalent in the general human population. Following pasteurization, it became an occupational disease primarily affecting packing house workers and individuals in the animal husbandry and health occupations. Worldwide, bovine brucellosis caused by *B. abortus* is the most prevalent of the brucellar infections (11). In cattle, the disease is transmitted through milk and contact with products of parturition. It produces last-trimester abortions, retained placentas, weak calves, metritis, mastitis, lameness, male genital lesions, hygroma, and synovitis. In horses, *B. abortus* causes fistulous withers, a chronic supraspinous bursitis with accompanying cellulitis and development of draining fistulas involving the withers. In humans, the disease is clinically similar to that of *B. melitensis*, though usually less severe.

B. suis (biovars 1 and 3) produces abortion at any stage of gestation, lesions in male genitalia, and lameness in swine and is considered second only to *B. melitensis* in clinical severity in humans. Unlike brucellosis in cattle, *B. suis* is transmitted venereally as well as by contact with infected fetuses, membranes, and fluids. *B. suis* infection has been reported to affect cattle having contact with infected swine in Florida and Latin America. A related organism, *B. suis* biovar 4, produces abortion, genital lesions in the male, and lameness in reindeer and caribou and clinical illness in humans.

Brucella canis is a disease of canids resulting in abortion, genital lesions in the male, and bone lesions. It is transmitted venereally and through contact with infected materials. Though rare in humans, it is occasionally transmitted to people assisting infected bitches at parturition. *Brucella ovis* is venereally transmitted in sheep; it primarily causes epididymitis and orchitis in rams but can cause abortion in ewes. The isolation of *Brucella neotomae* from desert wood rats (*Neotoma tepida*) in Utah was reported in 1965 (56): the organism has no known pathogenicity in any species. *B. neotomae* and *B. ovis* are not known to cause disease in humans. The most recent group of *Brucella* spp. to be discovered are those isolated from marine mammals. This group of as yet unnamed brucellae represent a truly emerging infectious condition, about which little is known.

COOPERATIVE STATE-FEDERAL BRUCELLOSIS ERADICATION PROGRAM

To understand the current Ozark Mountain emergence of brucellosis in wildlife, one must appreciate the 65-year effort to eliminate bovine brucellosis from the United States and the current status of the program. In the early 1900s, brucellosis was considered the most economically devastating disease of livestock, and in 1934, after many states had initiated brucellosis control efforts in cattle, the Cooperative State-Federal Brucellosis Eradication Program was begun. The eradication effort has used quarantine, test and slaughter, calfhood vaccination, and, recently, adult vaccination to accomplish its goal. In 1934, nearly 50% of the 12,000 cattle herds tested had seropositive or suspect animals (18). Two years later, 14 to 15% of all cattle tested were positive for brucellosis (62). In 1957, there still remained over 123,000 known brucellosis-affected cattle and bison herds in the United States (24). By August 1994, only 198 *Brucella*-affected herds remained (22), and regulatory veterinarians were predicting completion of the eradication effort within 4 years. As of December 1999, only seven known affected herds remain; two of the affected herd units have been depopulated and three are scheduled for depopulation (V. Ragan, personal communication).

Nationwide efforts to control swine brucellosis were begun in the late 1950s, at which time the disease was prevalent in domestic herds. Currently, swine brucellosis has been eliminated from domestic herds in the United States except for those that have contact with feral swine. These are generally small backyard operations where captured feral swine have been introduced into the herd or other contact with feral swine has occurred.

Since the eradication effort in the United States has focused primarily on bovine brucellosis, most of the work in wildlife has been to determine what, if any, wildlife reservoirs of *B. abortus* may exist. This work has consisted of serologic and bacteriologic surveys in numerous species of native and exotic, captive and free-ranging animals. While several animals of different species have been found with serologic evidence of exposure to the organism, few have been culture positive (13, 37). Most of these surveys were conducted at a time when brucellosis was widespread in cattle, and the infrequent occurrence of the organism in wildlife was considered atypical and self-limiting. To date, there is no known significant reservoir of *B. abortus* infection in wildlife in the United States other than in bison and elk in the greater Yellowstone area (GYA). Similarly, *B. suis*, though known to occasionally infect a variety of carnivores and ungulates, has no large reservoir of infection in nature outside of feral swine (biovars 1 and 3) and caribou (biovar 4).

The remainder of this chapter will deal with the scientific and, when applicable, socioeconomic facts surrounding each of four situations which are the main emerging disease issues concerning brucellosis in wildlife: *B. abortus* in elk and bison in the GYA, *B. suis* in feral swine, *B. suis* biovar 4 in caribou, and *Brucella* spp. in marine mammals.

B. ABORTUS IN BISON AND ELK IN THE GYA

Background

The GYA is a somewhat arbitrarily defined area of land including YNP, Grand Teton National Park (GTNP), the NER, and adjacent portions of Montana, Wyoming, and Idaho (Fig. 1). It is estimated that there are currently 120,000 elk and 3,000 bison in the GYA. There are approximately 2,500 bison in YNP and 400 in GTNP. The bison in GTNP, often referred to as the Jackson herd, migrate to the NER during winter, where they feed on pelleted hay dispersed for the elk. While there have been a few migrations of bison from YNP bison to the Jackson herd, the occurrence is rare and the herds are considered separate. The bison in YNP are not artificially fed. Bison migrations outside YNP boundaries have occurred with increased frequency since 1980, usually during the winter. When the bison migrate onto private land, they pose a potential threat of spreading brucellosis to domestic cattle, thereby putting the state's class free status at risk. Loss of a state's class free status automatically results in imposed testing requirements for interstate shipment of cattle. Because of the risk of transmission to cattle, YNP bison have been hazed, shipped to slaughter, and shot when they have left the park boundaries. This is the crux of the issue that has elevated this Ozark peak to national prominence. In the winter of 1996–1997, when the bison population was near an all-time high and a period of warm wet weather was followed by intense cold, resulting in icy conditions on much of the traditional winter range inside YNP, over 1,100 bison migrated out of the park and were shot or sent to slaughter, awakening national awareness of the problem.

Of the elk in the GYA, over 25,000 are artificially fed during the winter on the NER operated by the U.S. Fish and Wildlife Service and on 22 feedgrounds operated by the state of Wyoming. Most authorities believe that brucellosis is largely maintained in elk by the practice of artificial feeding due to increased risk of exposure to infected aborted fetuses and vaginal discharge in elk congregated on feedgrounds. Supporting evidence for this conclusion comes from serologic surveys that show the highest seroprevalence of disease (range, 16 to 60%) in feedground elk (26) with much lower prevalence (2.2% or less) in nonfed GYA elk (57). While surveys of nonfed elk in Idaho have not found brucellosis, the disease was recently diagnosed in elk on a GYA feedground in the southeastern part of the state. A survey of over 6,000 nonfed elk from the northern Yellowstone herd in the 1960s showed a seroprevalence of 1.7% (50); two additional surveys of northern Yellowstone elk conducted in the 1990s showed eight (1.2%) of 721 elk (43) and three (2.1%) of 143 animals (1) to

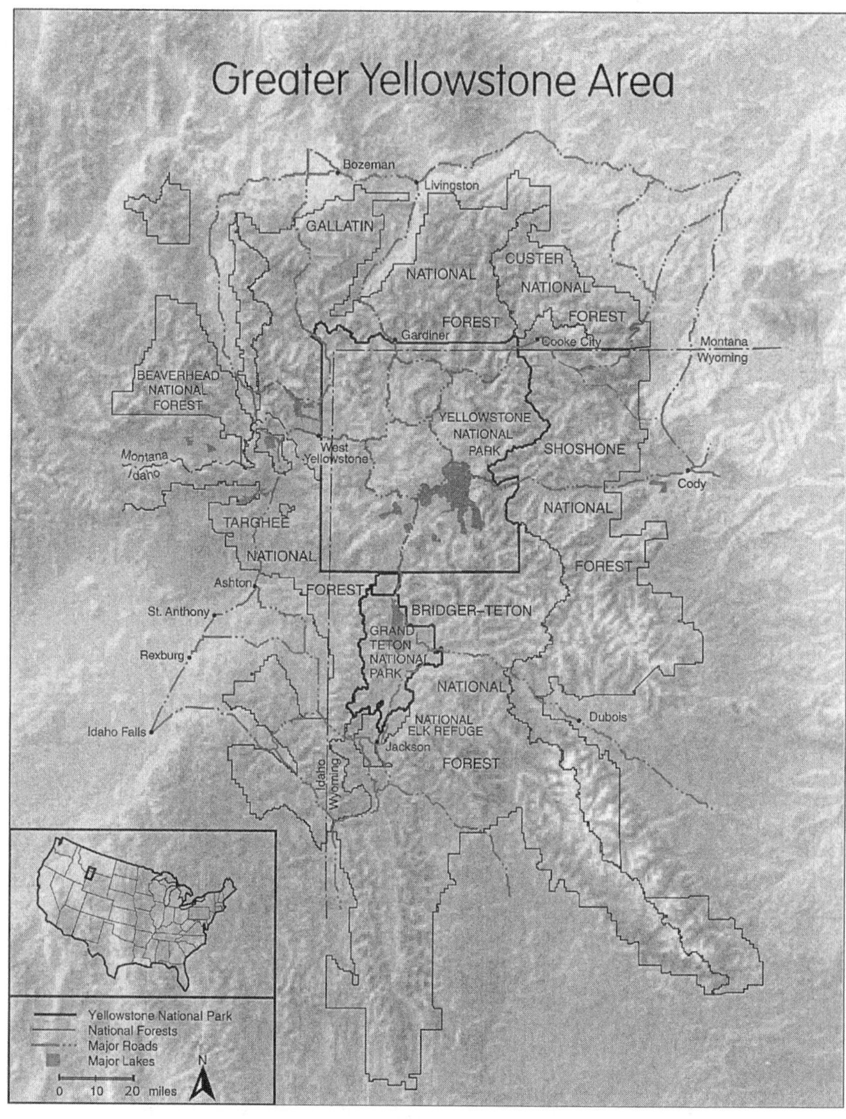

Figure 1. Map of the GYA. Courtesy of the National Park Service.

be seropositive. When elk are tested from other western states or from areas of Wyoming or Montana that are remote from the GYA, the prevalence of brucellosis is extremely low or zero (31, 57). The few seropositive animals in the northern herd and in other non-feedground areas of the

GYA may be migrants that have previously frequented feedgrounds. Nevertheless, eliminating the feedgrounds is not likely to be a politically accomplishable goal in the near future. Many of the feedgrounds were started in the early part of the century to alleviate ranchers' problems of winter feed predation by elk. Winter feeding also limits winter kill, resulting in the maintenance of a high number of animals for hunting, which is an important source of income for the state of Wyoming.

While state and federal agencies have been vigilant and largely successful in preventing spread of brucellosis from GYA wildlife to cattle on adjacent properties, there have been outbreaks of the disease in Wyoming cattle that epidemiologic evidence indicated were probably from GYA elk or, less likely, bison (8). Reported human cases of brucellosis among hunters in the GYA have been surprisingly few. An investigation by the Greater Yellowstone Interagency Brucellosis Committee, an interagency committee established to facilitate the development and implementation of brucellosis management plans for wildlife in the GYA, found that in Montana, only two cases have been reported to public health officials that could be traced to exposure to GYA wildlife.

Disease in Bison

Mohler first reported serologic evidence of brucellosis in bison that had aborted in YNP in 1917 (36). It has been proposed that brucellosis was probably transmitted to bison from cattle located in YNP in the early 1900s to provide milk for park employees (32). Conceivably, horses with fistulous withers or cow milk fed to bison calves could also have carried the infection. The recent isolation of biovar 2 from a YNP bison (Rhyan et al., unpublished data) and biovar 4 from GYA elk in Idaho (D. R. Ewalt, unpublished data) and Wyoming (53) as well as the repeated isolation of biovar 1 from GYA bison and elk suggest there was more than a single source of *Brucella* infection for GYA wildlife.

In 1930, *B. abortus* was isolated from the epididymis of male bison in YNP (12). Serologic surveys have demonstrated a seroprevalence of between 40 and 60% in YNP bison (8) and 77% in GTNP bison (63). However, abortions in GYA bison were not reported again until 1993, when Williams and coworkers described metritis and the isolation of *B. abortus* biovar 1 from a female bison from the Jackson herd that had recently aborted (63). The following year, the organism was recovered from an aborted bison fetus found inside YNP (45). The lack of reported abortions for most of the century led some scientists to speculate that the disease in this closed herd had ceased to produce many abortions (34). It was theorized that the infection was transmitted through the milk from dam to calf, acting as a calfhood vaccine to protect the animal from abortion during adult life. Since 1995, however, because of ongoing research in YNP and closer monitoring of the bison, five additional abortions due to *B. abortus* have been confirmed. In addition, a 2-week-old bison calf died, and large numbers of brucellae were cultured from numerous of its tissues (Rhyan et al., unpublished). Microscopic lesions, immunohistochemical staining (Fig. 2), and bacteriologic results from bison abortions are indistinguishable from those of cattle.

Additional studies on YNP bison since 1995 have shown the correlation of serologic test results and culture results to be similar to that in cattle (47). Utilizing a rigorous tissue collection and processing protocol, investigators found 12 of 26 seropositive bison and none of 8 seronegative or suspect bison to be positive on culture. Supramammary, iliac, and retropharyngeal lymph nodes were the sites from which the organism was most frequently isolated (Rhyan et al., unpublished). Interestingly, in this small sample of bison, the organism was not cultured from the mammary gland, whereas in cattle, the mammary gland is often infected. An ongoing prospective study of 40 radiocollared YNP female bison and their calves has found *B. abortus* in a single milk sample from a seropositive dam. In this study, the animals are immobilized and blood and milk specimens and swabs are collected in fall, winter, and immediately after calving. In addition, vaginal transmitters are installed in pregnant females in winter, allowing investigators to identify the site and time of parturition.

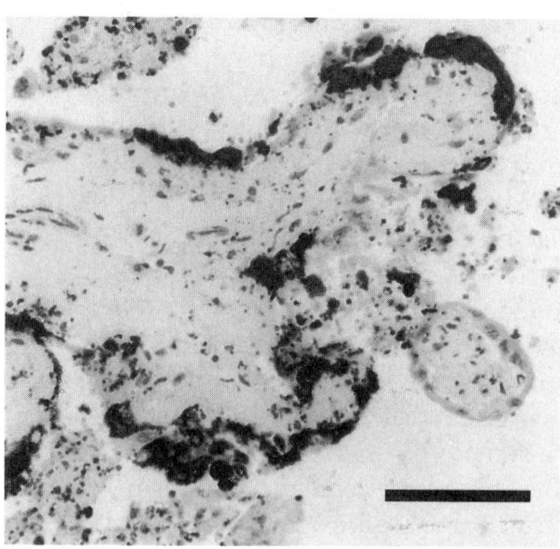

Figure 2. Placenta of aborted bison fetus from YNP. Note the immunohistochemical staining of brucellae in the trophoblastic epithelium. Immunoperoxidase stain was used. Bar = 100 μm.

Preliminary results indicate seroconversion in animals ranging in age from less than 6 months to 10 years and reproductive failure (diagnosed pregnant in fall or winter, and aborting or not having a calf at the first observation following calving) in the first gestation following seroconversion. Calves born to seropositive dams have either had an initial titer to *B. abortus* that became negative prior to a year of age or maintained a high titer which persisted through calfhood. One calf with high serologic titers was blood culture positive at 8 months of age. Serologic titers in seropositive adult animals have remained surprisingly consistent, with little variation over the 4-year duration of the study.

Results of experimental infection studies indicate the pathogenesis of brucellosis in bison is similar to that of cattle (3, 15). The organism appears somewhat more pathogenic in bison than in cattle. When given the same challenge inoculum as cattle, more bison abort and the abortions occur sooner than in cattle (S. C. Olsen, unpublished data). Additionally, results of safety studies on *B. abortus* vaccine strains 19 (14) and RB51 (41) in pregnant bison suggest that vaccinates are more likely to abort or shed organisms than cattle given the same dose. Venereal transmission is not thought to occur in bison, though only a single limited study has addressed the question (46). Transmission has occurred between captive bison and domestic cattle in nature (19) and experimentally (15).

An intriguing question about brucellosis in bison concerns the higher seroprevalence of disease in males than females. In a captive herd of approximately 3,500 animals that was being closely monitored, 80 primarily 3-year-old bulls seroconverted, but only four cows seroconverted (44). Management practices could not account for the disparity. Subsequent observation of YNP bison at calving, however, provided a likely explanation and insight into disease transmission (2). When calving season begins, the bison cows and calves congregate in nursery groups. Bull calves up to 2 years of age are still present with the nursery groups, while older bulls generally are not. Parturition often occurs with other animals present and early in the season generates a great deal of interest among surrounding bison. Other bison sniff and lick the presented placental membranes and neonate throughout the birth process. Cows and calves frequently engage in this activity, but the most interested animals are the 2-year-old bulls. Following parturition, cows with attached placentas are often followed by young bulls who continue to sniff and lick the placenta and fluids. As calving proceeds, the fascination of bison with the birthing process diminishes. This behavioral trait provides a likely explanation for the higher seroprevalence in bulls than cows and suggests that a *Brucella* abortion or retained placenta prior to or early in the calving season could be a source of infection for numerous bison.

Disease in Elk

After the initial discovery of serologically positive elk on the NER in 1930, various surveys demonstrated antibodies to *B. abortus* in elk from YNP (60) and state feedgrounds (55). The organism was first recovered from aborted elk fetuses on the NER in 1969 (55). Experimental infection of elk confirmed the production of late-term abortions and stillborn calves by *B. abortus* (54). The infection was transmitted to susceptible cattle confined with the infected elk. Other lesions attributable to *B. abortus* in elk are hygromata (carpal bursitis) and arthritis. Transmission is by ingestion of or contact with contaminated birth products; venereal transmission is not thought to occur in elk.

BRUCELLOSIS IN FERAL SWINE

Background

Serologic evidence of brucellosis in feral swine was first detected in Hawaii in 1962 (39), and confirmatory isolations of *B. suis* first occurred in South Carolina in 1976 (64). Since then, serologic surveys in feral animals have found seroprevalence rates from 4 to 53%, and the overall incidence rate is considered to lie between 10 and 25% (10).

The successful eradication of brucellosis from most of the domestic swine industry and recent progress made under the accelerated pseudorabies eradication program in domestic animals have greatly elevated these diseases in feral swine as Ozark peaks. Other factors also have intensified the feral swine situation in many states. Chief among these is the tremendous range expansion of feral hogs that has occurred in the last decade. Traditionally limited to the southeastern states, Hawaii, and California, the range now extends into the central tier of states from Colorado and Kansas to Indiana and Ohio (25). Twenty-six states now host a population that exceeds 2 million animals. Texas alone has an estimated 1 million feral swine. The most frequently given reason for this dramatic range expansion is the deliberate release of captured wild hogs into new areas by hunting enthusiasts to establish new stock for hunting. Indeed, hunting the "poor man's griz" has become a burgeoning sport industry in some parts of the country. In Texas, some outfitters average over $500 per animal from hunters shooting wild hogs on their property (7). Other factors complicating the feral swine issue are the lack of regulations or loose enforcement of regulations pertaining to movement and testing of animals, state-to-state variation in agency responsibility, and negative impacts on native wildlife and ecosystems. Also, the high reproductive potential of feral swine (51) makes it difficult to control populations with traditional management techniques.

Disease in Feral Swine

B. suis biovars 1 and 3 produce disease in swine characterized by abortion at any stage of gestation, stillbirth, weakness, sterility, arthritis, and orchitis. It is transmitted venereally and by contact with or ingestion of products of parturition. Semen from infected boars can contain large numbers of brucellae and may widely contaminate the environment due to the large volume of ejaculate. The organism occasionally infects dogs and cattle in contact with infected hogs and can persist in the mammary gland of cattle for at least 2 years (17), posing a public health risk to individuals drinking unpasteurized milk. Neither abortions nor intraspecific transmission has been reported in *B. suis*-infected cattle.

B. suis infection in humans is usually a disease of packing plant workers, livestock owners, or, with increasing frequency, hunters, who typically develop night sweats 3 to 4 weeks after dressing a wild hog.

BRUCELLOSIS IN REINDEER AND CARIBOU

Background

As of 1997, there were 15 herds of reindeer totaling 20,000 animals on or near the Seward Peninsula of western Alaska. There are several hundred thousand free-ranging caribou in the western arctic herd which ranges across Alaska and northwestern Canada. Caribou are native to the entire circumpolar arctic, while the nonspecific domesticated reindeer were imported to North America from Siberia in 1891. In Alaska, reindeer are ranged on vast tracts of unfenced land allotted to herders by the federal government. While the animals are periodically caught and eartagged, they are otherwise free-ranging animals with little artificial feeding or intensive management. The animals are raised for antlers and meat, with most of the meat being consumed locally. There is demand for importation of Alaskan reindeer as breeding stock into the lower 48 states, where there is currently an estimated population of 4,000 reindeer. However, because of the prevalence of brucellosis in some Alaskan herds, concerns about the risk of importing infected animals into the lower 48 states may lead to increased restrictions on reindeer exports from Alaska. Brucellosis has not been diagnosed in reindeer in the lower 48 states.

Disease in Reindeer and Caribou

The first evidence of brucellosis in Alaska was the finding of positive serologic titers in native Americans as early as 1939. Subsequently, an organism resembling *B. suis* was isolated from blood and bone marrow from humans, and epidemiologic evidence suggested caribou or reindeer as a potential source of infection. In 1963, the organism was isolated from caribou (27); later studies determined isolates from humans and reindeer in Alaska, Canada, and Siberia to be the same organism, which was classified as *B. suis* biovar 4 (33). Caribou and reindeer are now considered the primary host of biovar 4, but the organism is also known to infect humans, dogs, foxes, wolves, bears, muskox, and moose. Transmission from infected reindeer to cattle in confinement was recently demonstrated experimentally (20).

Seroprevalence of brucellosis in reindeer and caribou tends to be cyclic, varying from less than 1% to as high as 24%. Transmission is by ingestion of or contact with infected products of parturition. Whether venereal transmission or transmission via the milk occurs is unknown. A unique finding reported by Vashkevich in 1978 (61) was the isolation of *B. suis* biovar 4 from different developmental stages of the reindeer warble fly (*Oedemagena tarandi*) and demonstration of transovarian transmission of brucellosis by the warble fly to naive reindeer. In reindeer and caribou, the disease produces abortion, retained placentas, lameness, sterility, orchitis, epididymitis, seminal vesiculitis, metritis, mastitis, nephritis, and bursitis (42, 52). In human populations, there is a seroprevalence of 5 to 20% where caribou and reindeer are hunted or raised, killed, and eaten (6).

BRUCELLOSIS IN MARINE MAMMALS

Background

Though currently the lowest of the *Brucella* infections on the "biopolitical heat" scale, this infection in marine mammals represents the one truly emerging disease or group of diseases. Previously unrecognized *Brucella* spp. were first isolated in 1994 from harbor seals (*Phoca vitulina*), harbor porpoises (*Phocoena phocoena*), and a common dolphin (*Delphinus delphis*) found dead on the Scottish coast (48). Also in 1994, a similar *Brucella* spp. was isolated from the aborted fetus of a captive bottlenose dolphin (*Tursiops truncatus*) in California (16). Subsequent serologic (Table 1) and bacteriologic (Table 2) surveys have demonstrated positive titers and *Brucella* spp. in numerous marine mammal species (29, 30, 49, 59; L. B. Forbes, O. Nielson, L. Measures, and D. R. Ewalt, submitted for publication; O. Nielsen, R. E. A. Stewart, K. Neilsen, and P. Duignan, submitted for publication; N. Thomas, unpublished data). Recent molecular characterization of the marine mammal isolates suggests that they are distinct from brucellae in terrestrial mammals and may, in keeping with historical precedent, deserve to be designated as new *Brucella* species (5).

Disease in Marine Mammals

The only evidence of pathogenicity of the marine mammal brucellae is the isolation of large numbers of organisms from subcutaneous lesions in harbor porpoises, common dolphins, and striped dolphins (*Stenella coeruleoalba*) (21) and the demonstration of two brucellar abortions in captive bottlenose dolphins in California (16, 35). Both abortion cases had suppurative placentitis; immunohistochemical staining identified brucellae in trophoblastic epithelium and in endothelium and leukocytes in blood vessels. Otherwise, the organism has usually been cultured in small or moderate numbers from a variety of grossly normal-appearing tissues, including spleen, blood, mammary gland, reproductive organs, lung, liver, and lymph nodes.

A fascinating discovery was made in a pathologic investigation of one culture-positive harbor seal from Puget Sound in Washington (23). Cultures from the animal resulted in light growth of a *Brucella* sp. from various lymph nodes and liver and heavy growth from the lung. Immunohistochemical staining of lung tissue demonstrated large numbers of brucellae in the intestine and uterus of lungworms (*Parafilaroides* spp.) located in the small airways of the infected seal. Light and electron microscopic studies with immunogold labeling demonstrated the intrauterine brucellae in the adult nematode to be located largely in the membranes separating and surrounding developing larvae (Fig. 3). Host tissues surrounding infected lungworms contained occasional organisms, usually within the cytoplasm of phagocytes. Also present in the lung were occasional microgranulomas containing degenerating nematodes and nematode larvae, mixed inflammatory cells, giant cells, and abundant brucellae in the inflammatory lesions. In more recent studies, *Brucella*-positive *Parafilaroides* spp. have been identified in two additional harbor seals from Puget Sound (J. C. Rhyan, unpublished data); a *Brucella* was isolated from one of the seals, and no cultures were taken from the other. One of the animals had been given an anthelmintic as part of a rehabilitation protocol a few weeks prior to euthanasia. Besides containing intact *Parafilaroides* with brucellae located in the gut and uterus, lung sections from this animal had numerous degenerating nematodes consisting of the remaining cuticles filled with sheets of brucellae. The degenerating

Table 1. Summary of reported positive *Brucella* serological results on various species of marine mammals

Species	Location[a]	Dates collected	Tests[b]	n	No. (%) positive[c]	Reference(s)[d]
Harbor seal (*Phoca vitulina*)	UK(S)	1991–1993	1, 2, 3	140	69 (49), 25 (18), 45 (32)	49
	UK(EW)	1989–1995	3, 8	12	1 (8)	29
	Washington	1994–1997	1, 4, 5, 6	646	102 (16)	30, A
	Can&US(Atl)	1987–1996	7	130	14 (11)	B
	Can(VanIs)	1992–1993	7	33	7 (21)	B
Ringed seal (*Phoca hispida*)	Norway	1992–1995	3	49	5 (10)	59
	Can Arctic	1984–1997	7	876	17 (2)	40, B
Harp seal (*Phoca groenlandica*)	Norway	1983–1996	3	811	15 (2)	59
	Can Arc&N	1988–1997	7	453	8 (2)	B
Grey seal (*Halichoerus grypus*)	UK(S)	1991–1993	1, 2, 3	31	10 (32), 4 (13), 7 (23)	49
	UK(EW)	1989–1995	3, 7	62	6 (10)	29
	Can(Atl)	1991–1997	7	255	10 (4)	B
Hooded seal (*Cystophora cristata*)	Norway	1991–1995	3	137	48 (35)	59
	Can(Atl)	1988–1997	7	204	10 (5)	B
California sea lion (*Zalophus californianus*)	Washington	1994–1995	1, 4, 5, 6	50	4 (8)	A
	California	1992–1997	1, 4, 5, 6	89	5 (6)	C
Atlantic walrus (*Odobenus rosmarus*)	Can Arctic	1984–1997	7	229	12 (5)	40, B
Killer whale (*Orcinus orca*)	UK(EW)	1989–1995	3, 7	1	1	29
Pilot whale (*Globicephala melas*)	UK(EW)	1989–1995	3, 7	1	1	29
Bottlenose dolphin (*Tursiops truncatus*)	UK(EW)	1989–1995	3, 7	1	1	29
	California[e]	Before 1995	1, 2, 5	121	3 (2)	35
Common dolphin (*Delphinus delphis*)	UK(S)	1991–1993	1, 2, 3	1	1	49
	UK(EW)	1989–1995	3, 7	29	9 (31)	29

Striped dolphin (*Stenella coeruleoalba*)	UK(EW)	1989–1995	3, 7	4	1 (25)	29
Harbor porpoise (*Phocoena phocoena*)	UK(S)	1991–1993	1, 2, 3	18	5 (28), 2 (11), 4 (22)	49
	UK(EW)	1989–1995	3, 7	35	11 (31)	29
	Washington	1998–1999	1, 4, 5, 6	16	5 (31)	A
Beluga whale (*Delphinapterus leucas*)	Can Arctic and Can(Atl)	1984–1997	7	488	28 (6)	B
Narwhal (*Monodon monocerus*)	Can Arctic	1984–1997	7	77	5 (6)	B
Fin whale (*Balaenoptera physalus*)	Norway	1983–1989	3	108	12 (11)	59
Minke whale (*Balaenoptera acutorostrata*)	Norway	1991–1996	3	216	17 (8)	59
Sei whale (*Balaenoptera borealis*)	Norway	1983–1989	3	49	7 (14)	59
Southern sea otter (*Enhydra lutris neieis*)	California	1992–1999	1, 2, 5, 6	25	1 (4)	D

[a] UK(S), United Kingdom (coasts of Scotland); UK(EW), United Kingdom (coasts of England and Wales); Can&US(Atl), Atlantic coast of Canada and United States; Can(VanIs), Vancouver Island, British Columbia, Canada; Norway, northern Norway coast and areas north and west; Can Arctic, Canadian Arctic; Can Arc&N, Canadian Arctic and Newfoundland; Can(Atl), Canadian Atlantic coast.

[b] 1, Rosebengal plate test or standard card test; 2, standard agglutination test (standard tube test); 3, indirect enzyme-linked immunosorbent assay (ELISA); 4, buffered antigen plate agglutination; 5, rivanol; 6, complement fixation; 7, competitive ELISA.

[c] Percent positive includes all positives plus weak positives; some investigators include positives plus suspects. Some investigators give prevalence for each test.

[d] A, D. M. Lambourn, unpublished data; B, Nielsen et al., submitted; C, F. M. D. Gulland, unpublished data; D, N. Thomas, unpublished data.

[e] Captive U.S. Navy animals, majority originally wild-caught in the Gulf of Mexico.

Table 2. Summary of reported positive *Brucella* culture results on various species of marine mammals

Species	Location[a]	Dates collected	No. of animals culture positive	Tissues positive[b]	Reference(s)[c]
Harbor seal (*Phoca vitulina*)	UK(S)	1993–1994	7	LN, Sp,TF	21, 28, 49
	Washington	1996–1998	7	LN, Sp, Lu, L, K, P, T, Th, Ts, E, F, UB	23, 30, A
Ringed seal (*Phoca hispida*)	Can Arctic	1995–1997	4	LN	B
Harp seal (*Phoca groenlandica*)	Can&US(Atl)	1999	2	LN, Lu	A, B
Grey seal (*Halichoerus grypus*)	UK(S)	1994	1	Lu, T	21, 28
Hooded seal (*Cystophora cristata*)	UK(S)	1994	1	LN, Sp	21, 28
Bottlenose dolphin (*Tursiops truncatus*)	California[d]	1992 and 1997	3	Ft, VD, Lu	16, 35, A
Common dolphin (*Delphinus delphis*)	UK(S)	1993	1	SC, TF	21, 28, 49
Striped dolphin (*Stenella coeruleoalba*)	UK(S)	1994–1995	2	SC, LN, Sp, M	21, 28
Atlantic white-sided dolphin (*Lagenorhynchus acutus*)	UK(S)	1994	1	LN, Sp	21, 28
Harbor porpoise (*Phocoena phocoena*)	UK(S&E)	1991–1994	4	SC, Sp, M, B, U, TF	21, 28, 49
Minke whale (*Balaenoptera acutorostrata*)	Norway	1995	1	Sp, L	9, 59
European otter (*Lutra lutra*)	UK(S)	1994	1	LN	21

[a] UK(S), United Kingdom (Scotland); UK(S&E), United Kingdom (Scotland and England); Can Arctic, Canadian Arctic; Can&US(Atl), Atlantic coast of Canada and United States.

[b] LN, lymph node(s); Sp, spleen; TF, thoracic fluid; Lu, lung; L, liver; K, kidney; P, pancreas; T, testis; Th, thymus; Ts, tonsil; E, eye; F, feces; UB, urinary bladder; Ft, fetus; VD, vaginal discharge; SC, subcutaneous lesion; M, mammary gland; B, blood; U, uterus.

[c] A, Ewalt, unpublished; B, Forbes et al., submitted.

[d] Captive U.S. Navy animals, the majority originally wild-caught in the Gulf of Mexico.

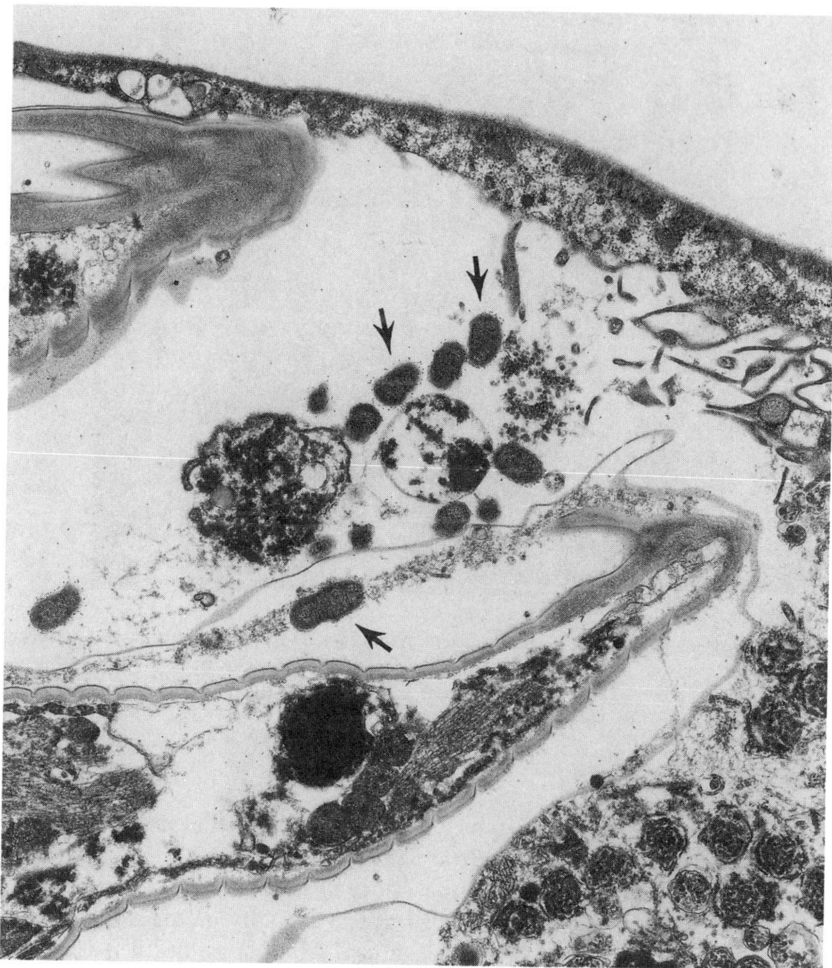

Figure 3. Electron photomicrograph of a developing *Parafilaroides* sp. larva in the uterus of an adult nematode. Note the immunogold labeling of brucellae adjacent to the larva (arrows).

nematodes were often surrounded by a zone of mixed inflammation containing large numbers of brucellae. In some sections of adult *Parafilaroides*, brucellar antigen could be identified within the developing larvae in the uterus (Fig. 4). Additionally, sections of the trachea of one infected seal contained neutrophils with intracytoplasmic brucellar antigen, and sections of seal intestine contained intraluminal nematode larvae with

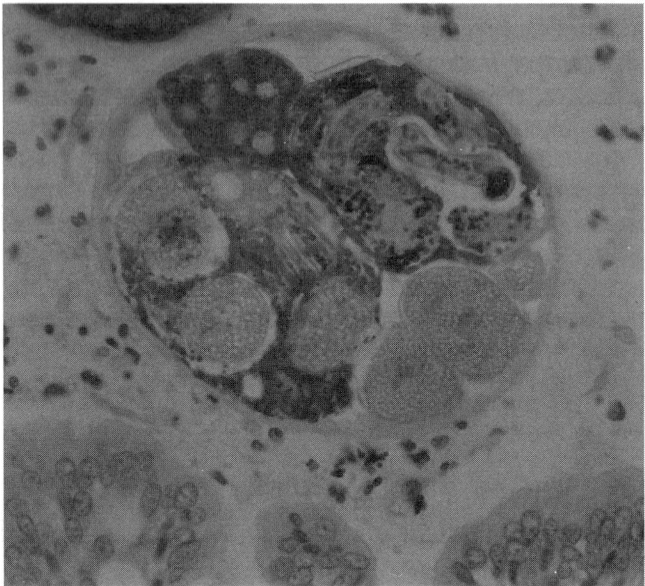

Figure 4. Cross section through an adult *Parafilaroides* sp. in the lung of a harbor seal. Note the marked immunohistochemical staining of brucellar antigen surrounding gonadal cells and developing larvae, in developing larvae, and in seal leukocytes adjacent to the nematode. Immunoperoxidase stain was used.

internal brucellar antigen. Bacteriologic culture of the feces of that animal resulted in heavy growth of a *Brucella* sp.

These findings suggest the possibility of *Parafilaroides* spp. as a vector of brucellosis between harbor seals. At the least, the nematode serves to amplify the brucellar infection in the seal's lungs, giving it a continual portal of exit via the mucociliary apparatus and fecal shedding. Fecal-feeding fish may then serve as vectors of the infection to the next hungry seal or conceivably may ingest *Brucella*-infected lungworm larvae and serve as intermediate hosts of the developing, infected *Parafilaroides* spp. The pathogenicity of this *Brucella* sp. for harbor seals is as yet undetermined.

Because of the presence of cattle in close proximity to the infected seal populations in Puget Sound and the possibility of sea gulls serving as mechanical vectors between the seal haul-out areas and cattle premises, a recent study examined the possibility of cattle infection by the seal isolate (Rhyan, unpublished). Investigators found that the organism could produce bovine abortion when 10^9 CFU were given intravenously. However, when inoculated intraconjunctivally at a similar dose, the seal isolate produced seroconversion but not abortion or persistence in tissues. From these findings, it is reasonable to conclude that the organism is not highly pathogenic for cattle but could result in short-term seroconversion detectable on standard cattle tests.

The zoonotic potential of the marine mammal brucellae is unknown. A single case of infection of a laboratory worker in the United Kingdom has been reported (4). The question of human health risk is an important one, as some Native American groups hunt and consume marine mammals, and there is potential risk of exposure to individuals involved in marine mammal research and rehabilitation.

FUTURE DIRECTIONS OF BRUCELLOSIS CONTROL AND ERADICATION IN WILDLIFE

The development of effective strategies to control and eradicate brucellosis from bison and elk in the GYA, from feral swine across the southern and central states, from reindeer and caribou in the arctic, and perhaps finally from marine mammals presents an unprecedented challenge. Even in the simplest situation, that of the bison in the GYA, the biopolitical complexity of disease management in public animals creates nearly innumerable hurdles that slow progress to a snail's pace. The intense public interest and multiagency involvement in each of these situations demand excellent research and accurate modeling before a plan can be developed and implemented.

With current and near-future technology, any disease control and eradication program will necessarily involve vaccination and other management strategies. These other management strategies are actions implemented to decrease the risk of intraspecific and interspecific disease transmission. The cattle brucellosis vaccine RB51 may be useful in bison and elk. Remaining research needs pertaining to the elk and bison in the GYA include additional studies on the efficacy of RB51 as a calfhood vaccine in bison and elk, safety and efficacy studies on RB51 as an adult vaccine in bison and elk, duration-of-immunity studies in bison and elk, and studies on the efficiency of remote vaccine delivery in the GYA. Of course, if RB51 does not prove safe or efficacious in bison or elk, then a great deal more work in vaccine development is needed before any vaccination program can be implemented. Additionally, work on newer, more efficacious, perhaps killed or subunit vaccines is needed. Also, development of safe and effective oral vaccines and species-specific oral delivery systems along with the development and validation of more accurate field diagnostic tests are needed in the long run.

If one assumes that long-term vaccination of YNP bison will reduce but not eliminate brucellosis from the herd, possible options for the final elimination of the disease should be investigated. Sterilization of seropositive animals may provide an alternative to testing and slaughter. This could involve surgical or immunocontraceptive techniques. Would 3 years of sterility for an infected female bison, perhaps with concurrent *Brucella* vaccination, allow her to clear the infection or protect her from abortion and/or shedding of organisms once she resumed reproduction? If so, perhaps an alternative to testing and slaughter could be developed for eliminating the last small percentage of seropositive animals. With the glacial speed of progress in this situation, research must begin early to examine the questions that will need answers in future years.

Research needed to address the feral swine brucellosis problem includes development of safe, more efficacious oral vaccines and vaccine delivery systems as well as development of better population control methods. Also, the development of reliable field diagnostic tests will greatly benefit any eradication effort.

Control and eradication of the disease in free-ranging caribou will also require the development of a highly efficacious vaccine and effective method of delivery. Additionally, improved serologic tests for reindeer and caribou are needed.

Regarding marine mammal brucellosis, the imminent research need is basic pathogenesis and epidemiology work to determine the characteristics of the infection in various species, the mechanisms of transmission, and the potential public health risks. Standard serologic tests used to detect antibodies to *B. abortus* have been shown to have poor sensitivity and specificity in some marine mammal species (35; Ewalt, unpublished); therefore, the development of more reliable diagnostic tests for use in marine mammals is needed. The finding of infected lungworms in harbor seals also beckons work to better determine the parasites' role in transmission and pathogenesis. That finding, combined with the transmission in reindeer via warble flies, invites additional work in other species to determine possible roles that parasites may play in the pathogenesis and epidemiology of brucellosis.

With the continuing emergence of this disease and these disease issues in free-ranging populations of wildlife, solutions can only be found based on good science, public education, and innovative, collaborative work between research scientists, regulatory and wildlife veterinarians, and wildlife managers.

REFERENCES

1. **Aguirre, A. A., D. E. Hansen, E. E. Starkey, and R. G. McLean.** 1995. Serologic survey of wild cervids for potential disease agents in selected national parks in the United States. *Prevent. Vet. Med.* **21:**313–322.
2. **Aune, K., T. Roffe, J. Rhyan, J. Mack, and W. Clark.** 1998. Preliminary results on home range, reproduction and behavior of female bison in northern Yellowstone National Park, p. 61–70. *In* L. Irby and J. Knight (ed.), *International Symposium on Bison Ecology and Management in North America.* Montana State University, Bozeman, Mont.
3. **Bevins, J. S., J. E. Blake, L. G. Adams, J. W. Templeton, J. K. Morton, and D. S. Davis.** 1996. The pathogenicity of *Brucella suis* biovar 4 for bison. *J. Wildl. Dis.* **32:**581–585.
4. **Brew, S. D., L. L. Perrett, J. A. Stack, A. P. MacMillan, and N. J. Staunton.** 1999. Human exposure to *Brucella* recovered from a sea mammal. *Vet. Rec.* **144:**483.
5. **Bricker, B. J., D. R. Ewalt, A. P. MacMillan, G. Foster, and S. Brew.** 2000. Molecular characterization of *Brucella* strains isolated from marine mammals. *J. Clin. Microbiol.* **38:**1258–1262.
6. **Brody, J. A., B. Huntley, T. M. Overfield, and J. Maynard.** 1966. Studies of human brucellosis in Alaska. *J. Infect. Dis.* **116:**263–269.
7. **Chambers, M.** 1999. Conserving a resource, p. 115–116. *In Proceedings of the 1999 National Feral Swine Conference.* Texas Animal Health Commission, Ft. Worth, Tex.
8. **Cheville, N. F., D. R. McCullough, and L. R. Paulson.** 1998. *Brucellosis in the Greater Yellowstone Area.* National Research Council, National Academy of Sciences, National Academy Press, Washington, D.C.
9. **Clavereau, C., V. Wellemans, K. Walravens, M. Tryland, J.-M. Verger, M. Grayon, A. Cloeckaert, J.-J. Letesson, and J. Godfroid.** 1998. Phenotypic and molecular characterization of a *Brucella* strain isolated from the minke whale (*Balaenoptera acutorostrata*). *Microbiology* **144:**3267–3273.
10. **Conger, T. H., E. Young, and R. A. Heckmann.** 1999. *Brucella suis* in feral swine, p. 98-107. *In Proceedings of the 1999 National Feral Swine Conference.* Texas Animal Health Commission, Ft. Worth, Tex.
11. **Corbel, M. J.** 1997. Brucellosis: an overview. *Emerg. Infect. Dis.* **3:**213–221.
12. **Creech, G. T.** 1930. *Brucella abortus* infection in a male bison. *N. Am. Vet.* **11:**35–36.
13. **Davis, D. S.** 1990. Brucellosis in wildlife, p. 322–334. *In* K. Neilsen and J. R. Duncan (ed.), *Animal Brucellosis.* CRC Press, Boca Raton, Fla.
14. **Davis, D. S., J. W. Templeton, T. A. Ficht, J. D. Huber, R. D. Angus, and L. G. Adams.** 1991. *Brucella abortus* in bison. II. Evaluation of strain 19 vaccination of pregnant cows. *J. Wildl. Dis.* **27:**258–264.
15. **Davis, D. S., J. W. Templeton, T. A. Ficht, J. D. Williams, J. D. Kopek, and L. G. Adams.** 1990. *Brucella abortus* in captive bison. I. Serology, bacteriology, pathogenesis, and transmission to cattle. *J. Wildl. Dis.* **26:**360–371.
16. **Ewalt, D. R., J. B. Payeur, B. M. Martin, D. R. Cummins, and W. G. Miller.** 1994. Characteristics of a *Brucella* species from a bottlenose dolphin (*Tursiops truncatus*). *J. Vet. Diagn. Investig.* **6:**448–452.
17. **Ewalt, D. R., J. B. Payeur, J. C. Rhyan, and P. L. Geer.** 1997. Brucella suis biovar 1 in naturally infected cattle: a bacteriological, serological, and histological study. *J. Vet. Diagn. Investig.* **9:**417–420.
18. **Fitch, C. P.** 1934. Report on the committee on Bang's disease project. *Proc. U. S. Livest. Sanit. Assoc.* **38:**311–317.
19. **Flagg, D. E.** 1983. A case history of a brucellosis outbreak in a brucellosis free state which originated in bison. *U.S. Anim. Health Assoc. Proc.* **87:**171–172.

20. **Forbes, L. B., and S. V. Tessaro.** 1993. Transmission of brucellosis from reindeer to cattle. *J. Am. Vet. Med. Assoc.* **203:**289–294.
21. **Foster, G., K. L. Jahans, R. J. Reid, and H. M. Ross.** 1996. Isolation of *Brucella* species from cetaceans, seals and an otter. *Vet. Rec.* **138:**583–586.
22. **Frye, G. H. and B. R. Hillman.** 1997. National Cooperative Brucellosis Eradication Program, p. 79-85. *In* E. T. Thorne, M. S. Boyce, P. Nicoletti, and T. J. Kreeger (ed.), *Brucellosis, Bison, Elk, and Cattle in the Greater Yellowstone Area: Defining the Problem, Exploring Solutions.* Greater Yellowstone Interagency Brucellosis Committee and Wyoming Game and Fish Department, Cheyenne, Wyo.
23. **Garner, M. M., D. M. Lambourn, S. J. Jeffries, P. B. Hall, J. C. Rhyan, D. R. Ewalt, L. M. Polzin, and N. F. Cheville.** 1997. Evidence of *Brucella* infection in *Parafilaroides* lungworms in a Pacific harbor seal (*Phoca vitulina richardsi*). *J. Vet. Diagn. Investig.* **9:**298–303.
24. **Gilsdorf, M. J.** 1998. Brucellosis in bison—case studies, p. 1–10. *In* L. Irby and J. Knight (ed.), *International Symposium on Bison Ecology and Management in North America.* Montana State University, Bozeman, Mont.
25. **Gipson, P. S., B. Hlavachick, and T. Berger.** 1998. Range expansion by wild hogs across the central United States. *Wildl. Soc. Bull.* **26:**279–286.
26. **Herriges, J. D., Jr., E. T. Thorne, and S. L. Anderson.** 1992. Vaccination to control brucellosis in free-ranging elk on western Wyoming feed grounds, p. 107–112. *In* R. D. Brown (ed.), *The Biology of Deer.* Springer-Verlag, New York, N.Y.
27. **Huntley, B. E., R. N. Philip, and J. E. Maynard.** 1963. Survey of brucellosis in Alaska. *J. Infect. Dis.* **112:**100–106.
28. **Jahans, K. L., G. Foster, and E. S. Broughton.** 1997. The characterization of *Brucella* strains isolated from marine mammals. *Vet. Microbiol.* **57:**373–382.
29. **Jepson, P. D., S. Brew, A. P. MacMillan, A. P. Baker, J. R. Barnett, J. Kirkwood, J. K. Kuiken, T. Robertson, I. R. Simpson, and V. R. Simpson.** 1997. Antibodies to *Brucella* in marine mammals around the coast of England and Wales. *Vet. Rec.* **141:**513–515.
30. **Lambourn, D., S. Jeffries, L. Polzin, D. Ewalt, and M. Garner.** 1998. Disease screening of harbor seals (*Phoca vitulina*) from Gertrude Island, Washington. *In Proceedings of 1998 Puget Sound Research Conference.* Puget Sound Water Quality Action Team, Seattle, Wash.
31. **McCorquodale, S. M., and R. F. Digiacomo.** 1985. The role of wild North American ungulates in the epidemiology of bovine brucellosis: a review. *J. Wildl. Dis.* **21:**351–357.
32. **Meagher, M., and M. E. Meyer.** 1994. On the origin of brucellosis in bison of Yellowstone National Park: a review. *Conserv. Biol.* **8:**645–653.
33. **Meyer, M. E.** 1966. Identification and virulence studies of *Brucella* strains isolated from Eskimos and reindeer in Alaska, Canada, and Russia. *Am. J. Vet. Res.* **27:**353–358.
34. **Meyer, M. E., and M. Meagher.** 1995. Brucellosis in free ranging bison (*Bison bison*) in Yellowstone, Grand Teton, and Wood Buffalo National Parks: a review. *J. Wildl. Dis.* **31:**579–598.
35. **Miller, W. G., L. G. Adams, T. A. Ficht, N. F. Cheville, J. P. Payeur, D. R. Harley, C. House, and S. H. Ridgway.** 1999. *Brucella*-induced abortions and infection in bottlenose dolphins (*Tursiops truncatus*). *J. Zoo Wildl. Med.* **30:**100–110.
36. **Mohler, J. R.** 1917. Abortion disease, p. 105-106. *In Annual Reports of the Department of Agriculture.* U.S. Department of Agriculture, Washington, D.C.
37. **Moore, C. G., and P. R. Schnurrenberger.** 1981. A review of naturally occurring *Brucella abortus* infections in wild mammals. *J. Am. Vet. Med. Assoc.* **179:**1105–1112.
38. **Murie, O.** 1951. *The Elk of North America.* Stackpole Books, Harrisburg, Pa., and Wildlife Management Institute, Washington, D.C.
39. **Nichols, L., Jr.** 1962. *Ecology of the Wild Pig.* Hawaii Division of Fish and Game PR Project W5-R-13.

40. **Nielsen, O., K. Nielsen, and R. E. A. Stewart.** 1996. Serologic evidence of *Brucella* spp. exposure in Atlantic walruses (*Odobenus rosmarus rosmarus*) and ringed seals (*Phoca hispida*) of arctic Canada. *Arctic* **49:**383–386.

41. **Palmer, M. V., S. C. Olsen, M. J. Gilsdorf, L. M. Philo, P. R. Clarke, and N. F. Cheville.** 1996. Abortion and placentitis in pregnant bison (*Bison bison*) induced by the vaccine candidate, *Brucella abortus* strain RB51. *Am. J. Vet. Res.* **57:**1604–1607.

42. **Rausch, R. L., and B. E. Huntley.** 1978. Brucellosis in reindeer, *Rangifer tarandus* L., inoculated experimentally with *Brucella suis*, type 4. *Can. J. Microbiol.* **24:**129–135.

43. **Rhyan, J. C., K. Aune, D. R. Ewalt, J. Marquardt, J. W. Mertins, J. B. Payeur, D. A. Saari, P. Schladweiler, E. J. Sheehan, and D. Worley.** 1997. Survey of free-ranging elk from Wyoming and Montana for selected pathogens. *J. Wildl. Dis.* **33:**290–298.

44. **Rhyan, J. C., S. D. Holland, T. Gidlewski, D. A. Saari, A. E. Jensen, D. R. Ewalt, S. G. Hennager, S. C. Olsen, and N. F. Cheville.** 1997. Seminal vesiculitis and orchitis caused by *Brucella abortus* biovar 1 in young bison bulls from South Dakota. *J. Vet. Diagn. Investig.* **9:**368–374.

45. **Rhyan, J. C., W. J. Quinn, L. S. Stackhouse, J. J. Henderson, D. R. Ewalt, J. B. Payeur, M. Johnson, and M. Meagher.** 1994. Abortion caused by *Brucella abortus* biovar 1 in a free-ranging bison (*Bison bison*) from Yellowstone National Park. *J. Wildl. Dis.* **30:**445–446.

46. **Robison, C. D., D. S. Davis, J. W. Templeton, M. Westhusin, W. B. Foxworth, M. J. Gilsdorf, and L. G. Adams.** 1998. Conservation of germ plasm from bison infected with *Brucella abortus*. *J. Wildl. Dis.* **34:**582–589.

47. **Roffe, T. J., J. C. Rhyan, K. Aune, L. M. Philo, D. R. Ewalt, T. Gidlewski, and S. G. Hennager.** 1999. Brucellosis in Yellowstone National Park bison: quantitative serology and infection. *J. Wildl. Manage.* **63:**1132–1137.

48. **Ross, H. M., G. Foster, R. J. Reid, K. L. Jahans, and A. P. MacMillan.** 1994. *Brucella* species infection in sea-mammals. *Vet. Rec.* **134:**359.

49. **Ross, H. M., K. L. Jahans, P. MacMillan, R. J. Reid, P. M. Thompson, and G. Foster.** 1996. *Brucella* species infection in North Sea seal and cetacean populations. *Vet. Rec.* **138:**647–648.

50. **Smith, B. L., and R. L. Robbins.** 1994. *Migrations and Management of the Jackson Elk Herd*, p. 38–39. *National Biological Service Resource Publication 199.* U.S. Department of the Interior, Washington, D.C.

51. **Taylor, R. B., E. C. Hellgren, T. M. Gabor, and L. M. Ilse.** 1998. Reproduction of feral pigs in southern Texas. *J. Mammol.* **79:**1325–1331.

52. **Tessaro, S. V., and L. B. Forbes.** 1986. *Brucella suis* biotype 4: a case of granulomatous nephritis in a barren ground caribou (*Rangifer tarandus groenlandicus* L.) with a review of the distribution of rangiferine brucellosis in Canada. *J. Wildl. Dis.* **22:**479–483.

53. **Thorne, E. T., J. D. Herriges, Jr., and A. D. Reese.** 1991. Bovine brucellosis in elk: conflicts in the greater Yellowstone area, p. 296–303. *In* A. G. Christensen, L. J. Lyon, and T. N. Lonner (ed.), *Proceedings of Elk Vulnerability—A Symposium*. Montana State University, Bozeman, Mont.

54. **Thorne, E. T., J. K. Morton, F. M. Blunt, and H. A. Dawson.** 1978. Brucellosis in elk. II. Clinical effects and means of transmission as determined through artificial infections. *J. Wildl. Dis.* **14:**280–291.

55. **Thorne, E. T., J. K. Morton, and G. M. Thomas.** 1978. Brucellosis in elk. I. Serologic and bacteriologic survey in Wyoming. *J. Wildl. Dis.* **14:**74–81.

56. **Thorpe, B. D., R. W. Sidwell, J. B. Bushman, K. L. Smart, and R. Moyes.** 1965. Brucellosis in wildlife and livestock of west central Utah. *J. Am Vet. Med. Assoc.* **146:**225–232.

57. **Toman, T. L., T. Lemke, L. Kuck, B. L. Smith, S. G. Smith, and K. Aune.** 1997. Elk in the greater Yellowstone area: status and management, p. 56-64. *In* E. T. Thorne, M. S. Boyce,

P. Nicoletti, and T. J. Kreeger (ed.), *Brucellosis, Bison, Elk, and Cattle in the Greater Yellowstone Area: Defining the Problem, Exploring Solutions.* Greater Yellowstone Interagency Brucellosis Committee and Wyoming Game and Fish Department, Cheyenne, Wyo.

58. **Traum, J.** 1914. *Report of the Chief of Animal Industry,* p. 30. U.S. Department of Agriculture, Washington, D.C.
59. **Tryland, M., L. Kleivane, A. Alfredsson, M. Kjeld, A. Arnason, S. Stuen, and J. Godfroid.** 1999. Evidence of *Brucella* infection in marine mammals in the North Atlantic Ocean. *Vet. Rec.* **144:**588–592.
60. **Tunnicliff, E. A., and H. Marsh.** 1935. Bang's disease in bison and elk in the Yellowstone National Park and on the National Bison Range. *J. Am. Vet. Med. Assoc.* **86:**745–752.
61. **Vashkevich, R. B.** 1978. The reindeer warble fly (*Oedemagena tarandi* L.) as a vector of *Brucella. Vopr. Prirodnoaei Ochagovosti Bolezneaei* **9:**119–131.
62. **Wight, A. E.** 1936. Progress of the Federal-State Bang's disease program. *Proc. U.S. Livest. Sanit. Assoc.* **40:**268–272.
63. **Williams, E. S., E. T. Thorne, S. L. Anderson, and J. D. Herriges, Jr.** 1993. Brucellosis in free-ranging bison (*Bison bison*) from Teton County, Wyoming. *J. Wildl. Dis.* **29:**118–122.
64. **Wood, G. W., J. B. Hendricks, and D. E. Goodman.** 1976. Brucellosis in feral swine. *J. Wildl. Dis.* **12:**579–582.

Emerging Diseases of Animals
Edited by C. Brown and C. Bolin

Chapter 9

Leptospirosis

Carole Bolin

Leptospirosis is a world-wide zoonosis caused by infection with any of the many pathogenic serovars of *Leptospira.* This genus of organisms is tremendously variable and contains in excess of 250 serovars, with pathogenic serovars belonging to seven different species and saprophytic serovars belonging to five species (11). The prevalence of individual pathogenic serovars varies in different locales, probably influenced by such factors as climate, indigenous fauna, and agricultural practices. Leptospirosis affects virtually all mammalian species. Human leptospirosis also is important and occurs endemically in the tropics and as epidemics in temperate climates. However, unlike other animal species, humans are not important in the maintenance of this disease in nature, and human-to-human transmission is rare.

Leptospirosis is an economically important livestock disease in many parts of the world and causes suffering of affected humans and animals. In livestock, the main sources of economic loss caused by leptospirosis include abortions, stillbirths, birth of weak neonates, animal deaths, loss of milk production, and costs of veterinary care, treatment, and vaccines. Animals that have been exposed to or infected with *Leptospira* are not eligible for import or export because of the risk of transmission of the disease. This can have important economic impact, particularly in the movement of semen and breeding stock across borders. Leptospirosis in companion animals such as dogs and horses results in loss of companionship, animal suffering, and a relatively high risk of transmission of the infection to owners. The main economic losses associated with human leptospirosis include medical care and loss of work time, which can be considerable in the more severe forms of leptospirosis. Among workers in

Carole Bolin • Animal Health Diagnostic Laboratory, Department of Veterinary Pathology, Michigan State University, East Lansing, MI 48824.

certain occupations, e.g., farmers, slaughterhouse workers, and sugarcane cutters, outbreaks of leptospirosis are common in some areas, and leptospirosis can significantly impair the workforce.

Leptospirosis was first described as a disease in humans by Weil in 1886 (65), and his name is often used to describe the severe human illness caused by some types of *Leptospira.* The causative organism was recognized as a spirochete in 1907, and rapidly thereafter, researchers in Japan documented the infectious nature of leptospirosis and identified rats as carriers of the infection (reviewed in references 22 and 28). These observations led to an understanding of the role of carrier animals in the transmission of leptospirosis. It was not until several years later that a broad range of wild and domestic animals were also shown to be carriers of leptospirosis (22). Many clinicians believe that leptospirosis is a single disease entity characterized by fever, jaundice, and renal failure in humans and abortions in domestic livestock. In fact, numerous clinical syndromes, some subtle and some dramatic, occur in domestic animals, wildlife, and human beings as a result of infection with this diverse group of pathogenic spirochetes. The host-serovar combination must always be considered when discussing the epidemiology, transmission, clinical signs, treatment, and control of leptospirosis in animals and humans.

Perhaps the most confusing facet of leptospirosis is the classification and nomenclature of the causative agent. Bacteriologists recognized early on that there were clinically and epidemiologically distinct types of leptospirosis. They realized that serologically distinct leptospires were associated with different clinical syndromes (32). The different leptospiral isolates were organized into serovars based on their antigenic constitution, and groups of related serovars were organized into serogroups. All the pathogenic serovars were clustered into a single species, *Leptospira interrogans,* and the saprophytic serovars were all members of *Leptospira biflexa.* Clinicians recognized that there were different types of leptospirosis and that it was important to identify the type of leptospire, and therefore the likely animal reservoir, to control outbreaks of leptospirosis in their patients. Therefore, the serologic classification of *Leptospira* became well recognized and remains very useful in understanding the epidemiology of leptospirosis. However, there is considerable overlap in the antigens between different serovars. This causes confusion in identification of isolates and in serologic diagnosis of leptospirosis in animals and humans. In addition, the methods used to identify isolates of *Leptospira* serologically (14) are complex, expensive, and time-consuming and often lead to equivocal results. A new genetic system of classification of the genus *Leptospira* has now been widely adopted, and the pathogenic leptospires have been classified into seven different species (11). The serovar concept is still useful, as in most cases, the serovars also are clearly distinguishable genetically. Therefore, leptospiral isolates can be identified using traditional serological techniques (14), monoclonal antibodies (55), or a host of genetic techniques, including restriction endonuclease analysis (55, 56), mapped-site restriction polymorphisms in 16S rRNA genes (46, 47), and others. It is important that bacteriologists, physicians, veterinarians, biologics manufacturers, and public health authorities make the transition to the new nomenclature to avoid confusion. As an example of the changes in nomenclature, the old and new names of the serovars that are common in North America are listed in Table 1; a full list is available elsewhere (11, 22).

EPIDEMIOLOGY

In particular regions, different leptospiral serovars are prevalent and are associated with one or more maintenance hosts, which serve as reservoirs of infection (Table 1). Maintenance hosts are often wildlife species and sometimes domestic animals and livestock. Each serovar behaves differently within its maintenance host species than it does in other, incidental ("accidental") host species. A maintenance host relationship is characterized by efficient transmission between animals, a relatively high incidence of infection (30 to 50%), production of chronic rather than acute disease, and persistent infection in the kidney. Diagnosis of maintenance host infections is often difficult because of a relatively low antibody response against the infecting serovar and the presence of few organisms in the tissues of infected animals. In contrast, an incidental host relation-

Table 1. Old and new names and the most common maintenance hosts of the pathogenic leptospires that are common in the United States and Canada

Previous name	New name (if different)	Maintenance host(s)
L. interrogans serovar canicola		Dogs
L. interrogans serovar pomona		Pigs, cattle, opossums, skunks
L. interrogans serovar grippotyphosa	*L. kirschneri* serovar grippotyphosa	Raccoons, muskrats
L. interrogans serovar hardjo type hardjo-bovis	*L. borgpetersenii* serovar hardjo	Cattle
L. interrogans serovar icterohaemorrhagiae		Rats
L. interrogans serovar bratislava		Pigs, mice?, horses?
L. interrogans serovar ballum	*L. borgpetersenii* serovar ballum	Mice

ship is characterized by relatively low susceptibility to infection but high pathogenicity for the host, production of acute disease, sporadic transmission within the host species, and a short renal phase of infection. Diagnosis of incidental host infections is less problematic because of a marked antibody response to infection and the presence of large numbers of organisms in the tissues of infected animals. Some serovar-host combinations cannot be neatly categorized as either maintenance or incidental host infections. For example, swine infected with serovar pomona behave as a host intermediate between the two forms, with the organism persisting in the kidney but the host showing a marked antibody response to infection. Human beings are susceptible to infection with most of the pathogenic serovars of *Leptospira* but are always incidental hosts and therefore not important reservoirs of the infection.

Transmission of leptospirosis among maintenance hosts is often direct and involves direct contact with infected urine, placental fluids, or milk. The infection can also be transmitted transplacentally or venereally in some host-serovar combinations. Infection of incidental hosts is more commonly indirect, by contact with areas contaminated with the urine of maintenance hosts. Environmental conditions are critical in determining the frequency of indirect transmission. Survival of leptospires is favored by moisture, moderately warm temperatures, and neutral or mildly stagnant water; survival is brief in dry soil or at temperatures below 10°C (8). Therefore, leptospirosis occurs most commonly in the spring, autumn, and early winter in temperate climates and during the rainy season in the tropics.

The prevalence of various serovars within a population is influenced by many factors. Shifts in management of livestock or in the maintenance hosts within an area can lead to shifts in the epidemiology of leptospirosis in those areas. For example, the growth of modern multisite and total confinement systems for raising swine has decreased the incidence of leptospirosis acquired from wildlife and cattle dramatically, but infection with serovars bratislava, pomona, and tarrasovi remains a problem under these management conditions. As another example, there has recently been a dramatic increase in the incidence of canine leptospirosis and a change in the serovars infecting dogs in North America. Serovars pomona, grippotyphosa, and bratislava are becoming more frequently recognized as causes of canine leptospirosis (1, 6, 7, 13, 26, 48) than serovars canicola and icterohaemorrhagiae which used to predominate. This change in the prevalence of serovars in North America follows many years of widespread use of canine vaccines containing serovars canicola and icterohaemorrhagiae, and probably also reflects increasing contact in suburban areas between dogs and skunks, raccoons, and opossums, which serve as maintenance hosts for serovars pomona and grippotyphosa (7). The changing epidemiology of canine and swine leptospirosis in North America, with changing serovar patterns and clinical signs, reflects the variable nature of leptospirosis within animal populations. Table 2 lists the most common known causes of leptospirosis in major livestock species and companion animals.

The overall incidence of leptospirosis in humans varies tremendously in different areas of the world. Clearly, the disease is more common in humans and animals in tropical and subtropical areas. In these environments, the climate is conducive to the survival of the organisms outside of the host. Less-developed nations also have a higher incidence of human leptospirosis because of factors such as poor sanitation, lack of protective clothing such as boots and footwear, and close association between animals and humans. Often, livestock are kept in and around the living quarters of persons in developing nations, which facilitates the transmission of leptospirosis. There also is a significant occupational association with the incidence of leptospirosis. Workers in contact with rodents, e.g., cane cutters, banana cutters, and sewer workers, have a significantly higher incidence of leptospirosis than others. Leptospirosis is often associated with certain seasons or with certain events in the lives of the persons within an area. Epidemics of leptospirosis are often associated with the harvest of certain crops, e.g., rice, when there is close contact between the farmers and rats that are attracted to the ripe grain. In all species, epidemics of leptospirosis are associated with flooding—presumably because of the increased exposure to water contaminated with animal urine and a concentration of rodents and other reservoir hosts because of the flooding. These factors have been shown to be important in a number of recent outbreaks of human leptospirosis (60, 69). Another significant risk factor for the occurrence of leptospirosis in humans is participation in certain recreational activities, e.g., swimming and rafting, in bodies of fresh water (2, 66).

Table 2. Most common serovars isolated from livestock and companion animal species[a]

Species	Most common serovars
Cattle	hardjo, pomona, grippotyphosa
Swine	bratislava, pomona, tarrasovi
Horses	pomona, bratislava, grippotyphosa
Dogs	grippotyphosa, pomona, bratislava, canicola, icterohaemorrhagiae
Sheep and goats	hardjo, pomona

[a]Not all serovars are prevalent in a particular species in a particular region of the world.

PATHOGENESIS

The key steps in the pathogenesis of leptospirosis are similar in all host-serovar combinations, and the early steps are similar in incidental and maintenance hosts. Leptospires invade the body after penetrating exposed mucous membranes or damaged skin. After a variable incubation period (4 to 20 days), leptospires circulate in the blood for up to 7 days. During this period, leptospires enter and replicate in many tissues, including the liver, kidneys, lungs, genital tract, and central nervous system. During the period of bacteremia and tissue colonization, the clinical signs of acute leptospirosis, which vary by serovar and host, occur. Agglutinating antibodies can be detected in serum soon after leptospiremia occurs, and the presence of antibodies coincides with clearance of the leptospires from the blood and most organs. As the organisms are cleared from the blood, liver, and kidney (exclusive of the tubules), the clinical signs of acute leptospirosis begin to resolve. Leptospires remain only in sites sequestered from the systemic antibody response, including the renal tubules and, in some instances, the eye and genital tract.

After development of antibody, the pathogenesis of leptospirosis in incidental and maintenance hosts begins to diverge. Leptospires often remain in the renal tubules of incidental hosts for only a short period of time and are shed in the urine for a few days to several weeks. In contrast, in maintenance hosts, organisms remain in the renal tubules, genital tract, and, less commonly, the central nervous system and eyes for months to years after infection. Leptospires are shed in the urine and genital secretions of these persistently infected animals. This feature of the disease in maintenance hosts is critical in these animals' serving as the reservoir of leptospirosis within an ecosystem.

In both incidental and maintenance hosts, if the animal or human is pregnant at the time of infection, localization and persistence of the organism in the uterus of pregnant animals may result in fetal infection, with subsequent abortion, stillbirth, birth of weak neonates, or birth of healthy but infected offspring. In general, abortions which occur with incidental host infection are associated with acute disease, whereas in maintenance hosts, the abortions or other reproductive sequelae may be delayed by several weeks or months following infection.

The central lesion of leptospirosis is damage to small blood vessels as a result of endothelial damage. Certain tissues seem to be more susceptible to the effects of this lesion, including the liver, kidney, lung, and placenta (22). The vascular lesions lead to leakage of red blood cells and plasma into the tissues. Tissue injury occurs secondary to hypoxia. In addition, the presence of large numbers of leptospires may lead to tissue damage due to the mechanical lesions of migration through tissues, the elaboration of as yet poorly defined toxins or other bacterial products, and apoptosis. Immunopathology almost certainly plays a role in some of the chronic or persistent lesions of leptospirosis.

Because of the similarity in the key steps in the pathogenesis of leptospirosis in all hosts and serovars, it is possible to discuss the general stages of the infection in more detail. The key stages are (i) entry of the organism, (ii) survival, growth, and bacteremia, (iii) production of antibody, and (iv) development of persistent infection (reviewed in reference 22).

Entry

The precise route and mode of entry of leptospires in natural infections are not clear. Leptospires are presumed to enter via the mucous membranes of the mouth, conjunctiva, or genital tract or through small abrasions of the skin. It seems unlikely that leptospires penetrate intact skin; water-softened skin may allow penetration. Fetal infection occurs by direct invasion of the placenta from mother to fetus at any stage of pregnancy in mammals. Leptospires can penetrate tissues by passing through or between cells; pathogenic strains penetrate more rapidly and in greater numbers than nonpathogenic strains. It is likely but not proven that leptospiral motility is important for the ability to penetrate tissues.

Survival, Growth, and Spread

Once leptospires enter the body, they come in contact with a variety of nonspecific host factors, some of which are conducive to leptospiral growth and others which act as host defense mechanisms. These include pH changes, changes in concentrations of electrolytes, iron sequestration, and the presence of fatty acids, pharmacologically active tissue peptides and mediators, neutrophils, and natural antibodies. The survival of leptospires in tissues in a naive animal is a result of their resistance to natural antibodies in tissue fluids and plasma. The natural antibody can act together with complement and lysozyme to lyse the leptospires; nonpathogenic leptospires are susceptible, and pathogenic serovars are resistant (22).

Unlike many bacteria, leptospires do not cause an acute inflammatory response to their presence in tissues. Individual organisms or clumps of leptospires are seen in tissues during natural or experimental infections with little evidence of inflammation or cellular infiltrate. This "stealth" property is undoubtedly important in the survival of the organisms in the host. The cellular infiltrate seen in liver and kidney lesions is most likely a response to tissue injury or the immune response rather than a response to the presence of the leptospires themselves.

Leptospires spread rapidly from their site of entry to the bloodstream, where they circulate to all organs. The organisms gain access to the bloodstream either by way of the lymphatics or by direct penetration of the blood vessels. Organisms which survive by evading phagocytosis grow exponentially in the bloodstream and tissues. Growth ceases with either the appearance of specific antibodies, which lead to rapid opsonization of the organisms and clearance, or the death of the host (22). Animals may also die of the effects of the lesions after leptospires have been cleared from the circulation and tissues (20, 21).

Development of Antibody and Bacterial Clearance

Clearance of leptospires from the bloodstream occurs rapidly following the development of specific antibodies against the leptospiral lipopolysaccharide (LPS) (4, 12, 23, 63). These antibodies are agglutinating, opsonic, and serovar associated; cross-reactions with closely related serovars occur, but almost exclusively within serogroups. Opsinized leptospires are cleared by cells of the reticuloendothelial system (19). Ingested leptospires can be observed in phagocytic vacuoles in Küpffer cells in the liver, macrophages in the spleen, and elsewhere (20). The mechanisms of killing and digestion of leptospires inside phagocytes are assumed to be the same as for other bacteria.

There is some question whether leptospires can survive inside mammalian cells. There is evidence that leptospires are readily digested inside phagocytes (4, 63). However, recently evidence has been presented that virulent, low-passage isolates of *Leptospira* invade epithelial cells, may survive for some time, and induce apoptosis in some cell types (39, 40). Once inside a phagocytic vacuole, leptospires lose their helical shape and are believed to be nonviable (23).

Persistence

In animals which do not die of acute leptospirosis, particularly in maintenance hosts, leptospires may persist in certain immunologically privileged sites. These tissues include the proximal renal tubules, brain, anterior chamber of the eye, and genital tract. Here, protected from antibody in the circulation, leptospires persist and grow. In the kidney, leptospires colonize the brush border of the renal tubular epithelium (37, 38). Maintenance hosts may eventually clear the organism from the kidney. This clearance is mediated by the production of local antibody within the urinary tract (36).

Persistence of the organism within the male and female genital tract can lead to infertility and venereal transmission. The precise location of the persistence within the genital tract is not known, but leptospires have been isolated from several places in the genital tract and accessory sex glands of persistently infected males and females (16–18). Mammalian fetuses appear to act in similar ways to adults; if they survive acute infection, they may be born persistently infected (9).

Persistence of the organism within the eye can lead to chronic inflammation and periodic uveitis, particularly in horses (10, 42). This leads to a debilitating condition of horses known as periodic ophthalmia or moon blindness, which is a leading cause of blindness in horses. The inflammation is thought to be immune mediated, and the pathogenic mechanisms will be described in the next section.

MECHANISMS OF DISEASE

Because the basic lesions and disease processes, in general, are the same regardless of the serovar of pathogenic *Leptospira* and the host, it is a reasonable hypothesis that similar pathogenic mechanisms may be involved, albeit to different extents in the various host-serovar infections. Studies to determine the nature of host specificity and the relative expression of various leptospiral virulence factors are needed to better understand these relationships.

It is not clear precisely how leptospires produce lesions and disease. Much of the evidence and suggested virulence factors are based on conjecture, and some early, nondefinitive studies. Part of the reason that virulence factors have not been clearly defined for *Leptospira* is that the organism cannot be manipulated genetically to create mutants lacking candidate virulence factors. In addition, investigators often study one particular serovar in one particular host (often not the natural host), and it is difficult to compare results between studies. However, several pathogenic mechanisms have been proposed to be important in the development of leptospirosis, including bacterial motility and adhesion, bacterial toxins, persistence of infection, and immune-mediated tissue damage (Table 3).

Table 3. Causes of tissue damage in acute and chronic leptospirosis

Type of disease	Cause(s) of tissue damage
Acute	Mechanical damage by leptospires migrating through tissues Toxins produced by *Leptospira* Hypoxia due to vascular lesions Apoptosis
Chronic	Immune response to Nleptospiral antigens

Motility and Adhesion

It seems logical, but has not been proven experimentally, that the corkscrew, progressive motility that is characteristic of leptospires would be important in the initial entry and spread of the organisms in host tissues. Leptospires burrow through and between cells in vitro and in vivo and are capable of progressive motility in extremely viscous environments (24, 57).

Leptospires adhere to the surface of mammalian cells, although the precise mechanisms are not clear. Leptospires must adhere in some way to the luminal surface of renal epithelial cells, and this ability is central to the way in which leptospires are transmitted and maintained within a population of animals. The adherence seems to cause little, if any, damage to the epithelial cells. It has been postulated, based on electron microscopic observations in experimental infections, that leptospiral attachment to the renal tubular cell surfaces is due to the merging of surface membranes through their lipids (34, 38). Leptospires do attach to fibroblasts, kidney cells, and endothelial cells in culture; virulent strains attach readily, and avirulent strains do not (3, 31, 57, 64).

Toxins

When animals are injected with large numbers of leptospires, they may die within 48 h, without bacteremia and the classical manifestations of acute leptospirosis. The animals display signs of toxicity, including fever, depression, widespread petechial hemorrhages, and toxic nephropathy, with very few organisms seen in tissues (21, 62). These observations are consistent with the effects of a toxin.

Leptospires have been reported to be cytotoxic in tissue culture, but these observations have been difficult to reproduce because of difference in cells, culture conditions, and serovars of *Leptospira* studied (27, 35, 41,

49, 53, 62, 68). The lipid moiety of a leptospiral glycolipoprotein has been shown to be responsible for membrane damage caused by leptospires cocultivated with fibroblasts in vitro (62, 64). The lipid moiety may damage the membrane by intercalation of the bacterial lipids into the host cell membrane, resulting in cell leakage. The role of this toxin in vivo is not clear.

Leptospires are known to produce phospholipases which can cause damage to erythrocyte and other cell membranes containing the specific phospholipid substrate. The presence of many such "hemolysin" genes in various leptospires suggests a role for these molecules in the survival, growth, or pathogenicity of these organisms. A clear role in pathogenesis has not been established, but they may be responsible for the holes in erythrocytes seen in electron micrographs of erythrocytes of calves infected with serovar pomona (22, 58, 59). The phospholipid composition of the erythrocytes of various animals differs, and the specificity of the substrates of the various leptospiral phospholipases (15, 33) may help explain the presence or absence of hemolysis in different host-serovar combinations. The sphingomyelinase-specific phospholipases have been studied extensively (50–52); it is not clear whether they have a significant role in pathogenesis.

Leptospiral LPS has been postulated to be important in the pathogenesis of leptospirosis. While leptospiral LPS has many features in common with the LPS of other gram-negative bacteria, it is relatively nontoxic to cells and animals. Leptospiral LPS can activate macrophages (30) and is a mitogen for B cells in mice (29). LPS is a potent antigen, and the presence of leptospiral LPS in tissues may induce immune-mediated tissue damage. LPS is postulated to be important in the development of the endothelial damage and platelet aggregation that lead to thrombocytopenia in acute leptospirosis (22).

Persistent Infection

As the persistence of leptospires in various tissues is critical to this disease, some consideration has been given to how the organism manages to persist after the appearance of specific antibodies. First, the sites of persistence are generally sequestered from the systemic antibody response. Second, as mentioned previously, leptospires in tissues are relatively noninflammatory and thus attract little attention from phagocytes. Third, leptospires, like other spirochetes, downregulate the production of outer membrane proteins when growing in vivo, perhaps contributing to decreased availability of antigens under these conditions (25).

Immune-Mediated Tissue Damage

Many of the lesions seen in chronic or persistent leptospirosis appear to be immune mediated. In the kidney, the leptospires themselves are not associated with significant tissue damage or inflammation; rather, the cellular infiltrate is associated with leptospiral antigens which remain in the tissues. The inflammatory infiltrate is predominantly lymphocytic, with fewer macrophages. Chronic lesions contain large numbers of mature plasma cells and may become organized into germinal centers. Recent studies (5) have shown that, in a model of chronic leptospirosis, the inflammatory infiltrate is associated with the localization of certain leptospiral antigens in the renal interstitium.

Autoimmunity may play a role in the kidney lesions found in dogs infected with serovar canicola (61) and in the eye lesions which can occur in horses subsequent to leptospirosis. There is evidence for an immunological cross-reaction between a protein epitope of the equine cornea and a leptospiral antigen which may lead to autoimmunity (43–45).

SUMMARY AND CONCLUSIONS

Leptospirosis is caused by a highly diverse group of organisms that infect virtually every mammalian species. This disease is maintained and circulated among populations of animals; humans are susceptible but are not important in the epidemiology of the disease. The prevalent serovars vary by geographic region, and therefore all efforts to diagnose and control leptospirosis within an area must be based on a thorough understanding of the serovars present and their maintenance hosts. Public and animal health authorities must always be alert to a changing epidemiologic pattern of leptospirosis within their area, as it may indicate the introduction of a new serovar or a changing pattern of disease with an established serovar.

The mechanisms by which *Leptospira* causes disease in animals and humans are not well documented. Careful, thorough studies using modern methods of investigation are needed to understand the interplay between bacterial and host factors responsible for the production of disease. Clear understanding of the role of various potential leptospiral virulence factors will help in the design of vaccines and other strategies to effectively control leptospirosis in populations of animals and humans at risk.

Whether leptospirosis should really be considered an emerging or reemerging disease is subject to debate. The number of reported cases is certainly increasing in many areas. In addition, several large, dramatic outbreaks of leptospirosis have received considerable press attention. However, it is not clear whether the incidence of disease is really increasing or whether the disease is merely being increasingly recognized. Human leptospirosis is a notifiable disease in most parts of the world, with the notable exception of the United States; reporting requirements for animal leptospirosis are highly variable. Therefore, true incidence figures are difficult to determine. It has been suggested, and empirical data provide support, that the best predictor of the reported incidence of leptospirosis within an ecosystem is the number of leptospirologists in the area! The difficulties in recognition and diagnosis of this infectious disease are legendary, and the availability of quality diagnostic support is crucial. Physicians, veterinarians, and public health authorities must work closely together to recognize, document, and control this important zoonotic disease.

REFERENCES

1. **Adin, C. A., and L. D. Cowgill.** 2000. Treatment and outcome of dogs with leptospirosis: 36 cases (1990–1998). *J. Am. Vet. Med. Assoc.* **216:**371-375.
2. **Anonymous.** 1997. Outbreak of leptospirosis among white-water rafters—Costa Rica, 1996. *Morb. Mortal. Wkly. Rep.* **46:**577–579.
3. **Ballard, S. A., M. Williams, B. Adler, and S. Faine.** 1986. Interactions of virulent and avirulent leptospires with primary cultures of renal epithelial cells. *J. Med. Microbiol.* **21:**59–67.
4. **Banfi, H., M. Cinco, M. Bellini, and M. R. Sorano.** 1982. The role of antibodies and serum complement in the interaction between macrophages and leptospires. *J. Gen. Microbiol.* **128:**813–816.
5. **Barnett, J., D. Barnett, C. Bolin, T. Summers, E. Wagar, N. Cheville, R. Hartskeerl, and D. Haake.** 1999. Expression and distribution of leptospiral outer membrane components during renal infection of hamsters. *Infect. Immun.* **67:**853–861.
6. **Birnbaum, N., S. C. Barr, S. A. Center, T. Schermerhorn, J. F. Randolph, and K. W. Simpson.** 1998. Naturally acquired leptospirosis in 36 dogs: serological and clinicopathological features. *J. Small Anim. Pract.* **39:**231–236.
7. **Bolin, C. A.** 1996. Diagnosis of leptospirosis: a reemerging disease of companion animals. *Semin. Vet. Med. Surg. (Small Anim.)* **11:**166–171.
8. **Bolin, C. A., and J. F. Prescott.** 1999. Leptospirosis, p. 352–357. *In* J. L. Howard and R. A. Smith (ed.), *Current Veterinary Therapy: Food Animal Practice 4*. W. B. Saunders Company, Philadelphia, Pa.
9. **Bolin, C. A., A. B. Thiermann, A. L. Handsaker, and J. W. Foley.** 1989. Effect of vaccination with a pentavalent leptospiral vaccine on *Leptospira interrogans* serovar hardjo type hardjo-bovis infection of pregnant cattle. *Am. J. Vet. Res.* **50:**161–165.
10. **Brem, S., H. Gerhards, B. Wollanke, P. Meyer, and H. Kopp.** 1998. Intraokularer Leptospirennachweis bei 4 Pferden mit recidiviertender Uveitis (ERU). [Intraocular leptospira isolation in 4 horses suffering from equine recurrent uveitis (ERU)]. *Berl. Münch. Tierärztl. Wochenschr.* **111:**415–417.
11. **Brenner, D. J., A. F. Kaufmann, K. R. Sulzer, A. G. Steigerwalt, F. C. Rogers, and R. S. Weyant.** 1999. Further determination of DNA relatedness between serogroups and

serovars in the family *Leptospiraceae* with a proposal for *Leptospira alexanderi* sp. nov. and four new *Leptospira* genomospecies. *Int. J. Syst. Bacteriol.* **49:**839–858.

12. **Cinco, M., E. Banfi, and M. R. Soranzo.** 1981. Studies on the interaction between macrophages and leptospires. *J. Gen. Microbiol.* **124:**409–413.
13. **Cole, J., L. Sangsteer, C. Sulzer, A. Pursell, and H. Ellinghausen.** 1982. Infections with *Encephalitizoon cuniculi* and *Leptospira interrogans,* serovars *grippotyphosa* and *ballum,* in a kennel of Foxhounds. *J. Am. Vet. Med. Assoc.* **180:**435–437.
14. **Dikken, H., and E. Kmety.** 1978. Serological typing methods of leptospires, p. 260-295. *In* T. Bergan and R. Norris (ed.), *Methods in Microbiology,* vol. 11. Academic Press, New York, N.Y.
15. **Dózsa, L., F. Kemenes, and T. Szent-Iványi.** 1960. Susceptibility of the red blood cells of ruminants with a four-chambered stomach (Pecora) to the haemotoxin of pathogenic leptospires. *Acta Vet. Acad. Sci. Hung.* **10:**35–44.
16. **Ellis, W. A., J. A. Cassells, and J. Doyle.** 1986. Genital leptospirosis in bulls. *Vet. Rec.* **118:**333.
17. **Ellis, W. A., P. J. McParland, D. G. Bryson, and J. A. Cassells.** 1986. Boars as carriers of leptospires of the Australis serogroup on farms with an abortion problem. *Vet. Rec.* **118:**563.
18. **Ellis, W. A., J. G. Songer, J. Montgomery, and J. A. Cassells.** 1986. Prevalence of *Leptospira interrogans* serovar *hardjo* in the genital and urinary tracts of non-pregnant cattle. *Vet. Rec.* **118:**11–13.
19. **Faine, S.** 1964. Reticuloendothelial phagocytosis of virulent leptospires. *Am. J. Vet. Res.* **25:**830–835.
20. **Faine, S.** 1957. Virulence in Leptospira. I. Reactions of guinea-pigs to experimental infection with Leptospira icterohaemorrhagiae. *Br. J. Exp. Pathol.* **38:**1–7.
21. **Faine, S.** 1957. Virulence in Leptospira. II. The growth in vivo of virulent Leptospira icterohaemorrhagiae. *Br. J. Exp. Pathol.* **38:**8–14.
22. **Faine, S., B. Adler, C. Bolin, and P. Perolat.** 1999. *Leptospira and Leptospirosis,* 2nd ed. Medisci Press, Melbourne, Australia.
23. **Faine, S., A. Shahar, and M. Aronson.** 1964. Phagocytosis and its significance in leptospiral infection. *Aust. J. Exp. Biol. Med. Sci.* **42:**579–588.
24. **Greenberg, E. P., and E. Canale-Parola.** 1977. Relationship between cell coiling and motility of spirochetes in viscous environments. *J. Bacteriol.* **131:**960–969.
25. **Haake, D. A., E. M. Walker, D. R. Blanco, C. A. Bolin, J. N. Miller, and M. A. Lovett.** 1991. Changes in the surface of *Leptospira interrogans* serovar *grippotyphosa* during in vitro cultivation. *Infect. Immun.* **59:**1131–1140.
26. **Harkin, K., and C. L. Gartrell.** 1996. Canine leptospirosis in New Jersey and Michigan: 17 cases (1990–1995). *J. Am. Anim. Hosp. Assoc.* **32:**495–501.
27. **Harrington, D. D., and S. D. Sleight.** 1966. *Leptospira pomona* in tissue culture: preliminary study. *Am. J. Vet. Res.* **27:**249–256.
28. **Ido, Y., R. Hoki, H. Ito, and H. Wani.** 1917. The rat as a carrier of *Spirochaeta icterohaemorrhagiae,* the causative agent of Weil's disease (spirochaetosis icterohaemorrhagica). *J. Exp. Med.* **26:**341–353.
29. **Isogai, E., H. Isogai, N. Fujii, and K. Oguma.** 1990. Biological effects of leptospiral lipopolysaccharide on mouse B, T and NK cells. *Jpn. J. Vet. Sci.* **52:**923–930.
30. **Isogai, E., H. Isogai, N. Fujii, and K. Oguma.** 1990. Macrophage activation by leptospiral lipopolysaccharide. *Zentbl. Bakteriol.* **273:**200–208.
31. **Ito, T., and R. Yanagawa.** 1987. Leptospiral attachment to extracellular matrix of mouse fibroblast (L929) cells. *Vet. Microbiol.* **15:**89–96.
32. **Kaneko, R., and S. Morihana.** 1921. Untersuchungen über die Identität der *Spirochaeta icterohaemorrhagiae* (Inada und Ito) und der *Spirochaeta icterogenes* (Uhlenhuth und

Fromme) und über das Verhalten der *Spirochaeta hebdomadis,* des Erregers des Siebentagfiebers ("Nanukayami"), gegenüber der *Spirochaeta icterogenes. Z. Immunitätsforsch. Exp. Ther. I Orig.* **31:**201–221.

33. **Kasarov, L. B.** 1970. Degradation of the erythrocyte phospholipids and haemolysis of the erythrocytes of different animal species by Leptospirae. *J. Med. Microbiol.* **3:**29–37.
34. **Kefford, B., and K. C. Marshall.** 1984. Adhesion of *Leptospira* at a solid-liquid interface: a model. *Arch. Microbiol.* **138:**84–88.
35. **Knight, L. L., N. G. Miller, and R. J. White.** 1973. Cytotoxic factor in the blood and plasma of animals during leptospirosis. *Infect. Immun.* **8:**401–405.
36. **Leonard, F. C., P. J. Quinn, W. A. Ellis, and K. O'Farrell.** 1993. Association between cessation of leptospiruria in cattle and urinary antibody levels. *Res. Vet. Sci.* **55:**195–202.
37. **Marshall, R. B.** 1976. The route of entry of leptospires into the kidney tubule. *J. Med. Microbiol.* **9:**149–152.
38. **Marshall, R. B.** 1974. Ultrastructural changes in renal tubules of sheep following experimental infection with *Leptospira interrogans* serotype *pomona. J. Med. Microbiol.* **7:**505–508.
39. **Merien, F., G. Baranton, and P. Perolat.** 1997. Invasion of Vero cells and induction of apoptosis in macrophages by pathogenic *Leptospira interrogans* are correlated with virulence. *Infect. Immun.* **65:**729–738.
40. **Merien, F., J. Truccolo, Y. Rougier, G. Baranton, and P. Perolat.** 1998. In vivo apoptosis of hepatocytes in guinea pigs infected with *Leptospira interrogans* serovar icterohaemorrhagiae. *FEMS Microbiol. Lett.* **169:**95–102.
41. **Miller, R. E., N. G. Miller, and R. J. White.** 1966. Growth of *Leptospira pomona* and its effect on various tissue culture systems. *J. Bacteriol.* **92:**502–509.
42. **Morter, R. L., R. D. Williams, H. Bolte, and M. J. Freeman.** 1969. Equine leptospirosis. *J. Am. Vet. Med. Assoc.* **155:**439–442.
43. **Parma, A. E., S. I. Cerone, and S. A. Sansinanea.** 1992. Biochemical analysis by SDS-PAGE and western blotting of the antigenic relationship between Leptospira and equine ocular tissues. *Vet. Immunol. Immunopathol.* **33:**179–185.
44. **Parma, A. E., C. G. Santisteban, J. S. Villalba, and R. A. Bowden.** 1985. Experimental demonstration of an antigenic relationship between *Leptospira* and equine cornea. *Vet. Immunol. Immunopathol.* **10:**215–224.
45. **Parma, A. E., M. E. Sanz, P. M. Lucchesi, J. Mazzonelli, and M. A. Petruccelli.** 1997. Detection of an antigenic protein of *Leptospira interrogans* which shares epitopes with the equine cornea and lens. *Vet. J.* **153:**75–79.
46. **Perolat, P., F. Grimont, B. Regnault, P. A. D. Grimont, E. Fournie, H. Thevenet, and G. Baranton.** 1990. rRNA gene restriction patterns of leptospira: a molecular typing system. *Res. Microbiol.* **141:**159–171.
47. **Perolat, P., I. Lecuyer, D. Postic, and G. Baranton.** 1993. Diversity of ribosomal DNA fingerprints of *Leptospira* serovars provides a database for subtyping and species assignation. *Res. Microbiol.* **144:**5–15.
48. **Rentko, V. T., N. Clark, L. A. Ross, and S. H. Schelling.** 1992. Canine leptospirosis: a retrospective study of 17 cases. *J. Vet. Intern. Med.* **6:**235–244.
49. **Rose, G. W., W. C. Eveland, and H. C. Ellinghausen.** 1966. Mechanisms of tissue cell penetration by *Leptospira pomona*: active penetration studies *in vitro. Am. J. Vet. Res.* **27:**1461–1471.
50. **Segers, R. P., J. A. van Gestel, G. J. van Eys, B. A. van der Zeijst, and W. Gaastra.** 1992. Presence of putative sphingomyelinase genes among members of the family *Leptospiraceae. Infect. Immun.* **60:**1707–1710.
51. **Segers, R. P. A. M.** 1991. Thesis. Rijksuniversiteit te Utrecht, Utrecht, The Netherlands.
52. **Segers, R. P. A. M., A. van der Drift, A. de Nijs, P. Corcione, B. A. van der Zeijst, and**

W. Gaastra. 1990. Molecular analysis of a sphingomyelinase C gene from *Leptospira interrogans* serovar hardjo. *Infect. Immun.* **58:**2177–2185.

53. **Stalheim, O. H. V.** 1967. A toxic factor in *Leptospira pomona. Proc. Soc. Exp. Biol. Med.* **126:**412–415.
54. **Terpstra, W. J., H. Korver, G. J. Schoone, J. van Leeuwen, C. E. Schönemann, S. de Jonge-Aglibut, and A. H. J. Kolk.** 1987. Comparative classification of leptospira serovars of the Pomona group by monoclonal antibodies and restriction-endonuclease analysis. *Zentbl. Bakteriol. Mikrobiol. Hyg. Ser. A* **266:**412–421.
55. **Thiermann, A. B., A. L. Handsaker, S. L. Moseley, and B. Kingscote.** 1985. New method for classification of leptospiral isolates belonging to serogroup pomona by restriction endonuclease analysis: serovar *kennewicki. J. Clin. Microbiol.* **21:**585–587.
56. **Thiermann, A. B., and R. LeFebvre.** 1989. Restriction endonuclease analysis and other molecular techniques in identification and classification of *Leptospira* and other pathogens of veterinary importance, p. 145-183. *In* B. Swaminathan and G. Prakash (ed.), *Nucleic Acids and Monoclonal Antibodies Probes.* Marcel Dekker, New York, N.Y.
57. **Thomas, D. D., and L. M. Higbie.** 1990. In vitro association of leptospires with host cells. *Infect. Immun.* **58:**581–585.
58. **Thompson, J. C.** 1986. Morphological changes in red blood cells of calves caused by Leptospira interrogans serovar pomona. *J. Comp. Pathol.* **96:**512–527.
59. **Thompson, J. C., and B. W. Manktelow.** 1986. Pathogenesis and red blood cell destruction in haemoglobinaemic leptospirosis. *J. Comp. Pathol.* **96:**529–540.
60. **Trevejo, R. T., Epidemiology Working Group—Ministry of Health Ecuador, Epidemiology Working Group—CDC, and C. A. Bolin.** 1998. Epidemic leptospirosis associated with pulmonary hemorrhage—Nicaragua, 1995. *J. Infect. Dis.* **178:**1457–1463.
61. **van den Ingh, T. S., and E. G. Hartman.** 1986. Pathology of acute *Leptospira interrogans* serotype *icterohaemorrhagiae* infection in the Syrian hamster. *Vet. Microbiol.* **12:**367–376.
62. **Vinh, T., B. Adler, and S. Faine.** 1986. Glycolipoprotein cytotoxin from *Leptospira interrogans* serovar *copenhageni. J. Gen. Microbiol.* **132:**111–123.
63. **Vinh, T., B. Adler, and S. Faine.** 1982. The role of macrophages in the protection of mice against leptospirosis: *in vitro* and *in vivo* studies. *Pathology* **14:**463–468.
64. **Vinh, T., S. Faine, and B. Adler.** 1984. Adhesion of leptospires to mouse fibroblasts (L929) and its enhancement by specific antibody. *J. Med. Microbiol.* **18:**73–85.
65. **Weil, A.** 1886. Ueber eine eigenthümliche, mit Milztumor, Icterus und Nephritis einhergehende, acute Infectionskrankheit. *Deutsche Arch. Klin. Med.* **39:**209.
66. **Wisconsin Outbreak Investigation Team, Wisconsin Department of Health; Illinois Outbreak Investigation Team, Springfield Department of Health and Illinois Department of Public Health; Council of State and Territorial Epidemiologists; Zoonotic Diseases Research Unit, Agricultural Research Service, U.S. Department of Agriculture; and Meningitis and Special Pathogens Branch, Division of Bacterial and Mycotic Diseases, National Center for Infectious Diseases, and EIS officers, Centers for Disease Control and Prevention.** 1998. Update: leptospirosis and unexplained acute febrile illness among athletes participating in triathlons—Illinois and Wisconsin, 1998. *Morb. Mortal. Wkly. Rep.* **47:**676.
67. **Wollanke, B., H. Gerhards, S. Brem, H. Kopp, and P. Meyer.** 1998. Intraokulare und Serumantikorpertiter gegen Leptospiren bei 150 wegen equiner rezidivierender Uveitis (ERU) vitrektomierten Pferden. *Berl. Münch. Tierärztl. Wochenschr.* **111:**134–139.
68. **Yam, P. A., N. G. Miller, and R. J. White.** 1970. A leptospiral factor producing a cytopathic effect on L cells. *J. Infect. Dis.* **122:**310–317.
69. **Zaki, S. R., and W. J. Shieh.** 1996. Leptospirosis associated with outbreak of acute febrile illness and pulmonary haemorrhage, Nicaragua, 1995. *Lancet* **347:**535–536.

Emerging Diseases of Animals
Edited by C. Brown and C. Bolin

Chapter 10

Enterohemorrhagic *Escherichia coli* in Ruminant Hosts

Barry G. Harmon, Cathy A. Brown, Michael P. Doyle, and Tong Zhao

Enterohemorrhagic *Escherichia coli,* or pathogenic Shiga toxin-producing *E. coli* (STEC), is distinguished from other pathogenic *E. coli* by its ability to produce Shiga toxins designated Stx1, Stx2, Stx2c, and Stx2e. These toxins are structurally similar to the Shiga toxin produced by *Shigella dysenteriae* type 1. Unlike the genes that encode Shiga toxin in *S. dysenteriae,* the toxins in *E. coli* are encoded on temperate lambdoid phages. Shiga toxins have a profound cytopathic effect on Vero cells, hence the synonym verotoxin-producing *E. coli* (56). Pathogenic STEC causes hemorrhagic colitis and hemolytic uremic syndrome in humans, rarely colitis in calves, and edema disease in swine (29, 39, 50, 55, 62, 73). *E. coli* serotype O157:H7 is considered the prototype of pathogenic enterohemorrhagic *E. coli. E. coli* O157:H7 is highly pathogenic, possesses numerous virulence factors, and is the most common cause of STEC disease in humans in many parts of the world (29, 56). *E. coli* O157:H7 is associated with large outbreaks of disease, whereas other STEC serotypes are more commonly associated with sporadic disease in humans, and these non-O157:H7 serotypes are believed to be less pathogenic. However, in recent years, non-O157:H7 serotypes have been identified as important causes of both sporadic cases and outbreaks of hemorrhagic colitis (42).

Barry G. Harmon • Department of Pathology, College of Veterinary Medicine, The University of Georgia, Athens, GA 30602. **Cathy A. Brown** • Athens Diagnostic Laboratory, College of Veterinary Medicine, The University of Georgia, Athens, GA 30602. **Michael P. Doyle and Tong Zhao** • Center for Food Safety and Quality Enhancement, College of Agriculture and Environmental Sciences, The University of Georgia, Griffin, GA 30223.

The pathogenic qualities of *E. coli* O157:H7 are determined by known and unknown virulence factors. The presence of the known virulence properties can be associated with severe symptoms in human patients infected with different STEC isolates (57, 58). In *E. coli* O157:H7, the genes for Shiga toxins are encoded on lambdoid phages. Attaching and effacing factors, intimin, translocated intimin receptor, and a necessary type III secretion system are encoded on the chromosome in a pathogenicity island named the locus of enterocyte effacement (LEE) (56). A 60-MDa plasmid (pO157) encodes a distinct enterohemolysin designated EHEC-Hly that is different from *E. coli* alpha hemolysin (5). Some *E. coli* O157:H7 strains can survive for extended periods in acidic conditions by activating amino acid decarboxylases and/or producing a stationary-phase sigma factor encoded by the *rpoS* regulon (1, 15). This regulon encodes a sigma factor that changes promoter recognition of RNA polymerase and modifies gene expression in response to stress (2, 47). The presence or absence of these virulence determinants, as well as some unknown factors, determines the pathogenic potential of a given STEC isolate.

PREVALENCE OF STEC IN DOMESTIC RUMINANTS

Contaminated beef and dairy products are important sources of food-borne STEC infections, causing hemorrhagic colitis and hemolytic uremic syndrome in humans (3, 9, 29, 68). From the available data, ruminants appear to have the highest prevalence of STEC among food-producing animals tested (6, 42, 58). Over 150 serotypes of STEC have been isolated from humans, and more than 200 serotypes of STEC have been isolated from animals and other sources (42, 48, 72).

While *E. coli* O157:H7 is the major cause of hemolytic uremic syndrome and hemorrhagic colitis in the United States and in some other parts of the world, its overall prevalence in individual cattle and meat samples is low relative to that of other STEC. A 1994 survey of Wisconsin dairy farms found a herd prevalence of *E. coli* O157:H7 of 7.1% and an animal prevalence of 1.8% (25). The prevalence of *E. coli* O157:H7 in weaned calves, however, may be as high as 5% (26, 74). However, 63% of 100 feedlots in 13 states of the United States had at least one *E. coli* O157:H7-positive fecal sample, and 1.8% of the total number of samples were positive (19). Therefore, *E. coli* O157:H7 is widely distributed throughout the cattle population, but fecal shedding of *E. coli* O157:H7 is only detected in a few individual cattle at a given time. In contrast, when all serotypes of STEC are included in surveys, the prevalence in cattle and other ruminants is much higher. On Ontario dairy farms, evidence of STEC was detected in

17% of 1,790 cattle tested in 1988 and in 45% of 1,478 cattle tested in 1992 (16, 70). In a German survey, STEC was isolated from 66% of sheep, 56.1% of goats, 21.1% of cattle, 0.1% of chickens, 7.5% of pigs, 13.8% of cats, and 4.8% of dogs tested (6). In a Canadian survey of ground and deboned raw beef, 15 to 40% of samples collected from 1987 to 1994 had evidence of STEC (42). Evidence of STEC was found in 23% of raw ground beef samples in Seattle, Washington, and in 66% of veal and 48% of lamb samples (60). In a Belgian study, STEC was detected in a large variety of meat samples, but the highest contamination was found in venison and beef (58). The large majority of STEC isolated in these studies was non-O157:H7 serotypes (6, 42, 58).

Non-O157:H7 STEC is increasingly being isolated from sporadic cases of hemorrhagic colitis, and non-O157:H7 STEC is responsible for at least six outbreaks of hemorrhagic colitis (42). Only a few of the STEC serotypes found in cattle have been associated with human disease. For example, serotypes O26:H11, O111:H^-, and O145:H^- are often associated with outbreaks or sporadic cases of hemorrhagic colitis in humans (42). Therefore, a large heterogeneous population of STEC is present in ruminants, and only a small subset of these are pathogenic for humans (42, 61). One possible explanation for the high incidence of human infection with *E. coli* O157:H7 and the relatively low incidence in the cattle population compared to other STEC serotypes is that *E. coli* O157:H7 is more pathogenic and more infectious for humans than the other serotypes.

PATHOGENICITY OF STEC FOR CATTLE

Although STEC is present in the gastrointestinal tract of ruminants and contamination from this source is a cause of food-borne illness in humans, STEC is not a major cause of disease in cattle. Natural disease associated with STEC has not been reported in adult cattle; however, hemorrhagic colitis and diarrhea with attaching and effacing lesions have been reported in young calves in association with naturally acquired and experimentally induced STEC infections (27, 40, 62).

The STEC serotypes most commonly isolated from calves with diarrhea include O111:H^-, O5:H^-, and O26:H^- (27, 40, 62). Even though attaching and effacing lesions were observed in the colons of calves infected with STEC, a large percentage of calves were concurrently infected with other intestinal pathogens, such as *Cryptosporidium* sp., rotavirus, coronavirus, enterotoxigenic *E. coli*, bovine viral diarrhea virus, and coccidia (40). Experimental infection requires large doses of STEC and does not always result in diarrhea. Even though focal attaching and effacing lesions can be produced experimentally, diarrhea may only occur in the youngest calves, colostrum-deprived calves, or calves coinfected with intestinal viruses (13, 50, 54, 73). Among 5- to 10-day-old calves infected with *E. coli* O26:H11, none developed any clinical disease, but focal attaching-effacing lesions were observed in the colon of some (50). In a survey of healthy calves and calves with diarrhea from 12 Michigan farms, *E. coli* strains with both the *eae* and *stx* genes were found in 20 of 101 healthy calves and in only 11 of 114 calves with diarrhea (37). Of 15 diarrheic calves that were necropsied, 5 had no lesions and 10 had lesions consistent with attaching and effacing lesions. However, 7 of the 15 were coinfected with other enteric pathogens (37). *E. coli* O157:H7 is not associated with natural disease in cattle of any age; however, hemorrhagic enterocolitis with attaching and effacing lesions can be produced in neonatal colostrum-deprived calves with challenge doses of 10^{10} CFU (20). Therefore, disease surveillance cannot be used to predict which cattle are shedding STEC. In fact, surveys indicate that most calves and adult cattle that shed STEC are healthy.

EMERGENCE OF STEC IN FOOD-PRODUCING RUMINANTS

It is virtually unknown under what circumstances or where in the environment pathogenic STEC emerged. In a retrospective study of 20 fecal *E. coli* isolates recovered from samples taken from calves in 1965, five possessed at least one STEC-associated virulence marker, either *stx-2*, *eaeA*, or enterohemolysin activity (8). STEC strains were present in the human population in the 1960s (8). It is believed that Shiga toxin-converting phages were responsible for the emergence of *E. coli* O157:H7 from less pathogenic O55:H7 and O127:H16 progenitors (61, 69). The accumulation of virulence factors into pathogenicity islands may have resulted from continued acquisition of genes from bacteriophages or plasmids that integrate at common integration sites such as the bacterial tRNA genes (14, 61). Another possibility is that virulence genes are grouped as a result of homologous recombination between related phage genomes carrying different virulence genes (61). The rumen has a high phage density, 3×10^9 to 1.6×10^{10} phages per ml of rumen fluid (45). It is not known if phage transfer takes place within the ruminant host. It is interesting that STEC of different serotypes but originating from the same population of sheep or cattle are similar with respect to *stx* genes (7). These data may suggest that Stx-encoding bacteriophages are spread among different serotypes of *E. coli* within an animal population, creating a genetically heterogeneous population of STEC (7). Levels of verotoxin in fecal samples from human patients infected with STEC remain high even after the strains causing the original infection can no longer be detected (61, 63). It is plausible, therefore, that the source of toxin could arise from phage transfer in vivo to commensal *E. coli* in the intestines in human patients, and similar transfer of toxin genes to other bacteria may occur in ruminant carriers.

The competitive and environmental selective pressures on *E. coli* in the ruminant host may play a role in selection of some virulence factors or other factors that give STEC a competitive advantage in the rather hostile environment of the rumen. It is possible that the high volatile fatty acid environment in the rumen could select acid-resistant STEC that would be better able to survive in acidic foods and survive passage through the stomach (22). Acid adaptation in the rumen could also select more acid-resistant *E. coli* that could pass through the acidic bovine abomasum and colonize the large intestines in greater numbers. However, sustained acid tolerance in *E. coli* O157:H7 does not require acid adaptation but can be induced by stationary phase, starvation, and other environmental stresses (1, 15, 53). This mechanism of acid tolerance is induced by the *rpoS* gene, encoding a sigma factor (1, 34, 46). *E. coli* strains vary in acid tolerance, and this variability is also present in STEC. Acid tolerance to volatile fatty acids is not unique to rumen-derived strains of *E. coli.* Acid tolerance is also found in *E. coli* isolated from porcine feces (63). Acid tolerance would be important for survival in the ruminant host as well as for passage through the human stomach to colonize the intestines of human patients. Acid tolerance may explain how such a low infective dose of *E. coli* O157:H7 can cause human illness. Whether other virulence attributes impart a competitive advantage to STEC in the ruminant gastrointestinal tract or elsewhere in the environment will require further investigation.

The emergence of antibiotic-resistant STEC in animal populations has been anticipated by some. In a longitudinal survey of *E. coli* O157:H7 prevalence in three dairy herds, almost all isolates from a farm that practiced heavy use of antibiotics exhibited resistance to at least one antibiotic, whereas only 15% of isolates from a farm that used antibiotics sparingly were resistant (65). However, neomycin sulfate was given to all calves from birth to 12 weeks of age on one farm, and none of the 17 isolates from this farm was resistant to neomycin (65). In a survey of isolates of *E. coli* O157:H7 from 56 human patients from 1984 to 1987, no antibiotic resistance was detected; however, 13 of 176 isolates from the same geographic area collected from 1989 to 1991 were resistant to streptomycin, sulfisoxazole, or tetracycline, antibiotics commonly used in animal feeds (44). Human patients on antimicrobial therapy could be at increased risk of infection with antibiotic-resistant STEC strains. Antibiotic-resistant STEC may have a competitive advantage in livestock treated with antibiotics, resulting in increased fecal shedding.

PROSPECTS FOR INTERVENTION TO CONTROL STEC SHEDDING IN CATTLE

Control procedures to reduce fecal contamination of beef and dairy products at the processing plant by improving sanitation are already being implemented. Effective control procedures to reduce fecal shedding on the farm and prior to slaughter are still under study.

Fecal shedding of STEC is transient in cattle. In studies where samples were collected from the same animals over time, cattle shed *E. coli* O157:H7 for 8 to 46 days in one study (68) and for less than 1 month in another study (4). In a longitudinal study, dairy heifers shed *E. coli* O157:H7 for 1 to 16 months, and *E. coli* O157:H7 persisted in a herd for at least 2 years (65). *E. coli* O157:H7 is most commonly isolated from weaned calves and dairy heifers from 3 months old to breeding age (26, 30, 31, 32, 35, 74). Heifers shed *E. coli* O157:H7 at levels ranging from 2×10^2 to 8.7×10^4 per g of feces (65).

To reduce the likelihood of contamination of food with STEC, procedures designed to reduce the carriage of STEC in cattle, sheep, and other ruminants have been proposed. Early on, there was interest in producing a vaccine against *E. coli* O157:H7 for cattle. However, adult cattle experimentally infected with 10^{10} CFU of *E. coli* O157:H7 shed *E. coli* O157:H7 in their feces for up to 100 days, and calves similarly infected shed bacteria for up to 144 days. Furthermore, prior exposure does not prevent reinfection and subsequent shedding (17). These findings suggest that acquired immunity may not affect subsequent fecal shedding. There is no evidence of prolonged mucosal colonization, and no gastrointestinal lesions are detected in experimental infections (11). *E. coli* O157:H7 is restricted to the gastrointestinal tract, being recovered from the cecum, colon, small intestines, rumen, reticulum, and omasum at necropsy (11, 17). Following infection with 10^{10} CFU of *E. coli* O157:H7, a few calves develop mild transient pyrexia and diarrhea within 24 h of inoculation but remain bright and alert with good appetites (11). *E. coli* O157:H7 shedding fluctuates from day to day and is highly variable for individual calves and cows (11, 17). These experiments indicate that *E. coli* O157:H7 is not invasive or pathogenic in weaned calves and adult cattle. Furthermore, calves and cattle are capable of shedding *E. coli* O157:H7 for extended periods even though there is no apparent long-term colonization of the intestinal mucosa. Repeated isolation of *E. coli* O157:H7 from the rumen suggests that the rumen may be a source of continued reintroduction of *E. coli* O157:H7 to the colon (11). In addition, prior exposure and even seroconversion does not protect against reinfection or correlate with subsequent shedding (17, 41). Furthermore, mucosal colonization does not appear to be a prerequisite for prolonged shedding. Therefore, traditional vaccine strategies are not likely to be successful in reducing fecal shedding of STEC in cattle.

Another possible approach to controlling fecal shedding is to identify specific farm management practices that would reduce contamination or spread of STEC among cattle. Several epidemiological studies relating shedding of *E. coli* O157:H7 to farm management practices and environmental conditions have been completed (4, 19, 25, 30, 35, 49, 65).

The prevalence of *E. coli* O157:H7 in cattle feces increases in warmer months (30, 36). *E. coli* O157:H7 appears to be disseminated throughout farms by horizontal spread, possibly from a single source (36, 65). On farms where *E. coli* O157:H7 is isolated from cattle, *E. coli* O157:H7 is also isolated from horses, sheep, ponies, milk filters, and flies (36). In a Wisconsin study, *E. coli* O157:H7 was isolated from flies, pigeons, and water (65). Specific isolates tend to predominate on given farms, and those same isolates may persist for up to 2 years on a given farm (36, 65). Contaminated drinking water likely serves as a common source for spread among livestock (25, 65). The intermittent and reoccurring shedding by cattle may result from reinfection from a contaminated environmental source, such as drinking water, rather than long-term colonization of the intestines (65). Fecal shedding occurs in a higher percentage of weaned heifer calves, 3 to 12 months old, compared to older or younger cattle (30, 31, 32, 35, 36, 52). In one study, heifers were more likely to test positive after being grouped in superhutches, suggesting that increased exposure or stress associated with grouping might be responsible for increased prevalence in calves of this age (65). In another study of 60 dairies, herds positive for *E. coli* O157:H7 were more likely to be smaller herds ($P = 0.08$). Positive herds were more likely to graze land irrigated with slurry ($P < 0.10$), and in those herds that did use slurry application, shorter intervals between application and grazing were associated with an increased prevalence of *E. coli* O157:H7 (30). The prevalence of *E. coli* O157:H7 was higher in herds that weaned calves abruptly than in herds that used a variety of methods to gradually wean calves (35). This same study found an increase in the prevalence of *E. coli* O157:H7 associated with monensin, lasalocid, or decoquinate use in heifer rations ($P = 0.10$), whereas no association could be found with *E. coli* O157:H7 prevalence and feeding of ionophores in another study of dairies (26). Likewise, no association with feeding of ionophores was shown in a study of *E. coli* O157:H7 prevalence in feedlots (19). In the feedlot study, no association of *E. coli* O157:H7 prevalence was found with feeding antibiotics, coccidiostats, soy meal, or urea; however, there was an increase in the likelihood of positive pens if barley was fed or if the cattle had been on feed for less than 20 days. In addition, heavier cattle (>700 lb) entering feed yards were less likely to have positive samples. Increased prevalence of *E. coli* O157:H7 shedding was observed in dairy heifers fed corn silage (35). Corn silage rations from 12 of 16 dairies were positive for *E. coli* O157:H7, and 5 had concentrations above 1,000 CFU/g (49). High moisture in feeds such as corn silage tends to support the growth of *E. coli.* High lactate concentrations and low propionate concentrations in experimentally mixed rations supported replication of *E. coli* (49). Supplementation of these feeds with lasalocid had no effect on replication of *E. coli* in feed (49). The above findings indicate that some farm management practices may influence STEC shedding. In particular, contaminated water sources may serve to spread STEC and to reintroduce STEC to animals. Reducing contamination of water sources and regular cleaning of watering tanks might reduce the dissemination of *E. coli* O157:H7. More studies are needed to determine the effects of weaning practices on shedding. Feeds could be a source of contamination, but this has not been clearly demonstrated. Therefore, with the exception of ensuring a clean water supply and improving overall sanitation, there is no clear indication that alteration of other management procedures would significantly reduce fecal shedding and horizontal spread of STEC on farms.

Ruminants such as sheep, cattle, and deer are pregastric fermentors that have enlarged pregastric nonsecretory rumens in which the passage of dietary plant material is delayed. The rumen is primarily an anaerobic environment where bacteria hydrolyse cellulose, arabinoxylins, and other plant polymers (24). Rumen anaerobes produce volatile fatty acids, acetate, propionate, and butyrate by dissimilation of pyruvate. Rumen anaerobic bacteria outnumber aerobes by 100-fold or more, and *E. coli* are typically recovered at populations of only 10^5 CFU/ml of rumen contents (24, 28).

Growth of *E. coli* and *Salmonella* spp. increases in the rumens of cattle and sheep that have been fasted (12, 28). Therefore, avoiding fasting, particularly before slaughter, might reduce the concentration of coliforms in the rumen and possibly in the feces at the time of slaughter. In vitro, *E. coli* O157:H7 growth is supported by rumen fluid from fasted cattle and grows poorly in rumen fluid from well-fed cattle (59). The inhibitory effect on coliform growth is attributed to a combination of low pH and high concentrations of volatile fatty acids (18, 67, 71). However, in two separate studies (18, 33), fasting did not result in a significant increase in fecal shedding or rumen growth of *E. coli* O157:H7 in calves that were experimentally inoculated with *E. coli* O157:H7. Fasting resulted in a marked increase in rumen pH and decrease in volatile fatty acids but no corresponding increase in rumen proliferation or fecal shedding in calves already shedding *E. coli* O157:H7 (33). However, when calves were fasted prior to experimental inoculation with *E. coli* O157:H7, subsequent shedding was increased (18). Cattle fed roughage for 4 days shed fewer *E. coli* biotype 1 populations than cattle that remained on a high-energy finishing ration (43). Cattle left on the high-energy diet shed 6.43 log *E. coli* CFU/g of feces, compared to 6.70 log *E. coli* CFU/g of feces for cattle switched to roughage. However, when cattle were fasted for 48 h, the cattle on roughage were shedding a higher concentration of *E. coli* (8.78 log *E. coli* CFU/g of feces) than cattle on a finishing ration (8.14 log *E. coli* CFU/g of feces). Fasting resulted in a significant increase in fecal shedding by both groups; however, the increase occurred slowly. Therefore, even if cattle on full feed shed fewer *E. coli* O157:H7 in the feces, this effect would have to be weighed against the increased likelihood of contaminating carcasses with feces from cattle placed on full feed before slaughter (43).

The effects of feeding high-energy grain diets or mostly forage diets on fecal shedding were investigated. A reduction in fecal shedding of *E. coli* was observed in cattle following a switch from feeding grain to hay. After switching the feed, the same cattle had a lower percentage of acid-resistant *E. coli* in the feces (21). Low rumen pH values and high volatile fatty acid concentrations in the rumen may induce acid tolerance in STEC, so that they survive passage through the acidic abomasum and proliferate in the colon. However, in studies where cattle were inoculated with *E. coli* O157:H7, cattle fed hay shed *E. coli* O157:H7 for an average of 39 to 42 days, compared to 4 days for cattle on a grain diet. Furthermore, acid resistance was similar for isolates taken from the feces of both groups of animals (38). Acid tolerance can be induced by less than favorable growth conditions, such as starvation and stationary phase (1, 15). Therefore, it is not surprising to find acid-resistant strains in the feces of cattle. From these data, it is not clear that feeding either grain or forage diets would significantly reduce fecal shedding in cattle.

Some feed substances such as whole cottonseed (26, 30) and soybean meal (19), are negatively associated with prevalence of *E. coli* O157:H7. Therefore, it may be possible to reduce shedding by feeding inhibitory feed substances if such substances can be identified. *E. coli* O157:H7 growth is inhibited by plant-derived coumarin aglycones (23). Some of the most numerous bacterial species in the rumen and colon, such as *Bacteroides* spp. and *Prevotella* spp. (66), can hydrolyze esculin to the aglycone esculetin, thus producing inhibitory aglycones in the gastrointestinal tract. It may be possible to alter the very competitive nature of the rumen to exclude or reduce *E. coli* O157:H7, particularly since coliforms are such a minor component of the rumen microflora. Rumen isolates of *Pseudomonas aeruginosa* inhibit the growth of *E. coli* O157:H7 in vitro (S. H. Duncan, C. J. Doherty, J. R. W. Govan, and C. S. Stewart, Abstr. 97th Gen. Meet. Am. Soc. Microbiol. 1997, abstr. N-233, p. 419). The major inhibitory factor under aerobic conditions is pyocyanin and under anerobic conditions appears to be fluorescein (24). Many of the gastrointestinal *E. coli* are colicinogenic, and some strains of *E. coli* O157:H7 are sensitive to colicins, whereas others are resistant (10, 24). Seventeen bovine fecal isolates of *E. coli* and one isolate of *Proteus mirabilis* were found to produce metabolites that inhibit growth of *E. coli* O157:H7 in vitro (75). Calves were inoculated with 10^{10} CFU of a mixture of these inhibitory bacteria and then inoculated with 10^{10} CFU of *E. coli* O157:H7. Five of six treated calves stopped shedding *E. coli* O157:H7 before the end of the study, whereas *E. coli* O157:H7 was cultured from the rumen or colon of all nine control calves. Only one treated calf at the end of the study, 27 to 35 days after inoculation, was still culture positive (75). The possibility of adding specific feed substances, feed additives, or competitive exclusion bacteria to the diets of cattle to reduce fecal shedding and rumen proliferation of STEC is promising but will require further study.

SUMMARY

Ruminants used as food are an important source of food-borne STEC infections in humans. Shedding of STEC is widespread in healthy cattle throughout the world. The populations of STEC in ruminants are heterogeneous, and only a subset of these populations are pathogenic for humans. If transfer of virulence genes by phages and other mobile DNA elements occurs with any frequency, we can expect new pathogens to emerge. Prevalence of *E. coli* O157:H7 has been associated with certain farm management and feeding practices; however, no cause-and-effect relationships have been established. Some preliminary results suggest that feed metabolites, feed additives, and probiotic bacteria can reduce shedding of *E. coli* O157:H7. More prospective studies that directly evaluate the effects of management, feed additives, and probiotics are needed. In addition, basic research on the mode of virulence factor transfer and the role of selective pressures in the ruminant on survival of STEC are needed if sound strategies for control are to be developed.

REFERENCES

1. **Arnold, K. W., and C. W. Kasper.** 1995. Starvation- and stationary-phase-induced acid tolerance in *Escherichia coli* O157:H7. *Appl. Environ. Microbiol.* **61:**2037–2039.
2. **Bearson, S., B. Bearson, and J. W. Foster.** 1997. Acid stress responses in enterobacteria. *FEMS Microbiol. Lett.* **147:**173–180.
3. **Bell, B. P., M. Goldoft, P. M. Griffin, M. A. Davis, D. C. Gordon, P. I. Tarr, C. A. Bartleson, J. H. Lewis, T. J. Barrett, J. G. Wells, R. Baron, and J. Kobayashi.** 1994. A multistate outbreak of *Escherichia coli* O157:H7-associated bloody diarrhea and hemolytic uremic syndrome from hamburgers: the Washington experience. *JAMA* **272:**1349–1353.
4. **Besser, T. E., D. D. Hancock, L. C. Pritchett, E. M. McRae, D. H. Rice, and P. I. Tarr.** 1997. Duration of detection of fecal excretion of *Escherichia coli* O157:H7 in cattle. *J. Infect. Dis.* **175:**726–729.
5. **Beutin, L., M. A. Montenegro, I. Orskov, F. Orskov, J. Prada, S. Zimmermann, and R. Stephan.** 1989. Close association of verotoxin (shiga-like toxin) production with enterohemolysin production in strains of *Escherichia coli*. *J. Clin. Microbiol.* **27:**2559–2564.
6. **Beutin, L., D. H. Geier, S. Steinruck, S. Zimmermann, and F. Scheutz.** 1993. Prevalence and some properties of verotoxin (Shiga-like toxin)-producing *Escherichia coli* in seven different species of healthy domestic animals. *J. Clin. Microbiol.* **31:**2483–2488.
7. **Beutin, L., D. Geier, S. Zimmermann, S. Aleksic, H. A. Gillespie, and T. S. Whittam.** 1997. Epidemiological relatedness and clonal types of natural populations of *Escherichia coli* strains producing Shiga toxins in separate populations of cattle and sheep. *Appl. Environ. Microbiol.* **63:**2175–2180.
8. **Beutin, L., and W. Muller.** 1998. Cattle and verotoxigenic *Escherichia coli* (VTEC), an old relationship? *Vet. Rec.* **142:**283–284.
9. **Boyce, T. G., D. L. Swerdlow, and P. M. Griffin.** 1995. *Escherichia coli* O157:H7 and the hemolytic-uremic syndrome. *N. Engl. J. Med.* **333:**364–368.
10. **Bradley, D. E., S. P. Howard, and H. Loir.** 1991. Colicinogeny of O157:H7 enterohemorrhagic *Escherichia coli* and the shielding of colicin and phage receptors by their O-antigenic side chains. *Can. J. Microbiol.* **37:**97–104.
11. **Brown, C. A., B. G. Harmon, T. Zhao, and M. P. Doyle.** 1997. Experimental *Escherichia coli* O157:H7 carriage in calves. *Appl. Environ. Microbiol.* **63:**27–32.
12. **Brownlie, L. E., and F. H. Grau.** 1967. Effect of food intake on growth and survival of salmonellas and *Escherichia coli* in the bovine rumen. *J. Gen. Microbiol.* **46:**125–134.
13. **Chanter, N., G. A. Hall, A. P. Bland, A. J. Hayle, and K. R. Parsons.** 1986. Dysentery in calves caused by an atypical strain of *Escherichia coli* (102–9). *Vet. Microbiol.* **12:**241–253.
14. **Cheetham, B. F., and M. E. Katz.** 1995. A role for bacteriophages in the evolution and transfer of bacterial virulence determinants. *Mol. Microbiol.* **18:**201–208.
15. **Cheville, A. M., K. W. Arnold, C. Burchrieser, C.-M. Cheng, and C. W. Kasper.** 1996. *rpoS* regulation of acid, heat, and salt tolerance in *Escherichia coli* O157:H7. *Appl. Environ. Microbiol.* **62:**1822–1824.
16. **Clark, R. C., J. B. Wilson, S. C. Read, S. Renwick, K. Rahn, R. P. Johnson, D. Alves, M. A. Karmali, H. Loir, S. A. McEwen, J. Spika, and C. L. Gyles.** 1994. Verotoxin-producing *Escherichia coli* O157:H7 (VTEC) in the food chain: preharvest processing perspectives, p. 17–24. *In* M. A. Karmali and A. G. Goglio (ed.), *Recent Advances in Verocytotoxin-Producing* Escherichia coli *Infections.* Elsevier Science B. V., Amsterdam, The Netherlands.
17. **Cray, W. C., and H. W. Moon.** 1995. Experimental infection of calves and adult cattle with *Escherichia coli* O157:H7. *Appl. Environ. Microbiol.* **61:**1586–1590.

18. **Cray, W. C., T. A. Casey, B. T. Bosworth, and M. A. Rasmussen.** 1998. Effect of dietary stress on fecal shedding of *Escherichia coli* O157:H7 in calves. *Appl. Environ. Microbiol.* **64:**1975–1979.
19. **Dargatz, D. A., S. J. Wells, L. A. Thomas, D. D. Hancock, and L. P. Garber.** 1997. Factors associated with the presence of *Escherichia coli* O157:H7 in feces of feedlot cattle. *J. Food Prot.* **60:**466–470.
20. **Dean-Nystrom, E. A., B. T. Bosworth, W. C. Cray, and H. W. Moon.** 1997. Pathogenicity of *Escherichia coli* O157:H7 in the intestines of neonatal calves. *Infect. Immun.* **65:**1842–1848.
21. **Diez-Gonzalez, F., T. R. Callaway, and J. B. Russell.** 1998. Grain feeding and the dissemination of acid-resistant *Escherichia coli* from cattle. *Science* **281:**1666–1668.
22. **Diez-Gonzalez, F., and J. B. Russell.** 1998. The ability of *Escherichia coli* O157:H7 to decrease its intracellular pH and resist the toxicity of acetic acid. *Microbiology* **143:**1175–1180.
23. **Duncan, S. H., H. J. Flint, and C. S. Stewart.** 1998. Inhibitory activity of gut bacteria against *Escherichia coli* O157:H7 mediated by dietary plant metabolites. *FEMS Microbiol. Lett.* **164:**283–288.
24. **Duncan, S. H., K. P. Scott, H. J. Flint, and C. S. Stewart.** 1999. Commensal-pathogen interactions involving *Escherichia coli* O157:H7 and the prospects for control, p. 71–89. *In* C. S. Stewart, and H. J. Flint (ed.), Escherichia coli *O157:H7 in Farm Animals.* CABI Publishing, New York, N.Y.
25. **Faith, N. G., J. A. Shere, R. Brosch, K. W. Arnold, S. E. Ansay, M.-S. Lee, J. B. Luchansky, and C. W. Kaspar.** 1996. Prevalence and clonal nature of *Escherichia coli* O157:H7 on dairy farms in Wisconsin. *Appl. Environ. Microbiol.* **62:**1519–1525.
26. **Garber, L. P. , S. J. Wells, D. D. Hancock, M. P. Doyle, J. Tuttle, J. A. Shere, and T. Zhao.** 1995. Risk factors for fecal shedding of *Escherichia coli* O157:H7 in dairy calves. *J. Am. Vet. Med. Assoc.* **207:**46–49.
27. **Goffaux, F., J. Mainil, V. Pirson, G. Charlier, P. Pohl, E. Jacquemin, and B. China.** 1997. Bovine attaching and effacing *Escherichia coli* possess a pathogenesis island related to the LEE of the human enteropathogenic *Escherichia coli* strain E2348/69. *FEMS Microbiol. Lett.* **154:**415–421.
28. **Grau, F. H., L. E. Brownlie, and E. A. Roberts.** 1968. Effect of some pre-slaughter treatments on the *Salmonella* populations in the bovine rumen and faeces. *J. Appl. Bacteriol.* **31:**157–163.
29. **Griffin, P. M., and R. V. Tauxe.** 1991. The epidemiology of infections caused by *Escherichia coli* O157:H7, other enterohemorrhagic *E. coli*, and the associated hemolytic uremic syndrome. *Epidemiol. Rev.* **13:**60–98.
30. **Hancock, D. D., T. E. Besser, M. L. Kinsel, P. I. Tarr, D. H. Rice, and M. G. Paros.** 1994. The prevalence of *Escherichia coli* O157:H7 in dairy and beef cattle in Washington State. *Epidemiol. Infect.* **113:**199–207.
31. **Hancock, D. D., T. E. Besser, D. H. Rice, D. E. Herriott, and P. I. Tarr.** 1997. Longitudinal study of *Escherichia coli* O157:H7 in fourteen cattle herds. *Epidemiol. Infect.* **118:**193–195.
32. **Hancock, D. D., D. H. Rice, D. E. Herriott, T. E. Besser, E. D. Ebel, and L. V. Carpenter.** 1997. Effects of farm manure handling practices on *Escherichia coli* O157:H7 prevalence in cattle. *J. Food Prot.* **60:**363–366.
33. **Harmon, B. G., C. A. Brown, S. Tkalcic, P. O. E. Mueller, A. Parks, A. V. Jain, T. Zhao, and M. P. Doyle.** 1999. Fecal shedding and rumen growth of *Escherichia coli* O157:H7 in fasted calves. *J. Food Prot.* **62:**574–579.
34. **Hengge-Aronis, R.** 1993. Survival of hunger and stress: the role of *rpoS* in early and stationary phase gene regulation in *E. coli*. *Cell* **72:**165–168.

35. **Herriott, D. E., D. D. Hancock, E. D. Ebel, L. V. Carpenter, D. H. Rice, and T. E. Besser.** 1998. Association of herd management factors with colonization of dairy cattle by Shiga toxin-positive *Escherichia coli* O157:H7. *J. Food Prot.* **61:**802–807.
36. **Heuvelink, A. E., F. L. A. M. van den Biggelaar, J. T. M. Zwartkruis-Nahuis, R. G. Herbes, R. M. Huyben, N. Nagelkerke, W. J. G. Melchers, L. A. H. Monnens, and E. De Boer.** 1998. Occurrence of verocytotoxin-producing *Escherichia coli* O157:H7 on Dutch dairy farms. *J. Clin. Microbiol.* **36:**3480–3487.
37. **Holland, R. E., R. A. Wilson, M. S. Holland, V. Yuzbasiyan-Gurkan, T. P. Mullaney, and D. G. White.** 1999. Characterization of *eae+ Escherichia coli* isolated from healthy and diarrheic calves. *Vet. Microbiol.* **66:**251–263.
38. **Hovde, C. J., P. R. Austin, K. A. Cloud, C. J. Williams, and C. W. Hunt.** 1999. Effect of cattle diet on *Escherichia coli* O157:H7 acid resistance. *Appl. Environ. Microbiol.* **65:**3233–3235.
39. **Imberechts, H., H. De Greve, and P. Lintermans.** 1992. The pathogenesis of edema disease in pigs: a review. *Vet. Microbiol.* **31:**221–233.
40. **Janke, B. H., D. H. Francis, J. E. Collins, M. C. Libal, D. H. Zeman, D. D. Johnson, and R. D. Neiger.** 1990. Attaching and effacing *Escherichia coli* infection as a cause of diarrhea in young calves. *J. Am. Vet. Med. Assoc.* **196:**897–901.
41. **Johnson, R. P., W. C. Cray, and S. T. Johnson.** 1996. Serum antibody responses of cattle following experimental infection with *Escherichia coli* O157:H7. *Infect. Immun.* **64:**1879–1883.
42. **Johnson, R. P., R. C. Clarke, J. B. Wilson, S. C. Read, K. Rahn, S. A. Renwick, K. A. Sandhu, D. Alves, M. A. Karmali, H. Lior, S. A. McEwen, J. S. Spika, and C. L. Gyles.** 1996. Growing concerns and recent outbreaks involving non-O157:H7 serotypes of verotoxigenic *Escherichia coli. J. Food Prot.* **59:**1112–1122.
43. **Jordon, D., and S. A. McEwen.** 1998. Effects of duration of fasting and short-term high-roughage rations on the concentration of *Escherichia coli* biotype 1 in cattle feces. *J. Food Prot.* **61:**531–534.
44. **Kim, H. H., M. Samadpour, L. Grimm, C. R. Clausen, T. E. Besser, M. Baylor, J. M. Kobayashi, M. A. Neill, F. D. Schoenknecht, and P. I. Tarr.** 1994. Characteristics of antibiotic-resistant *Escherichia coli* O157:H7 in Washington State, 1984–1991. *J. Infect. Dis.* **170:**1606–1609.
45. **Klieve, A. V., and R. A. Swain.** 1993. Estimation of ruminal bacteriophage numbers by pulsed-field gel electrophoresis and laser densitometry. *Appl. Environ. Microbiol.* **59:**2299–2303.
46. **Lange, R., and R. Hengge-Aronis.** 1991. Identification of a central regulator of stationary-phase gene expression in *Escherichia coli. Mol. Microbiol.* **5:**49–59.
47. **Loewen, P., and R. Hengge-Aronis.** 1994. The role of sigma factor σ^S in bacterial global regulation. *Annu. Rev. Microbiol.* **48:**53–80.
48. **Loir, H.** 1994. *Escherichia coli* O157:H7 and verotoxigenic *Escherichia coli* (VTEC). *Dairy Food Environ. Sanit.* **14:**378–382.
49. **Lynn, T. V., D. D. Hancock, T. E. Besser, J. H. Harrison, D. H. Rice, N. T. Stewart, and L. L. Rowan.** 1998. The occurrence and replication of *Escherichia coli* in cattle feeds. *J. Dairy Sci.* **81:**1102–1108.
50. **Mainil, J. G., C. J. Duchesnes, S. C. Whipp, L. R. M. Marques, A. D. O'Brien, T. A. Casey, and H. W. Moon.** 1987. Shiga-like toxin production and attaching effacing activity of *Escherichia coli* associated with calf diarrhea. *Am. J. Vet. Res.* **48:**743–748.
51. **McCann, M. P., J. P. Kidwell, and A. Matin.** 1991. The putative σ factor KatF has a central role in development of starvation-mediated general resistance in *Escherichia coli. J. Bacteriol.* **173:**4188–4194.

52. **Mechie, S. C., P. A. Chapman, and C. A. Siddons.** 1997. A fifteen month study of *Escherichia coli* O157:H7 in a dairy herd. *Epidemiol. Infect.* **118:**17–25.
53. **Miller, L. G., and C. W. Kasper.** 1994. *Escherichia coli* O157:H7 acid tolerance and survival in apple cider. *J. Food Prot.* **57:**460–464.
54. **Moxley, R. A., and D. H. Francis.** 1986. Natural and experimental infection with an attaching and effacing strain of *Escherichia coli* in calves. *Infect. Immun.* **53:**339–346.
55. **Nielsen, N. O., and R. E. Clugston.** 1971. Comparison of *E. coli* endotoxin shock and acute experimental edema disease in young pigs. *Ann. N. Y. Acad. Sci.* **176:**176–189.
56. **Paton, J. C., and A. W. Paton,** 1998. Pathogenesis and diagnosis of Shiga toxin-producing *Escherichia coli* infections. *Clin. Microbiol. Rev.* **11:**450–479.
57. **Pierard, D., D. Stevens, L. Moriau, H. Lior, and S. Lauwers.** 1997. Isolation and virulence factors of verotoxin-producing *Escherichia coli* in human stool samples. *Clin. Microbiol. Infect.* **3:**531–540.
58. **Pierard, D., L. Van Damme, L. Moriau, D. Stevens, and S. Lauwers.** 1997. Virulence factors of verotoxin-producing *Escherichia coli* isolated from raw meats. *Appl. Environ. Microbiol.* **63:**4585–4587.
59. **Rassmusen, M. A., W. C. Cray, T. A. Casey, and S. C. Whipp.** 1993. Rumen contents as a reservoir of enterohemorrhagic *Escherichia coli. FEMS Microbiol. Lett.* **114:**79–84.
60. **Samadpour, M., J. E. Ongerth, J. Liston, N. Tran, D. Nguyen, T. S. Whittam, R. A. Wilson, and P. I. Tarr.** 1994. Occurrence of Shiga-like toxin-producing *Escherichia coli* in retail fresh seafood, beef, lamb, pork, and poultry from grocery stores in Seattle, Washington. *Appl. Environ. Microbiol.* **60:**1038–1040.
61. **Saunders, J. R., M. J. Sergeant, A. J. McCarthy, K. J. Mobbs, C. A. Hart, T. S. Marks, and R. J. Sharp.** 1999. Genetic and molecular ecology of *Escherichia coli* O157:H7, p. 1–25. *In* C. S. Stewart and H. S. Flint (ed.), *Escherichia coli O157:H7 in Farm Animals.* CABI Publishing, New York, N.Y.
62. **Schooderwoerd, M., R. C. Clarke, A. A. van Dreumel, and S. A. Rawluk.** 1988. Colitis in claves: natural and experimental infection with a verotoxin-producing strain of *Escherichia coli* O111:NM. *Can. J. Vet. Res.* **52:**484–487.
63. **Scotland, S. M., B. Rowe, H. R. Smith, G. A. Willshaw, and R. J. Gross.** 1988. Vero cytotoxin-producing strains of *Escherichia coli* from children with haemolytic uraemic syndrome and their detection by specific DNA probes. *J. Med. Microbiol.* **25:**237–243.
64. **Scott, K. P., and H. J. Flint.** 1995. Transfer of plasmids between strains of *Escherichia coli* under rumen conditions. *J. Appl. Bacteriol.* **78:**189–193.
65. **Shere, J. A., K. J. Bartlett, and C. W. Kaspar.** 1998. Longitudinal study of *Escherichia coli* O157:H7 dissemination on four dairy farms in Wisconsin. *Appl. Environ. Microbiol.* **64:**1390–1399.
66. **Stewart, C. S., H. J. Flint, and M. P. Bryant.** 1997. The rumen bacteria, p. 10–72. *In* P. N. Hobson and C. S. Stewart (ed.), *The Rumen Microbial Ecosystem.* Blackie, London, United Kingdom.
67. **Wallace, R. G., M. L. Falconer, and P. K. Bhargava.** 1989. Toxicity of volatile fatty acids at rumen pH prevents enrichment of *Escherichia coli* by sorbitol in rumen contents. *Curr. Microbiol.* **19:**277–281.
68. **Wells, J. G., L. D. Shipman, K. D. Greene, E. G. Sowers, J. H. Green, D. N. Cameron, F. P. Downes, M. L. Martin, P. M. Griffin, S. M. Ostroff, M. E. Potter, R. V. Tauxe, and I. K. Wachsmuth.** 1991. Isolation of *Escherichia coli* serotype O157:H7 in a survey of dairy herds. *Appl. Environ. Microbiol.* **61:**1290–1293.
69. **Whittam, T.S.** 1995.Genetic population structure and pathogenicity in enteric bacteria, p. 217–246. *In* S. Baumberg, J. P. W. Young, E. M. H. Wellington, and J. R. Saunders (ed.), *Population Genetics of Bacteria.* Society for General Microbiology, Cambridge University Press, Cambridge, United Kingdom.

70. **Wilson, J. B., S. A. McEwen, R. C. Clarke, K. E. Leslie, R. A. Wilson, D. Walter-Toews, and C. L. Gyles.** 1992. Distribution and characteristics of verocytotoxigenic *Escherichia coli* isolated from Ontario dairy cattle. *Epidemiol. Infect.* **108:**423–439.
71. **Wolin, M. J.** 1969. Volatile fatty acids and the inhibition of *Escherichia coli* growth by rumen fluid. *Appl. Microbiol.* **17:**83–87.
72. **World Health Organization.** 1995. *Report of the WHO Working Group on Shiga-Like Toxin-Producing* Escherichia coli *(SLTEC), with Emphasis on Zoonotic Aspects.* World Health Organization, Geneva, Switzerland.
73. **Wray, C., I. Mclaren, and G. R. Pearson.** 1989. Occurrence of 'attaching and effacing' lesions in the small intestine of calves experimentally infected with bovine isolates of verocytotoxic *E. coli. Vet. Rec.* **125:**365–368.
74. **Zhao, T., M. P. Doyle, J. Shere, and L. Garber.** 1995. Prevalence of enterohemorrhagic *Escherichia coli* O157:H7 in a survey of dairy herds. *Appl. Environ. Microbiol.* **61:**1290–1293.
75. **Zhao, T., M. P. Doyle, B. G. Harmon, C. A. Brown, P. O. E. Mueller, and A. H. Parks.** 1998. Control of growth of enterohemorrhagic *Escherichia coli* O157:H7 in cattle by inoculation with probiotic bacteria. *J. Clin. Microbiol.* **36:**641–647.

Emerging Diseases of Animals
Edited by C. Brown and C. Bolin

Chapter 11

The Global Epidemiology of Multiresistant *Salmonella enterica* Serovar Typhimurium DT104

Dale Hancock, Thomas Besser, John Gay, Daniel Rice, Margaret Davis, and Clive Gay

Salmonellosis is among the most important food-borne diseases of human beings, resulting in millions of cases and hundreds of deaths annually in the United States (82). Furthermore, nontyphoidal salmonellosis has failed to respond to control measures that have drastically reduced typhoid in industrialized countries and successfully controlled certain other food-borne diseases. The putative reason for our continuing failure to control nontyphoidal salmonellosis is that the primary reservoir is external to humans, mainly in food-producing animals. A number of lines of evidence support this view for industrialized countries (4, 51).

During the past decade, considerable efforts have been made in meat processing to reduce the prevalence and load of pathogens in meat and poultry (107). Although some progress has been made in the United States in reducing pathogens in meat and poultry, the incidence of human salmonellosis has been reduced only marginally (25). In contrast to progress in food manufacturing, relatively few efforts have been undertaken in most countries to control the livestock reservoir of salmonellae. This has been due at least partly to the lack of understanding of the key elements of *Salmonella* ecology.

One aspect of *Salmonella* ecology that presents both an enigma and the prospect of control if it could be better understood is the periodic

Dale Hancock, John Gay, Daniel Rice, Margaret Davis, and Clive Gay • Field Disease Investigation Unit, College of Veterinary Medicine, Washington State University, Pullman, WA 99164-6610. **Thomas Besser** • Department of Veterinary Microbiology and Pathology, College of Veterinary Medicine, Washington State University, Pullman, WA 99164-6610.

emergence and dissemination of new clones across broad geographic areas with apparent displacement of previously existing clones. The most recent and best documented of these epidemic clones is a clone of *Salmonella enterica* serovar Typhimurium termed multiresistant DT104 (mr-DT104) that has rapidly disseminated around the world to become one of the most frequently isolated salmonellae from human and animal salmonellosis cases. The purpose of this chapter is to review the epidemiological features of the global dissemination of mr-DT104, to discuss the possible origins of this clone, to hypothesize about the possible mechanisms by which rapid global dissemination and displacement of other clones may have occurred, and to discuss possible means of control.

DEFINITION AND IDENTIFICATION OF mr-DT104

All members of the clone designated mr-DT104 are serotype Typhimurium or Typhimurium (Copenhagen). The appellation definitive type 104 (DT104) is derived from a pattern of susceptibility to a set of phages (2). However, not all DT104 serovar Typhimurium isolates are members of the mr-DT104 clone (23, 95). The other most common phenotypic identifier of mr-DT104 is resistance to ampicillin, chloramphenicol, streptomycin, sulfonamides, and tetracycline (ACSSuT). Resistance of mr-DT104 is not limited to these antimicrobials, and the characteristic resistance pattern is sometimes designated ACSSuT+ to reflect this. Glynn et al. (46) found that 91% of serovar Typhimurium isolates collected from human disease occurring after 1995 that were ACSSuT+ were DT104 or a related phage type. The clonal nature of mr-DT104 is well established by the genetics of antimicrobial resistance (6, 20, 23, 100) and by restriction fragment length polymorphism analyses (23, 95, 104).

GLOBAL DISSEMINATION OF mr-DT104

Reconstruction of the global dissemination of a *Salmonella* clone is difficult owing to the vagaries of diagnosis, surveillance, and subtyping. Only an estimated 1 in 38 human salmonellosis cases is diagnosed and reported in the United States (82), and in only a fraction of these is an isolate both subtyped and entered into a national or regional database. Subtyping of nontyphoidal salmonellae is rare in developing countries. Nevertheless, three important features of the dissemination of mr-DT104 can be inferred.

(i) By the early 1990s, mr-DT104 was undergoing rapid local dissemination in several countries in (at least) North America, Asia, and Europe (Table 1).

(ii) A period of intercontinental dissemination followed by latent local dissemination would have necessarily preceded the period of patent local dissemination. Given that the first isolates of serovar Typhimurium with resistotype ACSSuT+ and phage type 104 were reported in North America (46), Europe (118), and Asia (74) by the mid-1980s, one could reasonably hypothesize that the intercontinental dissemination of mr-DT104 began in the 1970s or early 1980s. The high degree of genetic relatedness among mr-DT104 isolates (100) suggests that the genetic origin of mr-DT104 did not long antedate the onset of global dissemination.

(iii) The rapid local dissemination phase of mr-DT104 appeared to be associated with a symmetrical decline in other serovar Typhimurium clones, consistent with clonal competition and displacement. By the mid-1990s, mr-DT104 had become the most common subtype of serovar

Table 1. Year of first known isolations, by species and country, of serovar Typhimurium with resistance pattern ACSSuT and phage type 104[a]

Country	Yr	Species	Reference
Hong Kong	1973–1982[b]	Human	74
United Kingdom	1984	Imported wild and exotic birds	56
United Kingdom	1984	Human	118
United States	1984–1985	Human	46
United Kingdom	1988	Bovine	56
France	1989	Bovine	7
Germany	1990	Bovine	95
United States (Northwestern)	1990	Bovine	This chapter
Germany	1991	Human	95
Italy (Sardinia)	1991	Human	104
Canada	1993–1996	Human	94
Belgium	1991–1993	Swine	58
Japan	1993	Human	59
Belgium	1993	Bovine	58
Israel	By 1994[c]	Human	85
Czech Republic	By 1991–1994	Human	66
Czech Republic	By 1991–1994	Unspecified animals	66
United Kingdom	By 1991–1995	Feline	131
Italy	By 1992–1996	Human	41
Denmark	1996	Swine	11

[a] Only first reported isolations in humans and domestic animals are listed.

[b] Data in this article were reported for a range of years, so that the specific year of the first confirmed serovar Typhimurium isolate with the ACSSuT+ resistance pattern and of phage type 104 could not be determined.

[c] Serovar Typhimurium with the ACSSuT+ resistance pattern and phage type 104 was known to be present in the specified year, but subtyping of any substantial number of serovar Typhimurium isolates before this year was not reported.

Typhimurium in (at least) England and Wales (56, 118), Scotland (75), France (23, 93), Germany (95), Belgium (58), the Czech Republic (66), Israel (85), Canada (94), and the northwestern United States (16, 17, 38). Yet the total number of serovar Typhimurium salmonellosis cases generally did not increase. For example, between 1980 and the late 1990s, the total number of reported human serovar Typhimurium salmonellosis cases remained roughly constant in the United States (46) and declined slightly in England and Wales (Public Health Laboratory Services website http://www.phls.co.uk/facts/salm.htm). In the northwestern United States, which has experienced a much higher fraction of serovar Typhimurium strains that are mr-DT104 than other regions of the United States, no evident increase in serovar Typhimurium-associated cases in either cattle or humans has occurred (16, 17, 38). The same general pattern has been reported in Germany (95), Austria (95), and Sardinia (104). At least two exceptions occurred to this pattern. In Israel (85) and in Italy (41), the dissemination of mr-DT104 was associated with increases in total serovar Typhimurium isolations and in the fraction of all human salmonellosis cases from which serovar Typhimurium was isolated.

This pattern of clonal dissemination with apparent displacement has been observed for other serovar Typhimurium clones, including DT12 (1, 132), DT204 (102, 115), DT193 (102), and DT10 (67). It has also been noted for other serovars, notably including a multiresistant clone of *Salmonella* serovar Wien, first noted in a nosocomial pediatric outbreak in Algeria that spread to a number of European countries during the 1970s (81, 88). If the pattern of these previous epidemic clones holds, mr-DT104 will eventually decline in terms of total numbers of isolates and relative to other serovar Typhimurium clones.

ORIGIN OF mr-DT104

Livestock Origin Hypothesis

Some authors have inferred a livestock origin for mr-DT104 (94, 95, 133), and at least one review implicates British cattle in particular (133). One must be careful, in this discussion, to distinguish an immediate source for human infection from a genetic origin of mr-DT104. By the time that mr-DT104 was recognized as an epidemic clone, it was already widely disseminated in a variety of domestic and wild animals, and cattle figured prominently in early publications as a source for human infection (118, 130). Clearly, cattle and other livestock are an important source of human infection with mr-DT104 via food-borne exposure (130), and direct contact with infected cattle imparts an unusual degree of risk for human infection (17, 130). However, no direct evidence exists for a livestock genetic origin of mr-DT104. The first isolations of the clone occurred in humans and imported birds several years prior to the first isolations in livestock (Table 1). In England and Wales, where subtyping of human and animal isolates was common during the 1980s, several dozen human mr-DT104 isolates were observed prior to the first livestock isolate (56, 118). This tends to refute the British cattle origin hypothesis and, more importantly, suggests that cattle were not the species in which the primary intercontinental dissemination occurred. In Germany the first isolates of mr-DT104 in cattle preceded the first isolates in humans (1990 and 1991, respectively) (95), and this has been taken as evidence of a cattle origin (94, 95). As noted earlier, however, mr-DT104 had already undergone wide dissemination by 1990.

The most compelling arguments for a livestock origin are based on deductive reasoning. Arguments along these lines are made explicitly by Angulo et al. (4) and appear to be implicit in other studies. Briefly, the reasoning proceeds as follows. Premise 1 is that no human reservoir for *Salmonella* spp. exists and livestock represent the main source for human infection. Premise 2 is that a large amount of antimicrobials are used in animal production for both therapeutic and growth promotion purposes. Premise 3 is that the clone mr-DT104 is resistant to more antimicrobials than other serovar Typhimurium subtypes and must have undergone continuing antimicrobial selection pressure, causing it to genetically emerge and disseminate. The conclusion is that the clone mr-DT104 must have a genetic origin in livestock.

Yet, as will be developed in the next few paragraphs, the premise that serovar Typhimurium lacks a human reservoir is valid only in the confines of industrialized countries—a restriction that is not relevant for a globally disseminated clone. The rapid global dissemination of mr-DT104 suggests that humans rather than livestock are a more likely vehicle for primary (intercontinental) dissemination of mr-DT104, and a greater potential exists for human-to-livestock transmission of serovar Typhimurium in industrialized countries than has been appreciated.

Salmonellae, Including Serovar Typhimurium, Are Endemic in Some Human Populations

The best evidence for a human reservoir for *Salmonella* spp. is the host-specific serovar Typhi that is endemic in some developing countries (32, 84, 98, 103, 113) and is transmitted by inadequate sewage and water hygiene as well as by direct contact (113). Transmission of nontyphoidal salmonellae via improper sewage handling (83) and inadequate water sanitation (48, 105) also occurs in developing countries. In one fecal culture survey of apparently healthy urban Indonesians, salmonellae were found in 8.7% of 464 individuals (49). Nosocomial salmonellosis associated with nontyphoidal salmonellae occurs in some developing countries (36, 64, 76, 101, 127).

Intense Selection Pressure in Human Populations

Multiresistant clones of serovar Typhi from Asia are common (32, 84, 98, 103, 113), including those with fluoroquinolone resistance (113). Multiresistant, regionally disseminated, nosocomial strains of serovar Typhimurium have also been described in developing countries (64, 124, 127), and the pace at which multiresistance has developed in serovar Typhimurium has been greater in developing countries than in industrialized ones. Isolates of serovar Typhimurium R-type ACSSuT+ were reported in India by 1976 (110), and a large majority of Typhimurium isolates from throughout India possessed this resistance pattern by 1989–1990 (86). Chloramphenicol resistance was common among Brazilian serovar Typhimurium isolates before this occurred in North America (8). Human-origin *Escherichia coli* isolates in some developing countries have been reported to have a much greater frequency of resistance than animal isolates from the same regions (65, 77), which suggests that antimicrobial use in humans accounts for most of the resistance selection pressure. Resistance in human commensal *E. coli* has been directly linked to antimicrobial usage patterns in one region of a developing country (13).

Human-Origin Salmonellae Disseminate Globally

Dissemination of multiresistant serovar Typhi from developing to industrialized countries occurs relatively frequently and provides a model for rapid intercontinental dissemination (84, 103). Most surveillance systems in industrialized countries do not discriminate between imported and domestic cases of human serovar Typhimurium salmonellosis. However, imported multiresistant serovar Typhimurium associated with human salmonellosis has been reported (24, 72, 80, 91, 109), as have cases of infection with other imported nontyphoidal serovars (53, 61, 114). In at least two of these reports, dissemination beyond the index case was reported (61, 72). Some of the early isolates of mr-DT104 in England and Wales were reported to have been associated with foreign travel (55).

Several Routes of Dissemination into Livestock Populations

Sewage transmission of *Salmonella* spp. from humans to livestock has been reported (61, 69, 99). Human sewage in industrialized countries is frequently contaminated with salmonellae (62, 70, 89), including serovar Typhimurium (69). Salmonellae are not dependably removed by sewage processing, and both treated effluent discharged into rivers (70) and processed sludge (89) are often contaminated with salmonellae. Birds and other animals that frequent sewage plants have an increased prevalence of colonization with salmonellae (22, 43, 69) and have been implicated in transmission to livestock (60, 99), including three incidents in which serovars and/or subtypes were newly introduced by this route (33, 61, 69). On farms, birds tend to congregate—often in great numbers—in feed bunks and feed storage areas, which they contaminate with feces. Feeds in commerce are often contaminated with a variety of salmonellae (37, 71, 78), including mr-DT104 (37), and birds are a major source of this contamination (37). Birds also are a likely source of infection of cats, in which mr-DT104 has been a frequent finding (131), and cats in turn have been implicated in dissemination of mr-DT104 to cattle herds (40).

Direct transmission of human fecal organisms to livestock also occurs, as illustrated by the long-term *Taenia saginata* cysticercosis problem in the northwestern United States (50, 137). Humans are the only definitive host for *T. saginata,* and human fecal contamination of cattle feed is thus necessary for cattle infection. The sustained rate of >1,000 cases of cysticercosis per year between 1984 and 1998 in northwestern United States cattle (D. D. Hancock, T. E. Besser, J. M. Gay, and C. C. Gay, unpublished data), along with the low sensitivity of detection of that condition and the presumed low prevalence of *T. saginata* in humans, indicates that large numbers of cattle in the region were exposed to human fecal contamination during the period of mr-DT104's dissemination. It seems notable that mr-DT104 spread through the northwestern United States earlier, and to a greater extent, than in any other region of the country (17).

Different Dissemination Patterns of Human and Bovine Host-Specific Salmonellae

Intercontinental and even intranational movements of livestock are much less than human traffic. It is instructive to consider that many years were required for serovar Dublin (the serogroup D *S. enterica* serovar adapted to cattle) to disseminate from the west to the east coast of the United States (19) and from southern Australia to Queensland (123). In contrast, the rapid global dissemination of serovar Typhi (the serogroup D *S. enterica* serovar adapted to humans) has been well described (84, 103). Although one cannot say in which species or country mr-DT104 genetically originated, it is clear that international dissemination is a characteristic feature of mr-DT104 and other serovar Typhimurium clones and that human traffic is more likely to be involved in this than that of livestock.

HYPOTHETICAL MODELS ACCOUNTING FOR EPIDEMIC *SALMONELLA* CLONES SUCH AS mr-DT104

A control strategy for an infectious agent is best developed with a model in mind that explains the ecological characteristics typical of the agent. In the case of mr-DT104 and several previous epidemic serovar Typhimurium clones, these ecological characteristics are rapid dissemination over great geographic distances, displacement of other serovar Typhimurium clones without an absolute increase in total serovar Typhimurium-associated cases, eventual decline of the clone so that it becomes less common relative to other serovar Typhimurium clones, and a high level of multiresistance to antimicrobials. Four models are discussed with respect to their ability to explain these features.

Model 1: the Enlarged Niche Hypothesis

Changes in habitat may provide an enlarged niche to allow expansion of a previously rare subtype. Rather than being associated with new dissemination, in this model the global epidemic is due to habitat changes occurring in multiple locations approximately concurrently. Under this hypothesis, one would expect that distinct but related subtypes would arise in different locations and that they would be the same subtypes extant in those regions (albeit more rare) prior to the enlarged-niche transition. Decline of a clone would occur only as a result of further habitat changes that tended to close or reduce the expanded niche. Available data indicate that this model explains the global epidemic of *S. enterica* serovar Enteritidis beginning in the 1970s. Rather than being a single clone, the global Enteritidis epidemic was produced by a cluster of related clones. Different but related phage types dominated on different continents (21). Enteritidis isolates collected before the onset of the epidemic were genetically indistinguishable from those collected after the onset in the same region (92), and it seems clear that this global epidemic was not caused by the global dissemination of a new Enteritidis clone. Changes in the poultry industry, such as reduced genetic variability of the birds, have been hypothesized as the cause of the epidemic (92). Recently, it was suggested that the eradication programs for *Salmonella* serovar Pullorum and *Salmonella* serovar Gallinarum—genetically and antigenically closely related to serovar Enteritidis—may have produced a niche for Enteritidis, previously a rodent-associated serovar (14). The ongoing increase in or stable level of serovar Enteritidis-associated disease in many countries is also consistent with this model, since decline would be expected to occur only if the inciting habitat change reverted to its former type or to a new form no longer favoring the epidemic clones. However, this model does not explain the global epidemic of mr-DT104, which was associated with dissemination of a new clone, nor does this model explain the multiresistance of mr-DT104 and other serovar Typhimurium isolates. In marked contrast to serovar Typhimurium isolates from poultry-origin disease outbreaks, serovar Enteritidis isolates are not typically multiresistant (85, 117).

Model 2: the Dissemination Fitness Hypothesis

Under this model, a clone emerges that is better able to disseminate than other subtypes and hence becomes more frequent relative to its competitors. Improved dissemination fitness could be due to physiological factors (e.g., better able to survive in a dried state), virulence factors, or antimicrobial resistance. Whatever factors are involved, the clone's improved ability to disseminate should result in an absolute increase in cases for the serovar due to its ability to disseminate to more potential hosts than displaced clones. Decline of the clone would occur only with the emergence of a clone with higher dissemination fitness, resulting in a still greater absolute increase in the number of cases.

Arguments against the dissemination fitness hypothesis for mr-DT104

While the dissemination fitness hypothesis is appealing, the lack of an increase in the total number of serovar Typhimurium cases associated with dissemination of mr-DT104 presents a seemingly fatal difficulty: it imposes the paradoxical constraint that dissemination fitness can be relative but not absolute. Logically, one would expect improved ability of a new clone to disseminate (say, via improved survival in foods or feeds) to result in an expansion of total cases via its dissemination to hosts that it would not otherwise have reached. The implication is that competition is confined to finite habitats that are already occupied by competing serovar Typhimurium clones. This will be the defining feature of the selection focus model discussed below.

Dissemination fitness via antimicrobial resistance

The dissemination fitness model could explain the multiple antimicrobial resistance that characterizes mr-DT104 and some other epidemic serovar Typhimurium clones. Under this hypothesis, use of antimicrobials in livestock, especially antimicrobials used in feeds at low concentrations for growth promotion, has selected for multiresistance by favoring survival and dissemination of increasingly multiresistant clones. Supporting this view is the argument that resistance genes and/or the clones possessing them would not survive and prosper in the absence of continuing antimicrobial selection pressure due to fitness costs associated with resistance (56). The hypothesis is compelling to the degree that it is sometimes stated as established fact (4, 27, 46), yet available data indicate that it is not valid for mr-DT104.

In the northwestern United States, the dissemination of mr-DT104 accounted for the only major changes in antimicrobial resistance of serovar Typhimurium of human and bovine origin over a period extending from the late 1980s to the late 1990s (38). By 1994, mr-DT104 represented over half of all serovar Typhimurium isolates from cattle in the northwest (17), yet, the dissemination of mr-DT104 and displacement of other serovar Typhimurium clones had no impact on the percentage of serovar Typhimurium isolates resistant to ampicillin, streptomycin, sulfonamides, and tetracycline, since the displaced clones were mainly R-type ASSuT (38). Under the hypothesis of antimicrobial resistance-driven dissemination, chloramphenicol and florfenicol would be the most obvious antimicrobials whose use would have favored mr-DT104 over ASSuT clones. However, chloramphenicol was banned from livestock use in 1983 in the United States, and florfenicol was not available until 1996. It is conceivable that dissemination was favored by resistance factors other than those of the ACSSuT clonal marker, but it has proven difficult to identify a candidate. Gentamicin and spectinomycin had moderate use in cattle on a prescription, therapeutic basis during the 1980s and 1990s. However, gentamicin resistance has not been a feature of mr-DT104 in the northwestern United States (38). And even though mr-DT104 isolates are resistant to spectintomycin, this was also common among bovine serovar Typhimurium isolates in 1988 and 1989, immediately prior to the first mr-DT104 isolation in cattle (Hancock et al., unpublished). In ruling out resistance to all the antimicrobials in common or even moderate use as factors that might have enhanced the dissemination fitness of mr-DT104 over contemporary competitors, one has gone far to refute the hypothesis.

Three additional lines of evidence tend to refute the antimicrobial resistance-driven dissemination hypothesis for mr-DT104. First, a large case-control study of factors associated with isolation of mr-DT104 on farms in Britain did not identify antimicrobial usage patterns as a risk factor (40). Second, the antimicrobials likely to favor dissemination of mr-DT104 over competing clones were banned from use for growth promotion in Britain after publication of the Swann report in 1969 (5), yet mr-DT104 disseminated throughout British livestock at least as quickly and to at least as great of an extent as in any region of the United States, where growth promotion usage continued. Finally, some epidemic serovar Typhimurium clones have not been multiresistant. Serovar Typhimurium strain DT10 is notable in that its pattern of rapid dissemination with clonal displacement in (at least) Canada in the 1970s (67) closely resembled that of mr-DT104 in many countries a decade later, yet about 90% of the phage type 10 isolates were sensitive to all antimicrobials tested. If multiresistance is not necessary for wide dissemination of a serovar Typhimurium clone, it is logically inconsistent to assume that a multiresistant clone that has become widely disseminated must have done so because of multiresistance. Recent research supports a view that genetic adaptations occur to reduce or eliminate the fitness cost of antimicrobial resistance and that these adaptations tend to preclude reversion to sensitive form (18, 108).

Two specific forms of the antimicrobial-driven dissemination hypothesis merit additional discussion.

Dissemination fitness via fluoroquinolone resistance

Incipient fluoroquinolone resistance, noted especially in Britain (119), could be taken as evidence that the main selective advantage determining dissemination of salmonellae in livestock is acquisition of additional antimicrobial resistance genes against those antimicrobials in widespread use in livestock. Tending to support this view are the findings that several different mutations in the *gyrA* gene of salmonellae account for fluoroquinolone resistance and that resistant strains are counter selected in the absence of fluoroquinolone use (44). The increasing frequency of fluoroquinolone resistance in mr-DT104 isolates would thus be explainable only on the basis of widespread—and independent—mutations leading to fluoroquinolone resistance at numerous locales (e.g., farms) where fluoroquinolones were in sustained use. Other data, however, contradict this. In a recent Danish outbreak involving mr-DT104 with reduced susceptibility to fluoroquinolones, the swine operations implicated as the source had not used fluoroquinolones for at least the previous year (87). The fluoroquinolone-resistant strain of mr-DT104 seemingly disseminated to and persisted in these Danish herds in the absence of fluoroquinolone selection pressure; indeed, the authors suggested that the offending strain might have been imported into Denmark. In Germany, the dissemination of mr-DT104 was associated with a sharp reduction in fluoroquinolone resistance among veterinary *Salmonella* isolates due to the displacement of DT204, which was commonly fluoroquinolone resistant (79). Together, these Danish and German data on fluoroquinolone resistance lend support to the selection focus model discussed below: clonal competition occurs in narrowly confined environments and is based on more than just antimicrobial resistance; dissemination from these foci is not dependent on antimicrobial selection pressure. It seems notable that, among human pathogens, the prevalence of fluoroquinolone resistance is much greater for organisms with a nosocomial epidemiology (47) and in developing countries (124) where hygienic controls on dissemination are relatively poor. Clonal dissemination of fluoroquinolone-resistant serovar Typhi has been described (32), as has intercontinental dissemination of nontyphoidal fluoroquinolone-resistant salmonellae (53, 103). Relevant to this argument, fluoroquinolones are licensed only for therapeutic use in livestock.

Dissemination fitness via florfenicol resistance

A recent hypothesis attributes the dissemination of mr-DT104 in France to florfenicol resistance (6, 7). In France, chloramphenicol was not prohibited for use in cattle until 1994, and florfenicol was licensed in 1995. Since the *floR* gene codes for resistance to both chloramphenicol and florfenicol, this would be expected to give mr-DT104 an advantage over clones lacking cross-resistance to florfenicol. Although this is an appealing hypothesis, it appears to be refuted by the data showing that, by 1994, mr-DT104 already constituted a majority of human and animal serovar Typhimurium isolates in France (23). Also, as mentioned above, mr-DT104 disseminated through the northwestern United States and became the dominant serovar Typhimurium clone during a period when neither chloramphenicol nor florfenicol was available for use in livestock.

Model 3: the Chance Dissemination Event Hypothesis

Under this model, chance contamination of a human food or animal feed results in broad dissemination over large geographic distances with exposure of many individuals, resulting in tremendous multiplication of the clone over a short time span. The relative frequency of subtypes under this model is simply a reflection of those that have undergone recent chance dissemination. Absolute decline in diagnostic isolation of the clone would require the absence of a long-term local reservoir. Relative decline would occur even faster due to the chance dissemination of other clones. In this model, competition never occurs among clones.

A number of salmonellosis outbreaks have occurred in humans over large geographical areas, associated with contaminated foods. These have included international outbreaks associated with serovar Agona in a savory snack (68), serovars Bovismorbificans and Stanley in sprouts (96), serovar Newport in alfalfa seeds used for sprouting (128), serovar Muenchen in orange juice (26), and serovar Anatum in powdered milk (120). A few cases of disseminated salmonellae have also been attributed to cattle feeds (3, 63), and not surprisingly given the relatively high rate of salmonella contamination in feeds, salmonellae have been identified in feeds in international commerce (39). Notably, mr-DT104 has been isolated from feeds (40, 71).

While this model could account for rapid dissemination over great geographic distances, it does not account for local displacement of other clones without an absolute increase in serovar Typhimurium cases. A contaminated food that was distributed over a sufficient area and to sufficient numbers of individuals to account for the global dissemination of mr-DT104 would seemingly have caused a massive increase in total serovar Typhimurium-related cases, as occurred for some of the serovars involved in the food-disseminated outbreaks. Also, the clonal decline that has occurred in the food-disseminated outbreaks has been more rapid than for mr-DT104 or previous epidemic serovar Typhimurium clones. It seems clear that the epidemic serovar Typhimurium clones have become entrenched within local reservoirs and that some reason other than lack of a reservoir accounts for the clonal decline. Finally, the multiresistance characteristic of most epidemic serovar Typhimurium clones has not been a regular feature of the salmonellae in food-disseminated outbreaks.

Model 4: the Selection Focus Hypothesis

Under this model, selection pressure is limited primarily to foci where salmonella infection is frequent, where a high potential for endemic salmonella transmission exists, and where competition among clones is high. In this model, selection foci are the ultimate, sustaining reservoir for pathogenic serovars such as serovar Typhimurium. The emergence of a new clone with displacement of other clones reflects competition limited to selection foci, even though these clones disseminate—but are not sustained indefinitely—outside the foci. Displacement occurs when a clone emerges that, under the conditions of selection foci, has a selective advantage over endemic clones. The multiresistance feature would be explainable in this model on the basis that selection foci would most likely occur in hospitals or their equivalent for livestock, where antimicrobial selection pressure via therapeutic use would often be intense.

Typhimurium livestock reservoir is highly heterogeneous

The best evidence for the selection focus model is the high degree of heterogeneity of serovar Typhimurium in the livestock population. The serovar can be found in only a minority of livestock operations, and multiple serovar Typhimurium clones, as required for clonal competition, occur only under the special circumstances of selection foci. Consider the distribution of serovar Typhimurium in cattle, the host species most often implicated in the dissemination of mr-DT104 (46, 94, 95). In the northwestern United States, mr-DT104 has mainly occurred in dairy herds (16, 17), yet neither that clone nor any other Typhimurium can be found in most dairy herds (J. M. Gay, C. C. Gay, T. E. Besser, and D. D. Hancock, unpublished data). Studies in dairy herds in other regions confirm the view that serovar Typhimurium can be isolated in only a small minority of them (73, 90, 112). Similarly, in a study of salmonellae in 100 feedlots in 13 states, salmonella isolation was common in randomly sampled, healthy, feedlot cattle, but serovar Typhimurium was an uncommon finding (42).

Yet locations exist where serovar Typhimurium is endemic and where there is a high potential for ongoing introduction of new serovar Typhimurium clones. By definition, these are selection foci. Human hospitals serve this role in some developing countries (36, 64, 76, 101, 127). The absence of a substantial human reservoir for *Salmonella* spp. in developed countries requires that animals serve as selection foci. Several feedlots with sustained, nosocomial, hospital pen salmonellosis associated with serovar Typhimurium have been observed in the northwestern United States (Gay et al., unpublished). Nosocomial salmonellosis is well documented in veterinary hospitals (54, 57, 121, 122, 125). Perhaps the best candidates for selection foci, however, are calf-raising operations. Calf raisers are specialized farms that receive newborn calves from a variety of sources and eventually resell them to cattle dealers, dairy herds, or feedlots. In addition to the risk imposed by receiving animals from many sources, some of the calves are typically received from livestock auctions, where they can be exposed to multiple serovar Typhimurium subtypes

(136). Not surprisingly, salmonellae of a variety of serovars, including multiple subtypes of serovar Typhimurium, are frequently found in calf-raising operations, and endemic transmission is favored by the susceptibility of neonates and the low level of hygiene (135, 136). Clinical disease associated with serovar Typhimurium and other salmonella serovars is a common problem and often results in intensive use of a diversity of therapeutic antimicrobial drugs (116).

Multiresistance consistent with selection focus model

As previously mentioned, the high level of antimicrobial resistance among most epidemic serovar Typhimurium clones leads one away from the expanded niche hypothesis and the chance dissemination hypothesis because salmonellae fitting those models have not tended to be multiresistant. Arguably, the selection focus model also does a better job of explaining the high level of multiresistance than does the model of dissemination fitness by antimicrobial resistance. Although very large amounts of antimicrobials are used in livestock production (4), much of this does not occur where it would be expected to select among competing serovar Typhimurium clones.

Multiple, competing serovar Typhimurium clones (or even one clone) exist in only a small fraction of farms, and only in unusual circumstances is there much opportunity for antimicrobial selection pressure. Most outbreaks associated with mr-DT104 have occurred in dairy farms (16, 40) and suckler (commercial beef) herds (40), which are among the least likely types of livestock operations to use subtherapeutic antimicrobials. Little opportunity exists for antimicrobial selection pressure via therapeutic antimicrobial use, since the duration of clinical cases associated with mr-DT104 is typically short and involves only a small fraction of the herd (40).

Only in selection foci does there seem to be the sustained opportunity for the clonal competition that appears to be the defining characteristic of the epidemiology of multiresistant serovar Typhimurium. It is erroneous to assume that this competition occurs solely on the basis of antimicrobial resistance. In the northwestern United States (38) and in Germany (79) mr-DT104 displaced serovar Typhimurium clones which were resistant to a similar or even longer list of commonly used antimicrobials, yet whatever factors are involved in clonal competition, the end result is to put successful competitors in environments—such as hospitals in developing countries and calf-raising operations in industrialized ones—where antimicrobial selection pressure is often intense. By analogy, methicillin-resistant *Staphylococcus aureus* (MRSA) has an increasing repertoire of antimicrobial resistance genes due to the persistence and dissemination of clones within and between human health care facilities (9, 45). This is so even though genetic factors other than antimicrobial resistance are involved in the epidemicity of clones (106, 129).

The key practical difference between the dissemination fitness model and the selection focus model—as well illustrated by methicillin-resistant *S. aureus*—is that hygienic measures rather than changes in antimicrobial use are the primary means of control (9, 34, 45). In the face of uncontrolled dissemination, no practical antimicrobial use policies would be effective in avoiding selection for antimicrobial resistance, and under good infection control practices, the opportunities for antimicrobial selection pressure are much reduced. To the degree that antimicrobial use is involved in the genesis and dissemination of epidemic clones, it is the small percentage of antimicrobials that are used within selection foci that are of concern, rather than the far greater total used within the community at large.

One cannot say with complete confidence that the selection focus model is the correct one for mr-DT104 and similar epidemic serovar Typhimurium clones, but the evidence points in that direction. The implications for control are considerable.

LESSONS FOR CONTROL

To date, the discussion of control efforts for mr-DT104 and similar clones has been limited primarily to debates on the use of antimicrobials in livestock, notably subtherapeutic antimicrobials. These discussions have often been formulated on the basis that development of antimicrobial resistance is a local phenomenon: local at the level of millions of farms in which antimicrobial resistance can be attributed to animals being fed subtherapeutic antimicrobials that develop resistant organisms from sensitive progenitors existing on those same millions of farms (26, 31, 46), and local in that internationally disseminated clones are assumed to account for a small minority of the total antimicrobial-resistant salmonellae in an industrialized country (31). These assumptions are not valid for mr-DT104. Being a clone, mr-DT104 could not have independently developed on thousands of farms in dozens of countries from sensitive progenitors; rather, it disseminated internationally and became the most common Typhimurium in a long list of countries. This pattern is not unique, except perhaps in extent, but appears to be common among serovar Typhimurium clones. The available evidence indicates that neither the intercontinental nor the local phase of dissemination of mr-DT104 was driven by selection pressure deriving from antimicrobial use in livestock. Thus, one cannot assume, even at the coarsest level of approximation, that a multiresistant Typhimurium found on a farm acquired its resistance genes on that farm (or even in the same country) or that it arrived on the farm because of antimicrobials in use there. Even if one agrees that greater control of antimicrobial use in livestock is needed (as we do), the disproportionate focus on this area has distracted from the broader discussions on control of epidemic salmonella clones that are needed. Those discussions should center on means to control dissemination.

At the international level, additional controls on dissemination may seem impractical. We already control intercontinental livestock movements to the point that they are rare, and we cannot practically collect fecal specimens from all international travelers. Yet where human-to-cattle transmission of fecal organisms regularly occurs—as in the northwestern United States—institution of controls would seem to be warranted. Also, given the evidence that birds transmit salmonellae from human sewage to livestock feeds and that this represents an important potential route for human-disseminated clones to spread to livestock populations, interventions aimed at preventing or discouraging bird contact with sewage plants and/or livestock feed should be considered.

At the national and regional levels, livestock movement likely constitutes the single most important cause of dissemination of mr-DT104 and other serovar Typhimurium clones. In particular, the exceptionally high risk of mr-DT104 infection associated with being a dealer herd found in one study (odds ratio, 14.25) (40) suggests that these herds play a key role in dissemination of mr-DT104 and that programs to more closely monitor them are worth considering. In the same study, the introduction of purchased animals was the second greatest risk factor for the occurrence of mr-DT104 on a farm (odds ratio, 2.51) (40). A 1996 survey of dairy herds in the United States found that only 15.3% of herds that had purchased pregnant heifers in the previous year provided for their isolation prior to mixing with the herd (28). Similarly, in poultry, much of the risk in an operation's serovar Typhimurium status has been attributed to the infection status in parent flocks and hygienic measures in hatcheries (29, 111). If farms are to become part of the integrated food safety system—extending from farm to consumer—a mandatory biosecurity program targeting serovar Typhimurium is the logical first step.

The Danish salmonella control program provides a more intensive example of a national control effort. Swine have been identified as the main source of human salmonella exposure in Denmark (10). This has led to a program in which meat juice serology is used to screen all Danish swine herds and identify those with suspected endemic salmonella infections (30). Suspect herds are evaluated with bacteriological sampling, and appropriate actions are taken (35). It is likely that this program of active surveillance, which is extended to a degree in other livestock species, has resulted in the much lower level of dissemination of mr-DT104 within Denmark than in most other European countries (10, 87). Notably, antimicrobials remain in common use in Danish livestock (87), and the apparent containment of the fluoroquinolone-resistant mr-DT104 strain found in two Danish swine herds was not due to any emergency program to restrict fluoroquinolone use. These herds were detected and dealt with as part of the routine surveillance program (87).

At the level of the individual herd, a number of measures are possible that would both reduce the risk of introduction of epidemic serovar Typhimurium clones and help eliminate serovar Typhimurium from herds in which it is endemic. The British case-control study (40) identified lack of isolation facilities on a farm, density of cats, and access of birds to feeds as risk factors for herd mr-DT104 infection status. In corroboration of one of these factors, an ongoing case-control study in the northwestern United States has found that herds infected with mr-DT104 more often have a common hospital-maternity pen than do control herds (Gay et al., unpublished). That is, salmonellosis cases are housed together with periparturient cows, which are particularly susceptible to infection with *Salmonella* spp. (97). In calf-raising operations, maintaining isolation between incoming calves both physically and virtually (treatment equipment, hands, air in enclosed buildings, etc.) goes far to prevent sustained horizontal transmission (52), as do certain other environmental health measures (134). Similarly, risk factors have been identified for the presence of salmonellae in swine herds (15) and for salmonellae in poultry hatcheries (12). The availability of such management tools at the level of the local operation is essential to make a mandatory national biosecurity program feasible and less onerous to producers.

CONCLUSION

The epidemiology of multiresistant serovar Typhimurium is not confined by national borders but is global in nature, as well illustrated by mr-DT104. Efforts by national governments aimed at preventing genetic emergence of new multiresistant clones are unlikely to be very effective since most of the world—notably developing countries, which account for much of the selection pressure for multiresistant serovar Typhimurium—falls outside the control of any single nation. A more promising approach would be the development of national programs aimed at reducing the potential for dissemination of epidemic serovar Typhimurium clones. The available evidence indicates that these efforts should center on selection foci, such as calf-raising operations, which serve as reservoirs of serovar Typhimurium and as sites for clonal competition among epidemic multiresistant strains.

REFERENCES

1. **Aarestrup, F. M., N. E. Jensen, and D. L. Baggesen.** 1997. Clonal spread of tetracycline-resistant *Salmonlla typhimurium* in Danish dairy herds. *Vet. Rec.* **140:**313–314.
2. **Anderson, E. S., L. R. Ward, and M. S. de Saxe.** 1977. Bacteriophage typing designations of *Salmonella typhimurium. J. Hyg.* **78:**297–300.
3. **Anderson, R. J., R. L. Walker, D. W. Hird, and P. C. Blanchard.** 1997. Case-control study of an outbreak of clinical disease attributable to *Salmonella menhaden* infection in eight dairy herds. *J. Am. Vet. Med. Assoc.* **210:**528–530.
4. **Angulo, F. J., R. V. Tauxe, and M. L. Cohen.** 1998. Significance and sources of antimicrobial-resistant nontyphoidal Salmonella infections in humans in the United States: the need for prudent use of antimicrobial agents, including restricted use of fluoroquinolones, in food animals. *Bov. Proc.* **31:**1–8.
5. **Anonymous.** 1969. *Report of the Joint Committee on the Use of Antibiotics in Animal Husbandry and Veterinary Medicine* (Swann Committee). Her Majesty's Stationery Office, London, United Kingdom.
6. **Arcangioli, M. A., S. Leroy-Setrin, J. L. Martel, and E. Chaslus-Dancla.** 1999. A new chloramphenicol and florfenicol resistance gene flanked by two integron structures in Salmonella typhimurium DT104. *FEMS Microbiol. Lett.* **174:**327–332.
7. **Arcangioli, M. A., S. Leroy-Setrin, J. L. Martel, and E. Chaslus-Dancla.** 2000. Evolution of chloramphenicol resistance, with emergence of cross-resistance to florfenicol, in bovine *Salmonella* Typhimurium strains implicates definitive phage type (DT) 104. *J. Med. Microbiol.* **49:**103–110.
8. **Asensi, M. D., A. P. Costa, E. Moura, E. M. dos Reis, and E. Hofer.** 1995. Lysotypes and plasmidial profile of *Salmonella* serovar typhimurium isolated from children with enteric processes in the cities of Rio de Janeiro, RJ, and Salvador, BA—Brazil. *Rev. Inst. Med. Trop. Sao Paulo* **37:**297–302.
9. **Ayliffe, G. A.** 1997. The progressive intercontinental spread of methicillin-resistant *Staphylococcus aureus. Clin. Infect. Dis.* **24**(Suppl. 1)**:**S74–S79.
10. **Baggesen, D. L., and H. C. Wegener.** 1994. Phage types of *Salmonella enterica* ssp. Enterica serovar typhimurium isolated from production animals and humans in Denmark. *Acta Vet. Scand.* **35:**349–354.
11. **Baggesen, D. L., and F. M. Aarestrup.** 1998. Characterisation of recently emerged multiple antibiotic-resistant *Salmonella enterica* serovar typhimurium DT104 and other multiresistant phage types from Danish pig herds. *Vet. Rec.* **143:**95–97.
12. **Bailey, J. S., R. J. Buhr, and N. A. Cox.** 1996. Effect of hatching cabinet sanitation treatments on *Salmonella* cross-contamination and hatchability of broiler eggs. *Poult. Sci.* **75:**191–196.
13. **Bartoloni, A., F. Cutts, S. Leoni, C. C. Austin, A. Mantella, P. Guglielmetti, M. Roselli, E. Salazar, and F. Paradisi.** 1998. Patterns of antimicrobial use and antimicrobial resistance among healthy children in Bolivia. *Trop. Med. Int. Health* **3:**116–123.

14. **Baulmer, A., B. M. Hargis, and R. M. Tsolis.** 1999. Tracing the origins of Salmonella outbreaks. *Science* **287:**50–52.
15. **Berends, B. R., H. A. Urlings, J. M. Snijders, and F. Van Knapen.** 1996. Identification and quantification of risk factors in animal management and transport regarding *Salmonella* spp. in pigs. *Int. J. Food Microbiol.* **30:**37–53.
16. **Besser, T. E., C. C. Gay, J. M. Gay, D. D. Hancock, D. Rice, L. C. Pritchett, and E. D. Erickson.** 1997. Salmonellosis associated with *S. typhimurium* DT104 in the USA. *Vet. Rec.* **140:**75.
17. **Besser, T. E., M. Goldoft, L. C. Pritchett, R. Khakhria, D. D. Hancock, D. H. Rice, J. M. Gay, W. Johnson, and C. C. Gay.** 2000. Multiresistant *Salmonella* Typhimurium DT104 infections of humans and domestic animals in the Pacific Northwest of the United States. *Epidemiol. Infect.* **124:**193–200.
18. **Bjorkman, J., D. Hughes, and D. I. Andersson.** 1998. Virulence of antibiotic-resistant *Salmonella typhimurium. Proc. Natl. Acad. Sci. USA* **95:**3949–3953.
19. **Blackburn, B. O., K. Sutch, and R. Harrington, Jr.** 1980. The changing distribution of *Salmonella dublin* in the United States. *Proc. Annu. Meet. U.S. Anim. Health Assoc.* **84:**445–451.
20. **Briggs, C. E., and P. M. Fratamico.** 1999. Molecular characterization of an antibiotic resistance gene cluster of *Salmonella typhimurium* DT104. *Antimicrob. Agents Chemother.* **43:**846–849.
21. **Buchrieser, C., R. Brosch, O. Buchrieser, A. Kristl, J. B. Luchansky, and C. W. Kaspar.** 1997. Genomic analyses of *Salmonella enteritidis* phage type 4 strains from Austria and phage type 8 strains from the United States. *Zentbl. Bakteriol. Mikrobiol. Hyg. Ser. A* **285:**379–388.
22. **Butterfield, J., J. C. Coulson, S. V. Kearsey, P. Monaghan, J. H. McCoy, and G. E. Spain.** 1983. The herring gull *Larus argentatus* as a carrier of salmonella. *J. Hyg. (Lond).* **91:**429–436.
23. **Casin, I., J. Breuil, A. Brisabois, F. Moury, F. Grimont, and E. Collatz.** 1999. Multidrug-resistant human and animal *Salmonella typhimurium* isolates in France belong predominantly to a DT104 clone with the chromosome- and integron-encoded beta-lactamase PSE-1. *J. Infect. Dis.* **179:**1173–1182.
24. **Centers for Disease Control.** 1982. Multiresistant salmonella and other infections in adopted infants from India. *Morb. Mortal. Wkly. Rep.* **31:**285–287.
25. **Centers for Disease Control and Prevention.** 1999. Incidence of foodborne illnesses: preliminary data from the foodborne diseases active surveillance network (FoodNet), United States, 1998. *Morb. Mortal. Wkly. Rep.* **48:**189–194.
26. **Centers for Disease Control and Prevention.** 1999. Outbreak of Salmonella serotype Muenchen infections associated with unpasteurized orange juice—United States and Canada, June 1999. *Morb. Mortal. Wkly. Rep.* **48:**582–585.
27. **Centers for Disease Control and Prevention.** 1999. *Salmonella* fact sheets, Office of Communications, Atlanta, Ga.
28. **Centers for Epidemiology and Animal Health.** 1996. *Biosecurity Practices of U.S. Dairy Herds.* U.S. Department of Agriculture, Animal and Plant Health Inspection Services, Centers for Epidemiology and Animal Health, Fort Collins, Co.
29. **Chriel, M., H. Stryhn, and G. Dauphin.** 1999. Generalised linear mixed models analysis of risk factors for contamination of Danish broiler flocks with *Salmonella typhimurium. Prev. Vet. Med.* **40:**1–17.
30. **Christensen, J., D. L. Baggesen, V. Soerensen, and B. Svensmark.** 1999. Salmonella level of Danish swine herds based on serological examination of meat-juice samples and Salmonella occurrence measured by bacteriological follow-up. *Prev. Vet. Med.* **40:**277–292.

31. **Cohen, M. L., and R. V. Tauxe.** 1986. Drug-resistant Salmonella in the United States: an epidemiologic perspective. *Science* **234:**964–969.
32. **Connerton, P., J. Wain, T. T. Hien, T. Ali, C. Parry, N. T. Chinh, H. Vinh, V. A. Ho, T. S. Diep, N. P. Day, N. J. White, G. Dougan, and J. J. Farrar.** 2000. Epidemic typhoid in Vietnam: molecular typing of multiple-antibiotic-resistant *Salmonella enterica* serotype Typhi from four outbreaks. *J. Clin. Microbiol.* **38:**895–897.
33. **Coulson, J. C., J. Butterfield, and C. Thomas.** 1983. The herring gull *Larus argentatus* as a likely transmitting agent of *Salmonella montevideo* to sheep and cattle. *J. Hyg. (Lond.)* **91:**437–443.
34. **Crowcroft, N. S., O. Ronveaux, D. L. Monnet, and R. Mertens.** 1999. Methicillin-resistant *Staphylococcus aureus* and antimicrobial use in Belgian hospitals. *Infect. Control Hosp. Epidemiol.* **20:**31–36.
35. **Dahl, J., A. Wingstrand, B. Nielsen, and D. L. Baggesen.** 1997. Elimination of *Salmonella typhimurium* infection by the strategic movement of pigs. *Vet. Rec.* **140:**679–681.
36. **Das, A. S., D. N. Mazumder, D. Pal, and U. K. Chattopadhyay.** 1996. A study of nosocomial diarrhea in Calcutta. *Indian J. Gastroenterol.* **15:**12–13.
37. **Davies, R. H., and C. Wray.** 1997. Distribution of salmonella contamination in ten animal feedmills. *Vet. Microbiol.* **57:**159–169.
38. **Davis, M. A., D. D. Hancock, T. E. Besser, D. H. Rice, J. M. Gay, C. Gay, L. Gearhart, and R. DiGiacomo.** 1999. Changes in antimicrobial resistance among *Salmonella enterica* serovar Typhimurium isolates from humans and cattle in the northwestern United States. *Emerg. Infect. Dis.* **5:**802–806.
39. **Eld, K., A. Gunnarsson, T. Holmberg, B. Hurvell, and M. Wierup.** 1991. Salmonella isolated from animals and feedstuffs in Sweden during 1983–1987. *Acta Vet. Scand.* **32:**2261–2277.
40. **Evans, S., and R. Davies.** 1996. Case control study of multiple resistant *Salmonella* Typhimurium DT104 infection in Great Britain. *Vet. Rec.* **139:**557–558.
41. **Fantasia, M., E. Filetici, S. Arena, and S. Mariotti.** 1998. Serotype and phage type distribution of salmonellas from human and non-human sources in Italy in the period 1973–1995. *Eur. J. Epidemiol.* **14:**701–710.
42. **Fedorka-Cray, P. J., D. A. Dargatz, L. A. Thomas, and J. T. Gray.** 1998. Survey of Salmonella serotypes in feedlot cattle. *J. Food Prot.* **61:**525–530.
43. **Fenlon, D. R.** 1983. A comparison of salmonella serotypes found in the feces of gulls feeding at a sewage works with serotypes present in the sewage. *J. Hyg. (Lond).* **91:**47–52.
44. **Giraud, E., A. Brisabois, J. L. Martel, and E. Chaslus-Dancla.** 1999. Comparative studies of mutations in animal isolates and experimental in vitro- and in vivo-selected mutants of *Salmonella* spp. suggest a counterselection of highly fluoroquinolone-resistant strains in the field. *Antimicrob. Agents Chemother.* **43:**2131–2137.
45. **Givney, R., A. Vickery, A. Holliday, M. Pegler, and R. Benn.** 1998. Evolution of an endemic methicillin-resistant *Staphylococcus aureus* population in an Australian hospital from 1967 to 1996. *J. Clin. Microbiol.* **36:**552–556.
46. **Glynn, M. K., C. Bopp, M. S. Dewitt, P. Dabney, M. Mokhtar, and F. J. Angulo.** 1998. Emergence of multidrug-resistant *Salmonella enterica* serotype Typhimurium DT104 infections in the United States. *N. Engl. J. Med.* **338:**1333–1338.
47. **Goldstein, F. W., and J. F. Acar.** 1995. Epidemiology of quinolone resistance: Europe and North and South America. *Drugs* **49**(Suppl. 2)**:**36–42.
48. **Gracey, M., P. Ostergaard, S. W. Adnan, and J. B. Iveson.** 1979. Fecal pollution of surface waters in Jakarta. *Trans. R. Soc. Trop. Med. Hyg.* **73:**306–308.
49. **Gracey, M., J. B. Iveson, and S. Suharyono.** 1980. Human salmonella carriers in a tropical urban environment. *Trans. R. Soc. Trop. Med. Hyg.* **74:**479–482.

50. **Hancock, D. D., S. E. Wikse, A. B. Lichtenwalner, R. B. Wescott, and C. C. Gay.** 1989. Distribution of cysticercosis in Washington. *Am. J. Vet. Res.* **50:**564–570.
51. **Hancock, D. D., T. V. Lynn, T. E. Besser, and S. E. Wikse.** 1997. Feasibility of preharvest food safety control. *Compend. Contin. Ed. Pract. Vet.* **19:**S200.
52. **Hardman, P. M., C. M. Wathes, and C. Wray.** 1991. Transmission of salmonellae among calves penned individually. *Vet. Rec.* **129:**327–329.
53. **Herikstad, H., P. Hayes, M. Mokhtar, M. L. Fracaro, E. J. Threlfall, and F. J. Angulo.** 1997. Emerging quinolone-resistant Salmonella in the United States. *Emerg. Infect. Dis.* **3:**371–372.
54. **Hird, D. W., M. Pappaioanou, and B. P. Smith.** 1984. Case-control study of risk factors associated with isolation of *Salmonella saintpaul* in hospitalized horses. *Am. J. Epidemiol.* **120:**852–864.
55. **Hogue, A., J. Akkina, F. Angulo, R. Johnson, K. Petersen, S. Saini, and W. Schlosser.** 1997. *Situation assessment: Salmonella typhimurium DT104.* U.S. Department of Agriculture, Food Safety and Inspection Service, Washington, D.C.
56. **Hollinger, K., C. Wray, S. Evans, S. Pascoe, S. Chappell, and Y. Jones.** 1998. *Salmonella typhimurium* DT104 in cattle in Great Britain. *J. Am. Vet. Med. Assoc.* **213:**1732–1733.
57. **House, J. K., R. C. Mainar-Jaime, B. P. Smith, A. M. House, and D. Y. Kamiya.** 1999. Risk factors for nosocomial Salmonella infection among hospitalized horses. *J. Am. Vet. Med. Assoc.* **214:**1511–1516.
58. **Imberechts, H., M. De Filette, C. Wray, Y. Jones, C. Godard, and P. Pohl.** 1998. *Salmonella typhimurium* phage type DT104 in Belgian livestock. *Vet. Rec.* **143:**424–425.
59. **Izumiya, H., K. Tamura, J. Terajima, and H. Watanabe.** 1999. *Salmonella enterica* serovar Typhimurium phage type DT104 and other multi-drug resistant strains in Japan. *Jpn. J. Infect. Dis.* **52:**133.
60. **Johnston, W. S., G. K. MacLachlan, and G. F. Hopkins.** 1979. The possible involvement of seagulls (Larus sp) in the transmission of salmonella in dairy cattle. *Vet. Rec.* **105:**526–527.
61. **Johnston, W. S., D. Munro, W. J. Reilly, and J. C. Sharp.** 1981. An unusual sequel to imported *Salmonella zanzibar. J. Hyg. (Lond.)* **87:**525–528.
62. **Johnston, W. S., G. F. Hopkins, G. K. Maclachlan, and J. C. Sharp.** 1986. Salmonella in sewage effluent and the relationship to animal and human disease in the north of Scotland. *Vet. Rec.* **119:**201–203.
63. **Jones, P. W., P. Collins, G. T. H. Brown, and M. Aitken.** 1982. Transmission of *Salmonella mbandaka* to cattle from contaminated feed. *J. Hyg. (Lond.)* **88:**2255–2263.
64. **Kariuki, S., O. Olsvik, E. Mitema, J. Gathuma, and N. Mirza.** 1993. Acquired tetracycline resistance genes in nosocomial *Salmonella typhimurium* infection in a Kenyan hospital. *East Afr. Med. J.* **70:**255–258.
65. **Kariuki, S., C. Gilks, J. Kimari, A. Obanda, J. Muyodi, P. Waiyaki, and C. A. Hart.** 1999. Genotype analysis of *Escherichia coli* strains isolated from children and chickens living in close contact. *Appl. Environ. Microbiol.* **65:**472–476.
66. **Karpiskova, R., and M. Mikulaskova.** 1995. Salmonella phage types distribution in the Czech Republic in 1991–1994. *Cent. Eur. J. Public Health* **3:**161–162.
67. **Khakhria, R., G. Bezanson, D. Duck, and H. Lior.** 1983. The epidemic spread of *Salmonella typhimurium* phage type 10 in Canada (1970–1979). *Can. J. Microbiol.* **29:**1583–1588.
68. **Killalea, D., L. R. Ward, D. Roberts, J. de Louvois, F. Sufi, J. M. Stuart, G. P. Wall, M. Susman, M. Schwieger, P. J. Sanderson, I. S. Fisher, P. S. Mead, O. N. Gill, C. L. Bartlett, and B. Rowe.** 1996. International epidemiological and microbiological study of outbreak of *Salmonella agona* infection from a ready to eat savory snack. I. England and Wales and the United States. *Br. Med. J.* **313:**1105–1107.

69. **Kinde, H., D. H. Read, A. Ardans, R. E. Breitmeyer, D. Willoughby, H. E. Little, D. Kerr, R. Gireesh, and K. V. Nagaraja.** 1996. Sewage effluent: likely source of *Salmonella enteritidis,* phage type 4 infection in a commercial chicken layer flock in southern California. *Avian Dis.* **40:**672–676.
70. **Kinde, H., M. Adelson, A. Ardans, E. H. Little, D. Willoughby, D. Berchtold, H. Read, R. Breitmeyer, D. Kerr, R. V. Tarbell, and E. D. Hughes.** 1997. Prevalence of Salmonella in municipal sewage treatment plant effluents in southern California. *Avian Dis.* **41:**392–398.
71. **Krytenburg, D., D. D. Hancock, D. H. Rice, T. E. Besser, C. C. Gay, and J. M. Gay.** 1998. *Salmonella enterica* in cattle feeds from the Pacific Northwest. *Anim. Feed Sci. Technol.* **75:**75–79.
72. **Lamb, V. A., C. G. Mayhall, A. C. Spadora, S. M. Markowitz, J. J. Farmer 3d, and H. P. Dalton.** 1984. Outbreak of *Salmonella typhimurium* gastroenteritis due to an imported strain resistant to ampicillin, chloramphenicol, and trimethoprim-sulfamethoxazole in a nursery. *J. Clin. Microbiol.* **20:**1076–1079.
73. **Lance, S. E., G. Y. Miller, D. D. Hancock, P. C. Bartlett, and L. E. Heider.** 1992. Salmonella infections in neonatal dairy calves. *J. Am. Vet. Med. Assoc.* **201:**864–868.
74. **Ling, J., P. Y. Chau, and B. Rowe.** 1987. Salmonella serotypes and incidence of multiply-resistant salmonellae isolated from diarrheal patients in Hong Kong from 1973–82. *Epidemiol. Infect.* **99:**295–306.
75. **Low, J. C., M. Angus, G. Hopkins, D. Munro, and S. C. Rankin.** 1997. Antimicrobial resistance of *Salmonella enterica* typhimurium DT104 isolates and investigation of strains with transferable apramycin resistance. *Epidemiol. Infect.* **118:**97–103.
76. **Mahajan, R., M. Mathur, A. Kumar, P. Gupta, M. M. Faridi, and V. Talwar.** 1995. Nosocomial outbreak of *Salmonella typhimurium* infection in a nursery intensive care unit (NICU) and pediatric ward. *J. Commun. Dis.* **27:**10–14.
77. **Mahipal, S., M. A. Chaudhry, J. N. S. Yadava, and S. C. Sanyal.** 1992. The spectrum of antibiotic resistance in human and veterinary isolates of *Escherichia coli* collected from 1984–86 in Northern India. *J. Antimicrob. Chem.* **29:**159–168.
78. **Malmqvist, M., K. G. Jacobsson, P. Haggblom, F. Cerenius, L. Sjoland, and A. Gunnarsson.** 1995. Salmonella isolated from animals and feedstuffs in Sweden during 1988–1992. *Acta Vet. Scand.* **36:**21–39.
79. **Malorny, B., A. Schroeter, and R. Helmuth.** 1999. Incidence of quinolone resistance over the period 1986 to 1998 in veterinary *Salmonella* isolates from Germany. *Antimicrob. Agents Chemother.* **43:**2278–2282.
80. **Matsushita, S., N. Konishi, M. Arimatsu, A. Kai, S. Yamada, S. Morozumi, H. Izumiya, J. Terajima, and H. Watanabe.** 1999. Drug-resistance and definitive type 104 of *Salmonella* serovar typhimurium isolated from sporadic cases in Tokyo, 1980–1998. *Kansenshogaku Zasshi* **73:**1087–1094.
81. **McConnell, M. M., H. R. Smith, J. Leonardopoulous, and E. S. Anderson.** 1979. The value of plasmid studies in the epidemiology of infections due to drug-resistant *Salmonella wien. J. Infect. Dis.* **139:**178–190.
82. **Mead, P. S., L. Slutsker, V. Dietz, L. F. McCaig, J. S. Bresee, C. Shapiro, P. M. Griffin, and R. V. Tauxe.** 1999. Food-related illness and death in the United States. *Emerg. Infect. Dis.* **5:**607–625.
83. **Melloul, A. A., and L. Hassani.** 1999. Salmonella infection in children from the waste-water-spreading zone of Marrakesh city. *J. Appl. Microbiol.* **87:**536–539.
84. **Mermin, J. H., J. M. Townes, M. Gerber, N. Dolan, E. D. Mintz, and R. V. Tauxe.** 1998. Typhoid fever in the United States, 1985–1994: changing risks of international travel and increasing antimicrobial resistance. *Arch. Intern. Med.* **158:**633–638.

85. **Metzer, E., V. Agmon, N. Andoren, and D. Cohen.** 1998. Emergence of multidrug-resistant *Salmonella enterica* serotype Typhimurium phage-type DT104 among salmonellae causing enteritis in Israel. *Epidemiol. Infect.* **121:**555–559.
86. **Mohan, V. P., K. B. Sharma, D. S. Agarwal, G. Purinima, and P. K. Pillai.** 1995. Plasmid profile and phage type of *Salmonella typhimurium* strains encountered in different regions of India. *Comp. Immunol. Microbiol. Infect. Dis.* **18:**283–290.
87. **Molbak, K., D. L. Baggesen, F. M. Aarestrup, J. M. Ebbesen, J. Engberg, K. Frydendahl, P. Gerner-Smidt, A. M. Petersen, and H. C. Wegener.** 1999. An outbreak of multidrug-resistant, quinolone-resistant *Salmonella enterica serotype typhimurium* DT104. *N. Engl. J. Med.* **341:**1420–1425.
88. **Nastasi, A., M. R. Villafrate, and C. Mammina.** 1990. Characterization of strains of *Salmonella enterica* subsp. enterica serovar Wien isolated in Italy: an epidemiological evaluation. *Microbiologica* **13:**317–321.
89. **Ottolenghi, A. C., and V. V. Hamparian.** 1987. Multiyear study of sludge application to farmland: prevalence of bacterial enteric pathogens and antibody status of farm families. *Appl. Environ. Microbiol.* **53:**1118–1124.
90. **Pacer, R. E., J. S. Spika, M. C. Thurmond, N. Hargrett-Bean, and M. E. Potter.** 1989. Prevalence of Salmonella and multiple antimicrobial-resistant Salmonella in California dairies. *J. Am. Vet. Med. Assoc.* **195:**59–63.
91. **Patton, W. N., G. M. Smith, M. J. Leyland, and A. M. Geddes.** 1985. Multiply resistant *Salmonella typhimurium* septicaemia in an immunocompromised patient successfully treated with ciprofloxacin. *J. Antimicrob. Chemother.* **16:**667–669.
92. **Pignato, S., A. Nastasi, C. Mammina, M. Fantasia, and G. Giammanco.** 1996. Phage types and ribotypes of *Salmonella enteritidis* in southern Italy. *Zentralbl. Bakteriol. Milcrobiol. Hyg. Ser. A* **283:**399–405.
93. **Poirel, L., M. Guibert, S. Bellais, T. Naas, and P. Nordmann.** 1999. Integron- and carbenicillinase-mediated reduced susceptibility to amoxicillin-clavulanic acid in isolates of multidrug-resistant *Salmonella enterica* serotype Typhimurium DT104 from French patients. *Antimicrob. Agents Chemother.* **43:**1098–1104.
94. **Poppe, C., N. Smart, R. Khakhria, W. Johnson, J. Spika, and J. Prescott.** 1998. *Salmonella typhimurium* DT104: a virulent and drug-resistant pathogen. *Can. Vet. J.* **39:**559–565.
95. **Prager, R., A. Liesegang, W. Rabsch, B. Gericke, W. Thiel, W. Voigt, R. Helmuth, L. Ward, and H. Tschape.** 1999. Clonal relationship of *Salmonella enterica* serovar typhimurium phage type DT104 in Germany and Austria. *Zentbl. Bakteriol. Mikrobiol. Hyg. Ser. A* **289:**399–414.
96. **Puohiniemi, R., T. Heiskanen, and A. Siitonen.** 1997. Molecular epidemiology of two international sprout-borne *Salmonella* outbreaks. *J. Clin. Microbiol.* **35:**2487–2491.
97. **Radostits, O. M., D. C. Blood, and C. C. Gay.** 1994. *Veterinary Medicine: A Textbook of the Diseases of Cattle, Sheep, Pigs, Goats, and Horses,* 8th ed., p. 730–745. Baillière Tindall, London, United Kingdom.
98. **Rao, P. S., V. Rajashekar, G. K. Varghese, and P. G. Shivananda.** 1993. Emergence of multidrug-resistant *Salmonella typhi* in rural southern India. *Am. J. Trop. Med. Hyg.* **48:**108–111.
99. **Reilly, W. J., G. I. Forbes, G. M. Paterson, and J. C. Sharp.** 1981. Human and animal salmonellosis in Scotland associated with environmental contamination, 1973–1979. *Vet. Rec.* **108:**553–555.
100. **Ridley, A., and E. J. Threlfall.** 1998. Molecular epidemiology of antibiotic resistance genes in multiresistant epidemic *Salmonella typhimurium* DT 104. *Microb. Drug Resist.* **4:**113–118.

101. **Riley, L. W., B. S. Ceballos, L. R. Trabulsi, M. R. Fernandes de Toledo, and P. A. Blake.** 1984. The significance of hospitals as reservoirs for endemic multiresistant *Salmonella typhimurium* causing infection in urban Brazilian children. *J. Infect. Dis.* **150:**236–241.
102. **Rowe, B., E. J. Threlfall, L. R. Ward, and A. S. Ashley.** 1979. International spread of multiresistant strains of *Salmonella typhimurium* phage types 204 and 193 from Britain to Europe. *Vet. Rec.* **105:**468–469.
103. **Rowe, B., L. R. Ward, and E. J. Threlfall.** 1997. Multidrug-resistant *Salmonella typhi:* a worldwide epidemic. *Clin. Infect. Dis.* **24**(Suppl. 1)**:**S106–S109.
104. **Rubino, S., E. Muresu, M. Solinas, M. Santona, B. Paglietti, A. Azara, A. Schiaffino, A. Santona, A. Maida, and P. Cappuccinelli.** 1999. IS200 fingerprint of *Salmonella enterica* serotype Typhimurium human strains isolated in Sardinia. *Epidemiol. Infect.* **120:**215–222.
105. **Saidi, S. M., Y. Iijima, W. K. Sang, A. K. Mwangudza, J. O. Oundo, K. Taga, M. Aihara, K. Nagayama, H. Yamamoto, P. G. Waiyaki, and T. Honda.** 1997. Epidemiological study on infectious diarrheal diseases in children in a coastal rural area of Kenya. *Microbiol. Immunol.* **41:**773–778.
106. **Sanches, I. S., Z. C. Saraiva, T. C. Tendeiro, J. M. Serra, D. C. Dias, and H. de Lencastre.** 1998. Extensive intra-hospital spread of a methicillin-resistant staphylococcal clone. *Int. J. Infect. Dis.* **3:**26–31.
107. **Sandrou, D. K., and I. S. Arvanitoyannis.** 1999. Implementation of hazard analysis critical control point in the meat and poultry industry. *Food Rev. Int.* **15:**265–308.
108. **Schrag, S. J., V. Perrot, and B. R. Levin.** 1997. Adaptation to the fitness costs of antibiotic resistance in *Escherichia coli. Proc. R. Soc. Lond. B* **264:**1287–1291.
109. **Seyfarth, A. M., H. C. Wegener, and N. Frimodt-Moller.** 1997. Antimicrobial resistance in *Salmonella enterica* subsp. enterica serovar typhimurium from humans and production animals. *J. Antimicrob. Chemother.* **40:**67–75.
110. **Sharma, K. B., M. B. Bhat, A. Pasricha, and S. Vaze.** 1979. Multiple antibiotic resistance among salmonellae in India. *J. Antimicrob. Chemother.* **5:**15–21.
111. **Skov, M. N., O. Angen, M. Chriel, J. E. Olsen, and M. Bisgaard.** 1999. Risk factors associated with *Salmonella enterica* serovar typhimurium infection in Danish broiler flocks. *Poult. Sci.* **78:**848–854.
112. **Smith, B. P., L. Da Roden, M. C. Thurmond, G. W. Dilling, H. Konrad, J. A. Pelton, and J. P. Picanso.** 1994. Prevalence of salmonellae in cattle and in the environment on California dairies. *J. Am. Vet. Med. Assoc.* **205:**467–471.
113. **Tarr, P. E., L. Kuppens, T. C. Jones, B. Ivanoff, P. G. Aparin, and D. L. Heymann.** 1999. Considerations regarding mass vaccination against typhoid fever as an adjunct to sanitation and public health measures: potential use in an epidemic in Tajikistan. *Am. J. Trop. Med. Hyg.* **61:**163–170.
114. **Terajima, J., A. Nakamura, and H. Watanabe.** 1998. Epidemiological analysis of *Salmonella enterica* Enteritidis isolates in Japan by phage-typing and pulsed-field gel electrophoresis. *Epidemiol. Infect.* **120:**223–229.
115. **Threlfall, E. J., L. R. Ward, and B. Rowe.** 1978. Epidemic spread of a chloramphenicol-resistant strain of *Salmonella typhimurium* phage type 204 in bovine animals in Britain. *Vet. Rec.* **103:**438–440.
116. **Threlfall, E. J., B. Rowe, J. L. Ferguson, and L. R. Ward.** 1985. Increasing incidence of resistance to gentamicin and related aminoglycosides in *Salmonella typhimurium* phage type 204c in England, Wales and Scotland. *Vet. Rec.* **117:**355–357.
117. **Threlfall, E. J., B. Rowe, and L. R. Ward.** 1993. A comparison of multiple drug resistance in salmonellas from humans and food animals in England and Wales, 1981 and 1990. *Epidemiol. Infect.* **111:**189–197.

118. **Threlfall, E. J., J. A. Frost, L. R. Ward, and B. Rowe.** 1994. Epidemic in cattle and humans of *Salmonella typhimurium* DT 104 with chromosomally integrated multiple drug resistance. *Vet. Rec.* **134:**577.
119. **Threlfall, E. J., F. J. Angulo, and P. G. Wall.** 1998. Ciprofloxacin-resistant *Salmonella typhimurium* DT104. *Vet. Rec.* **142:**255.
120. **Threlfall, E. J., L. R. Ward, M. D. Hampton, A. M. Ridley, B. Rowe, D. Roberts, R. J. Gilbert, P. Van Someren, P. G. Wall, and P. Grimont.** 1998. Molecular fingerprinting defines a strain of *Salmonella enterica* serotype Anatum responsible for an international outbreak associated with formula-dried milk. *Epidemiol. Infect.* **121:**289–293.
121. **Tillotson, K., C. J. Savage, M. D. Salman, C. R. Gentry-Weeks, D. Rice, P. J. Fedorka-Cray, D. A. Hendrickson, R. L. Jones, W. Nelson, and J. L. Traub-Dargatz.** 1997. Outbreak of *Salmonella infantis* infection in a large animal veterinary teaching hospital. *J. Am. Vet. Med. Assoc.* **211:**1554–1557.
122. **Timoney, J. F., H. C. Neibert, and F. W. Scott.** 1978. Feline salmonellosis: a nosocomial outbreak and experimental studies. *Cornell Vet.* **68:**211–219.
123. **Trueman, K. F., R. J. Thomas, A. R. MacKenzie, L. E. Eaves, and P. F. Duff.** 1996. *Salmonella dublin* infection in Queensland dairy cattle. *Aust. Vet.* **74:**367–369.
124. **Turnidge, J.** 1995. Epidemiology of quinolone resistance, Eastern hemisphere. *Drugs* **49** Suppl. 2:43–47.
125. **Uhaa, I. J., D. W. Hird, D. C. Hirsh, and S. S. Jang.** 1988. Case-control study of risk factors associated with nosocomial *Salmonella krefeld* infection in dogs. *Am. J. Vet. Res.* **49:**1501–1505.
126. **Vahaboglu, H., L. M. Hall, L. Mulazimoglu, S. Dodanli, I. Yildirim, and D. M. Livermore.** 1995. Resistance to extended-spectrum cephalosporins, caused by PER-1 beta-lactamase, in *Salmonella typhimurium* from Istanbul, Turkey. *J. Med. Microbiol.* **43:**294–299.
127. **Vahaboglu, H., S. Dodanli, C. Eroglu, R. Ozturk, G. Soyletir, L. Yildirim, and V. Avkan.** 1996. Characterization of multiple-antibiotic-resistant *Salmonella typhimurium* strains: molecular epidemiology of PER-1-producing isolates and evidence for nosocomial plasmid exchange by a clone. *J. Clin. Microbiol.* **34:**2942–2946.
128. **Van Beneden, C. A., W. E. Keene, R. A. Strang, D. H. Werker, A. S. King, B. Mahon, K. Hedberg, A. Bell, M. T. Kelly, V. K. Balan, W. R. MacKenzie, and D. Fleming.** 1999. Multinational outbreak of *Salmonella enterica* serotype Newport infections due to contaminated alfalfa sprouts. *JAMA* **281:**158–162.
129. **van der Zee A., H. Verbakel, J. C. van Zon, I. Frenay, A. van Belkum, M. Peeters, A. Buiting, and A. Bergmans.** 1999. Molecular genotyping of *Staphylococcus aureus* strains: comparison of repetitive element sequence-based PCR with various typing methods and isolation of a novel epidemicity marker. *J. Clin. Microbiol.* **37:**342–349.
130. **Wall, P. G., D. Morgan, K. Lamden, M. Ryan, M. Griffin E. J. Threlfal, L. R. Ward, and B. Rowe.** 1994. A case control study of infection with an epidemic strain of multiresistant *Salmonella typhimurium* DT104 in England and Wales. *Commun. Dis. Rep.* **14;4:**R130–5.
131. **Wall, P. G., E. J. Threlfall, L. R. Ward, and B. Rowe.** 1996. Multiresistant *Salmonella typhimurium* DT104 in cats: a public health risk. *Lancet* **348:**471.
132. **Wegener, H. C., D. L. Baggesen, and K. Gaarslev.** 1994. *Salmonella typhimurium* phage types from human salmonellosis in Denmark 1988–1993. *APMIS* **102:**521–525.
133. **World Health Organization.** 1997. Multi-drug resistant *Salmonella typhimurium.* World Health Organization Fact Sheet 139. World Health Organization, Geneva, Switzerland.
134. **Wray, C., J. N. Todd, and M. Hinton.** 1987. Epidemiology of *Salmonella typhimurium* infection in calves: excretion of *S. typhimurium* in the faeces of calves in different management systems. *Vet. Rec.* **121:**293–296.

135. **Wray, C., N. Todd, I. McLaren, Y. Beedell, and B. Rowe.** 1990. The epidemiology of Salmonella infection of calves: the role of dealers. *Epidemiol. Infect.* **105:**295–305.
136. **Wray, C., N. Todd, I. M. McLaren, and Y. E. Beedell.** 1991. The epidemiology of Salmonella in calves: the role of markets and vehicles. *Epidemiol. Infect.* **107:**521–525.
137. **Yoder, D. R., E. D. Ebel, D. D. Hancock, and B. A. Combs.** 1994. Investigation of an outbreak of bovine cysticercosis in an Idaho feedlot. *J. Am. Vet. Med. Assoc.* **205:**45–50.

Emerging Diseases of Animals
Edited by C. Brown and C. Bolin

Chapter 12

Bartonellosis

Craig E. Greene and Duncan C. Krause

The genus *Bartonella* is a consolidation of vector-borne hemotropic bacteria that were once members of three genera, *Bartonella, Grahamella,* and *Rochalimaea* (4) (Table 1). Carrión's disease, the first identified bartonellosis, is a hemolytic, vasculoproliferative disorder of people in the Andes Mountains of Peru, Colombia, and Ecuador. Oroya fever refers to the acute bacteremic and hemolytic form of this disease, usually observed in visitors to the region who have never been exposed. Verruga peruana is the form of the disease more frequently observed in the resident population and is characterized by benign cutaneous vascular lesions. *Bartonella bacilliformis,* the causative agent, is transmitted by the sand fly (*Lutzomyia*), and humans are the only known hosts. Trench fever is a multisystemic disease characterized by bacteremia, endocarditis, lymphadenomegaly, and vasculoproliferative lesions and was identified in military personnel stationed in Europe in World War I. The disease, caused by *Bartonella* (*Rochalimaea*) *quintana,* went largely ignored until later in the 20th century, when it was found as a complicating infection of immune-compromised people causing a condition termed bacillary angiomatosis (BA) (29, 32). Two new pathogenic species, *Bartonella henselae* and *Bartonella elizabethae,* were also isolated from BA patients (13, 51). *B. henselae* was subsequently also detected in more immune-competent patients with cat scratch disease (CSD) (59) and is now thought to be the leading cause of CSD. *B. henselae* is harbored by cats and transmitted by the cat flea *Ctenocephalides felis,* while *B. elizabethae* is likely harbored by rodents and transmitted by their fleas. A number of genetic variants of *B. henselae* have been detected among infected

Craig E. Greene • Departments of Small Animal Medicine and Medical Microbiology, College of Veterinary Medicine, University of Georgia, Athens, GA 30602. **Duncan C. Krause** • Department of Microbiology, University of Georgia, Athens, GA 30602.

Table 1. Clinically important members of the genus *Bartonella*[a]

Organism	Classical disease(s)	Geographic occurrence	Vector(s)	Reservoir host	Incidental host(s)	Clinical features in incidental host
B. bacilliformis	Oroya fever; verruga peruana	Andes Mountains	Sand fly (*Lutzomyia*)		Human	Hemolytic anemia, fever, asymptomatic bacteremia, oroya fever, indolent angiomatous skin lesions (verruga peruana)
B. quintana[b]	Trench fever	Focal World War I worldwide	Body louse (*Pediculus humanus*)	Human	Human	Bacteremia, localized tissue infection (angiomatosis, peliosis, granulomatous and pyogenic inflammation, lymphadenitis)
B. henselae[b]	Cat scratch disease	Worldwide	Cat flea (*Ctenocephalides felis*), tick?	Cat	Human	Same as for *B. quintana*
B. clarridgeiae	Cat scratch disease	Worldwide?	Cat flea (*C. felis*)?	Cat	Human	Same as for *B. quintana*
B. elizabethae[b]		Unknown	Flea (*Xenopsylla cheopis*)	Rat	Human	Endocarditis, bacteremia
B. koehlerae	Unknown	Unknown	Unknown	Cat	Unknown	
B. vinsonii subsp. *berkhoffii*[b]		Unknown	Brown dog tick (*Rhipicephalus sanguineus*)	Unknown	Dog, coyote	Endocarditis, bacteremia in splenectomized dog
B. vinsonii subsp. *arupensis*[b]		Texas	Unknown	Mouse?	Human	Bacteremia, fever

[a] Modified from reference 6 with permission of the publisher. The genus *Bartonella* also includes other organisms of uncertain pathogenicity: *B. vinsonii* (Canadian vole agent), *B. tribocorum* (wild rat isolate), *B. alsatica* (wild rabbit isolate); and former *Grahamella* species (*B. talpae, B. peromysci, B. grahamii, B. taylorii,* and *B. doshiae*) that infect small mammals, fish, and birds. Numerous other untyped species of *Bartonella* have been isolated from domestic and wild carnivores and herbivores.

[b] Formerly classified in the genus *Rochalimaea.*

people and cats (3, 53, 54). A closely related species, *Bartonella clarridgeiae*, constitutes approximately 10% of *Bartonella* isolates from clinically healthy cats and has been isolated from some people with CSD (22, 38). A recently discovered species, *Bartonella koehlerae*, was likewise isolated from cats and may be flea transmitted (15), but its role as a human pathogen is uncertain. Dogs and coyotes have been infected with *Bartonella vinsonii* subsp. *berkhoffii*, which is likely transmitted by ticks (50). *Bartonella vinsonii* subsp. *arupensis* was isolated from a cattle rancher in Texas, and similar isolates were found in naturally infected mice in the same region (57). Numerous other domestic and wild carnivores and herbivores have been found to be infected with members of the genus *Bartonella* whose species have not yet been determined. A number of hosts have been shown to be coinfected, and animals and people within the same household may have different species infecting them (7, 21, 35, 36). New *Bartonella* spp. as well as pathogenic and reservoir relationships with other hosts and their vectors will undoubtedly be described in coming years.

PATHOGENESIS AND EPIDEMIOLOGY OF DISEASE

Host Specificity

Bartonella spp. show a certain degree of host specificity. This may relate to inherent host immune defenses, organism adaptations, or vector preferences. Levels of bacteremia in reservoir hosts are noticeably higher than in inadvertent hosts. Those species of *Bartonella* studied to date are intraerythrocytic in their reservoir hosts, and this adaptation to host erythrocytes may afford them a protected existence and facilitate transmission via suitable blood-sucking vectors. A high prevalence and degree of *B. henselae* bacteremia in the healthy cat population indicates that the organism has adapted well to its reservoir host. This subclinical infection in cats can lead to inadvertent transmission to people, to other cats during blood transfusions, or among cats comingling in the presence of suitable vectors.

Pathogenesis

Erythrocyte invasion is a strategy utilized by *Bartonella* spp. The selective advantage of an intraerythrocytic presence for *Bartonella* spp. has not been clearly defined, but one can make a strong case for several benefits. Erythrocyte invasion might protect the bacteria from host immune defenses, facilitate acquisition of heme or iron, or enhance transmission to or survival in the flea vector. The pathogenic relation of *Bartonella* spp. with erythrocytes has been most thoroughly studied in *Bartonella bacilliformis* (47), which upon reaching the blood invades and ultimately lyses most circulating erythrocytes, resulting in anemia, malaise, fever, splenomegaly, and lymphadenomegaly. However, hemolysis has not been shown to be a significant manifestation of infections involving the other species in the genus *Bartonella*. *B. henselae* exhibits an affinity for erythrocytes, but in a host-dependent manner, and apparently without significant lysis of infected cells (45). Cats are apparently the major reservoir hosts for this agent and have high levels of intraerythrocytic bacteremia with no overt hemolysis. By electron microscopy, bacteria are noted within feline erythrocytes but not extracelluarly or on the cell surface during persistent bacteremia (33). Studies by confocal microscopy of feline erythrocytes infected in vitro revealed epicellular bacteria (45), probably representing a transient state in erythrocyte invasion, which has been reported to require several hours for *B. henselae* (45). In contrast to the chronic bacteremia common in infected cats, blood cultures are rarely positive in immune-competent humans, where *B. henselae* appears to localize primarily in regional lymph nodes. In immunodeficient people, bacteremia is detectable, although still at much lower levels than are seen in cats. The existence of an intraerythrocytic phase in humans has yet to be determined.

Bartonella Virulence Determinants

The characterization of potential virulence factors in *Bartonella* spp. has been well elucidated. An invasion-associated locus (Ial) has been described in *Bartonella bacilliformis* (48), including the gene for (di)nucleoside polyphosphate hydrolase, which may function in intracellular survival (10). A similar locus has been described in *B. henselae.* Both *B. bacilliformis* and *B. henselae* secrete a protein that deforms human erythrocyte membranes, yet *B. henselae* does not bind human erythrocytes (28). *B. henselae* invasion of human umbilical vein endothelial cells in vitro appears to involve bacterial aggregation on the cell surface, followed by internalization by a novel cellular structure termed the invasome (14). *B. henselae* produces type 4-like fimbriae, which are thought to participate in adherence to and invasion of human epithelial cells. Fimbria expression is phase variable and correlates with invasion of cultured human epithelial cells (2), but the role, if any, of fimbriae in feline erythrocyte invasion is not known. Finally, contact-dependent hemolytic activity has been described recently for *B. bacilliformis* that is distinct from erythrocyte-deforming activity (25).

B. henselae colonizes human vascular endothelial cells, and infections in immunocompromised individuals are often accompanied by vascular skin tumors resembling Kaposi's sarcoma. The association of *Bartonella* with angiogenic lesions of both Carrión's disease and BA, as well as the regression of these lesions with antimicrobial therapy, strongly supports a role for these organisms in triggering neovascularization. Significantly, *Bartonella* spp. are phylogenetically related to *Agrobacterium* and *Rhizobium* spp., which are associated with nodule or tumor formation in plants. *B. bacilliformis* releases an angiogenic factor that is responsible for the vascular proliferation seen in Carrión's disease, and related species may do similarly (46).

Transmission

Any means of mechanical or biologic transmission of blood is a potential means of spreading *Bartonella* infections. *B. henselae* has been cultured and detected by PCR from fleas feeding on naturally infected cats (11, 32). Infection has also been transmitted via fleas from naturally infected to laboratory-reared specific-pathogen-free (SPF) cats (11). Other blood-sucking arthropods such as ticks may be involved, as infected cats have high levels of bacteremia. However, in the absence of arthropod vectors, infection is not transmitted between cats (11, 18). *B. henselae* has been found in dissected flea midguts and can be cultured from feces of fleas being fed infected blood by artificial means (26). Replication of the organism in the fleas was also demonstrated. Cats have been infected by intradermal inoculation of feces obtained from fleas that were feeding on bacteremic cats (16). Hypothetically, transmission of *B. quintana* or *B. henselae* infections to humans occurs when arthropod feces containing the organisms is transmitted transcutaneously or transmucosally at the site of an abrasion or bite or scratch injury. In the case of cats, flea feces may contaminate claws or teeth, resulting in inoculation into people or other cats simultaneous with a bite or scratch.

Immunity

Experimentally infected cats develop an increased antibody titer to the organism but, despite this titer rise, may have variable degrees and durations of bacteremia (18, 52, 58). High-dose antimicrobial therapy may facilitate the clearance of the bacteremia, and once they have cleared the infection, cats are immune to rechallenge by intradermal inoculation of cultured organisms. The route of rechallenge is important in determining the outcome, since transmission between cats using transfused infected blood has not shown protection (34, 37).

Immunity to *Bartonella* appears to be species or strain specific. In naturally exposed cats, simultaneous infection with more than one species has been demonstrated. In experimental challenge experiments, successive infection with closely related *B. henselae* strains produces immunity, whereas heterologous strain-challenged cats are not protected (58). Bacteremia in some cats may be inconsistently present when monitored for many months. In untreated cats, relapsing bacteremia was universal among *B. clarridgeiae*-infected cats and less prevalent (22 to 36%) in cats infected with *B. henselae* variants (58).

Geographic Distribution

The distribution of feline bartonellosis has been studied by serologic and cultural studies in various countries. The prevalence of infection has ranged from approximately 6 to 80% when either method is used and varies among areas by the selected population of cats that is tested. Prevalence of infection is highest with increasing temperature and humidity, and in the United States, the highest prevalence data are from warmer moist regions and lowest in the dry regions (30). Other prevalence studies of feline bartonellosis have shown greater risk factors for infection, including stray cats, young cats (<1 year), and flea infestation (12). No correlation could be found between *Bartonella* infection and antibody prevalence to feline immunodeficiency virus. In other studies, seroprevalence of bartonellosis was highest with lack of cattery hygiene and greater flea infestation (17). Seropositivity to *B. vinsonii* (*B. berkhoffii*) in ill dogs was highest in dogs from rural environments, those exposed to cattle, and those heavily infested with ticks (50).

Clinical Features

Humans

Carrión's disease is characterized by a hemolytic anemia, which is a unique manifestation caused by *B. bacilliformis*. Other signs in people with bartonellosis caused by other species are recurrent fever, bacteremia with or without endocarditis, disseminated vascular lesions characterized as bacillary angiomatosis, and visceral bacillary peliosis. These forms of the disease are most common in immunocompromised hosts. In more immune-competent hosts, localization of infection to the lymphoid tissues results in classical CSD. This syndrome is associated with marked regional lymphadenomegaly, fever, and rarely further systemic dissemination of infection. Even then, people with CSD may have cranial manifestations of conjunctivitis (Parinaud's oculoglandular syndrome), meningitis, encephalitis, or neuroretinitis.

Cats

The pathogenic nature of *Bartonella* spp. for cats is still in question. There are reports of lesions in cats that resemble those seen with human bartonellosis, such as persistent lymphadenomegaly with argyrophilic staining bacteria (31) or peliosis hepatis (9). Self-limiting febrile illness and transient nonlocalizing neurologic dysfunction developed in two cats experimentally infected with *B. henselae* by blood transfusion (34). Cats experimentally inoculated with *B. henselae* or *B. clarridgeiae* by intravenous transfusion of blood from infected donors had unremarkable clinical signs and gross necropsy lesions despite persistent infection and bacteremia (19, 37). Lymph node hyperplasia and multifocal inflammatory lesions were seen in various parenchymal organs. With intradermal inoculations, localized papule formation and regional lymphadenomegaly are consistent but transient phenomena (18, 52). Intradermal inoculation of an apparently more pathogenic strain, using cultured organisms, infected cat blood, or flea feces, resulted in more systemic signs of infection (49). These signs included fever, lethargy, anorexia, and gross inflammation at the inoculation sites, which in some cats required supportive force feeding and fluid therapy (49). Reproductive failure but lack of perinatal transmission was observed in bacteremic SPF female cats that were bred with uninfected males (20). The male cats did not become infected.

Most studies of naturally infected cats indicate that infection is subclinical despite the persistent bacteremia. Some naturally infected cats developed brief periods of pronounced fever following routine surgical procedures that resolved without antibiotic therapy. Uveitis has also been found to result from *Bartonella* infection in cats (43). In general, *B. henselae* appears to be a subtle pathogen in its reservoir feline host in that clinical signs may only develop in infected cats during periods of stress or potentially in conjunction with concurrent diseases. The role of this organism in causing feline diseases will be difficult to determine in the natural setting because of the high percentage of chronically bacteremic healthy cats.

Dogs

B. vinsonii subsp. *berkhoffii,* the cause of endocarditis and bacteremia in dogs, has been associated with clinical illness (8, 40). Clinical signs in affected dogs have been typical of endocarditis, including lethargy, inappetence, shifting leg lameness, and heart murmur. Some of the dogs develop granulomatous lymphadenitis, infiltrative myocarditis, or granulomatous infiltration in other organs (6). A very similar clinical syndrome associated with endocarditis can be produced by a closely related alphaproteobacterium, and without proper genetic analysis of the isolate, the disease may be mistaken for bartonellosis (5).

Other animals

Numerous isolations of *Bartonella* spp. have been made from other domestic and wild animals and their ticks (23, 24, 27). Experimental inoculations of these various isolates into other animal hosts have been performed. In a majority of cases, clinical signs have been minimal although brief bacteremias are produced. At present, the clinical significance in other animal hosts is unknown.

DIAGNOSIS

Serologic testing for antibodies to *B. henselae* provides useful epidemiologic information relative to disease prevalence but is of limited usefulness in documenting actively infected cats. The indirect fluorescent antibody procedure, the reference standard, is only accurate to the genus level. Furthemore, cross reactions in human sera have been demonstrated for antibodies to *B. quintana, B. henselae, Coxiella burnetii,* and *Chlamydia* spp. In experimentally infected SPF cats, the magnitude and duration of immunoglobulin M or immunoglobulin G response to the organism are highly variable among individual cats although in individual cats, an anamnestic response is usually noted on rechallenge infection (18). Very high antibody titers (>256) seem to correlate with bacteremia in both naturally and experimentally infected cats, but infection can only be confirmed by detecting the organism.

Blood or tissue culture is one of the most clinically useful tools to document active infection. Due to higher levels of bacteremia (>100 CFU/ml) in cats, *B. henselae* is easier to isolate from feline blood or tissue specimens than those of humans. *B. henselae* bacteremia can persist in clinically healthy cats for months to sometimes years. The presence of bacteremia can be intermittent or recurrent in some cats, making multiple sampling necessary. Blood for culture should be taken in an aseptic manner and placed into sterile EDTA tubes or directly into centrifugation-lysis tubes. Blood samples can be frozen and thawed prior to culture to produce erythrocyte lysis, which improves the sensitivity of isolation and rapidity of colony formation. Colonies usually appear within 1 to 3 weeks on solid medium containing sheep, rabbit, or human blood-based agar or chocolate and charcoal-yeast extract agar. Incubation is at 37°C at high humidity in a 5% CO_2-enriched atmosphere. Species identification and strain recognition of isolated *Bartonella* spp. generally require genetic methods.

Gross pathologic changes are not observed in infected cats with the exception of lymphadenomegaly in some experimentally infected cats. Reactive follicular hyperplasia is the typical histologic feature in these lymph nodes. Multifocal granulomatous inflammation is a characteristic lesion that has been observed in infected cats and dogs. Findings are generally not definitive without special staining to identify the organism. Warthin-Starry silver staining is needed to detect the organism within tissues by light microscopy. Immunohistochemistry may be beneficial for specific identification of the organism in tissues.

Treatment

In vitro testing generally shows susceptibility of *Bartonella* to beta-lactams, chloramphenicol, tetracyclines, macrolides, aminoglycosides, and rifampin (55). The clinical response and clearance of bacterial infection are more pronounced in humans with bacteremia rather than localized lymphadenitis or multifocal granulomatous disease. This difference may relate to the inability of the antimicrobials to penetrate pyogranulomatous lesions or reverse the immunopathologic consequences of infection. Antimicrobial efficacy in eliminating *B. henselae* bacteremia in cats has not been clearly established. Several of the antibiotics considered efficacious for treating people have been reported to be ineffective in the cat, even with a prolonged duration of therapy. However, dose ranges and combination therapies have not been properly evaluated. Both treatment successes and failures have been reported with amoxicillin, doxycycline, and enrofloxacin in both naturally and experimentally *B. henselae*-infected bacteremic cats. Some reduction in bacteremia has been found with doxycycline and erythromycin at the recommended dosages (52). In other studies, enrofloxacin has been reported as being effective (39). On the contrary, when used at higher than recommended dosages for one week, doxycycline, amoxicillin, and amoxicillin-clavulanate were effective in suppressing bacteremia in infected cats (18). In experimental studies, rifampin has been effective in clearing bacteremia alone and in combination with doxycycline (6). Because of the uncertainty of their efficacy, 2 to 4 weeks of treatment are recommended and follow-up cultures should always be obtained to ensure that treatment is effective. Cultures should be taken at least 3 weeks after the antimicrobial therapy has been discontinued. Even then, periodic blood culture monitoring may be advised until sequential negative results are obtained. Although of minor clinical consequence to the cat, persistence of *Bartonella* bacteremia is an important concern for the household of an immune-compromised individual.

Prevention

Vaccination of cats would seem desirable to decrease potential human exposure to *Bartonella.* Cats that are treated and recover from *Bartonella* bacteremia following experimental intradermal infections with cultured organisms appear to be resistant to homologous strains on rechallenge (1, 18, 52). With heterologous reinfection, a protective response is not found (58). Unfortunately, a multivalent or genetically engineered vaccine will be necessary to provide protection against the diverse strains of *Bartonella* that infect cats. Furthermore, immunity to reinfection does not occur with intravenous transfusion of blood from infected cats (34), although this may not reflect a natural route of infection. Cats in households where immune-deficient people are present should be screened for infection by blood culture, and any positive cat should be treated. Strict flea control measures should be used under such circumstances. Cats and dogs should be screened for *Bartonella* infections prior to being used as blood donors.

SUMMARY AND CONCLUSION

Arthropod-borne infectious agents develop a complex relationship in concert with their vector and reservoir host. In general, reservoir hosts have few clinical signs despite organism proliferation, while inadvertent hosts suffer from more severe illness. *Bartonella* spp. undoubtedly have a long evolutionary history because of the numerous reservoir hosts and corresponding host-adapted *Bartonella* spp. that have been identified. Although human bartonelloses have been recognized as clinical entities for most of the 20th century, many of their underlying organisms have only been identified in recent years. Using newer cultural and genetic detection methods, other clinical manifestations and affected hosts have been identified. Undoubtedly, the species of *Bartonella* as well as the host range of and diseases caused by these species will continue to expand at a rapid rate. Additional study of genetic variants within given *Bartonella* species will be useful in future epidemiologic studies. This will help determine reservoir hosts, incriminate responsible vectors, and elucidate means of transmission.

REFERENCES

1. **Abbott, R. C., B. B. Chomel, R. W. Kasten, K. A. Floyd-Hawkins, Y. Kikuchi, J. E. Koehler, and N. C. Pedersen.** 1997. Experimental and natural infection with *Bartonella henselae* in domestic cats. *Comp. Immunol. Microbiol. Infect. Dis.* **20:**41–51.
2. **Batterman, H. J., J. A. Peek, J. S. Loutit, S. Falkow, and L. S. Tompkins.** 1995. *Bartonella henselae* and *Bartonella quintana* adherence to and entry into cultured human epithelial cells. *Infect. Immun.* **63:**4553–4556.
3. **Bergmans, A. M. C., J. F. P. Schellekens, J. D. A. van Embden, and L. M. Schouls.** 1996. Predominance of two *Bartonella henselae* variants among cat-scratch disease patients in the Netherlands. *J. Clin. Microbiol.* **34:**254–260.
4. **Birtles, R. J., T. G. Harrison, N. A. Saunders, and D. H. Molyneux.** 1995. Proposals to unify the genera *Grahamella* and *Bartonella,* with descriptions of *Bartonella talpae* comb. nov., *Bartonella peromysci* comb. nov., and three new species, *Bartonella grahamii* sp. nov., *Bartonella taylorii* sp. nov., and *Bartonella doshiae* sp. nov. *Int. J. Syst. Bacteriol.* **45:**1–8.
5. **Breitschwerdt, E. B., C. E. Atkins, T. T. Brown, D. L. Kordick, and P. S. Snyder.** 1999. *Bartonella vinsonii* subsp. *berkhoffii* and related members of the alpha subdivision of the Proteobacteria in dogs with cardiac arrhythmias, endocarditis, or myocarditis. *J. Clin. Microbiol.* **37:**3618–3626.
6. **Breitschwerdt, E. B., and C. E. Greene.** 1998. Bartonellosis, p. 337–343. *In* C. E. Greene (ed.), *Infectious Diseases of the Dog and Cat.* W. B. Saunders Co., Phildelphia, Pa.
7. **Breitschwerdt, E. B., B. C. Hegarty, and S. I. Hancock.** 1998. Sequential evaluation of dogs naturally infected with *Ehrlichia canis, Ehrlichia chaffeensis, Ehrlichia equi, Ehrlichia ewingii,* or *Bartonella vinsonii. J. Clin. Microbiol.* **36:**2645–2651.
8. **Breitschwerdt, E. B., D. L. Kordick, D. E. Malarkey, B. Keene, T. L. Hadfield, and K. Wilson.** 1995. Endocarditis in a dog due to infection with a novel *Bartonella* subspecies. *J. Clin. Microbiol.* **33:**154–160.
9. **Brown, P. J., J. P. Henderson, P. Galloway, H. Odair, and J. M. Wyatt.** 1994. Peliosis hepatis and telangiectasis in 18 cats. *J. Small Anim. Pract.* **35:**73–77.
10. **Cartwright, J. L., P. Britton, M. F. Minnick, and A. G. McLennan.** 1999. The Ia1A invasion gene of *Bartonella bacilliformis* encodes a (di)nucleoside polyphosphate hydrolase of the MutT motif family and has homologs in other invasive bacteria. *Biochem. Biophys. Res. Commun.* **256:**474–479.
11. **Chomel, B., R. W. Kasten, K. Floyd-Hawkins, B. Chi, K. Yamamoto, J. Roberts-Wilson, A. N. Gurfield, R. C. Abbott, and N. C. Pedersen.** 1996. Experimental transmission of *Bartonella henselae* by the cat flea. *J. Clin. Microbiol.* **34:**1952–1956.
12. **Chomel, B. B., R. C. Abbott, R. W. Kasten, K. A. Floyd-Hawkins, P. H. Kass, C. A. Glaser, N. C. Pederson, and J. E. Koehler.** 1995. *Bartonella henselae* prevalence in domestic cats in California: risk factors and association between bacteremia and antibody titers. *J. Clin. Microbiol.* **33:**2445–2450.
13. **Daly, J. S., M. G. Worthington, D. J. Brenner, C. W. Moss, D. G. Hollis, R. S. Weyant, A. G. Steigerwalt, R. E. Weaver, M. I. Daneshvar, and S. P. O'Connor.** 1993. *Rochalimaea elizabethae* sp. nov. isolated from a patient with endocarditis. *J. Clin. Microbiol.* **31:**872–881.
14. **Dehio, C., M. Meyer, J. Berger, H. Schwarz, and C. Lanz.** 1997. Interaction of *Bartonella henselae* with endothelial cells results in bacterial aggregation on the cell surface and the subsequent engulfment and internalization of the bacterial aggregate by a unique structure, the invasome. *J. Cell Sci.* **110:**2141–2154.
15. **Droz, A., B. Chi, E. Horn, A. G. Steigerwalt, A. M. Whitney, and D. J. Brenner.** 1999. *Bartonella koehlerae* sp. nov., isolated from cats. *J. Clin. Microbiol.* **37:**1117–1122.

16. **Foil, L., E. Andress, R. L. Freeland, A. F. Roy, R. Rutledge, P. C. Triche, and K. L. O'Reilly.** 1998. Experimental infection of domestic cats with *Bartonella henselae* by inoculation of *Ctenocephalides felis* (Siphonaptera: Pulicidae) feces. *J. Med. Entomol.* **35:**625–628.
17. **Foley, J. E., B. Chomel, Y. Kikuchi, K. Yamamoto, and N. C. Pedersen.** 1998. Seroprevalence of *Bartonella henselae* in cattery cats: association with cattery hygiene and flea infestation. *Vet. Q.* **20:**1–5.
18. **Greene, C. E., M. McDermott, P. H. Jameson, C. L. Atkins, and A. M. Marks.** 1996. *Bartonella henselae* infection in cats: evaluation during primary infection, treatment, and rechallenge infection. *J. Clin. Microbiol.* **34:**1682–1685.
19. **Guptill, L., L. Slater, C. Wu, T. Lin, L. Glickman, D. Welch, and H. Hogenesch.** 1997. Experimental infection of young cats with the zoonotic organism *Bartonella henselae. J. Infect. Dis.* **176:**206–216.
20. **Guptill, L., L. Slater, C.-C. Wu, T.-L. Lin, L. T. Glickman, D. F. Welch, J. Tobolski, and H. Hogenesch.** 1998. Evidence of reproductive failure and lack of perinatal transmission of *Bartonella henselae* in experimentally infected cats. *Vet. Immunol. Immunopathol.* **65:**177–189.
21. **Gurfield, A. N., H.-J. Boulouis, B. B. Chomel, R. Heller, R. W. Kasten, K. Yamamoto, and Y. Piemont.** 1997. Coinfection with *Bartonella clarridgeiae* and *Bartonella henselae* and with different *Bartonella henselae* strains in domestic cats. *J. Clin. Microbiol.* **35:**2120–2123.
22. **Heller, R., M. Artois, V. Xemar, D. DeBriel, H. Gehin, B. Jaulhac, H. Monteil, and Y. Piemont.** 1997. Prevalence of *Bartonella henselae* and *Bartonella clarridgeiae* in stray cats. *J. Clin. Microbiol.* **35:**1327–1331.
23. **Heller, R., M. Kubina, P. Mariet, P. Riegel, G. Delacour, C. Dehio, F. Lamarque, R. Kasten, H.-J. Boulouis, H. Monteil, B. Chomel, and Y. Piemont.** 1999. *Bartonella alsatica* sp. nov, a new *Bartonella* species isolated from the blood of wild rabbits. *Int. J. Syst. Bacteriol.* **49:**283–288.
24. **Heller, R., P. Riegel, Y. Hansmann, G. Delacour, D. Bermond, C. Dehio, F. Lamarque, H. Monteil, B. Chomel, and Y. Piemont.** 1998. *Bartonella tribocorum* sp. nov., a new *Bartonella* species isolated from the blood of wild rats. *Int. J. Syst. Bacteriol.* **48:**1333–1339.
25. **Hendrix, L. R.** 2000. Contact-dependent hemolytic activity distinct from deforming activity of *Bartonella bacilliformis. FEMS Microbiol. Lett.* **182:**119–124.
26. **Higgins, J. A., S. Radulovic, D. C. Jaworski, and A. F. Azad.** 1996. Acquisition of the cat scratch disease agent *Bartonella henselae* by cat fleas (Siphonaptera: Pulicidae). *J. Med. Entomol.* **33:**490–495.
27. **Hofmeister, E. K., C. P. Kolbert, A. S. Abdulkarim, J. M. H. Magera, M. K. Hopkins, J. R. Uhl, A. Ambyaye, S. R. Telford III, F. R. Cocerill III, and D. H. Persing.** 1998. Cosegregation of a novel *Bartonella* species with *Borrelia burgdorferi* and *Babesia microti* in *Peromyscus leucopus. J. Infect. Dis.* **177:**409–416.
28. **Iwaki-Egawa, S., and G. M. Ihler.** 1997. Comparison of the abilities of proteins from *Bartonella bacilliformis* and *Bartonella henselae* to deform red cell membranes and to bind to red cell ghost proteins. *FEMS Microbiol. Lett.* **157:**201–217.
29. **Jackson, L. A., D. H. Spach, D. A. Kippen, N. K. Sugg, R. L. Regnery, M. H. Sayers, and W. E. Stamm.** 1996. Seroprevalence to *Bartonella quintana* among patients at a community clinic in downtown Seattle. *J. Infect. Dis.* **173:**1023–1026.
30. **Jameson, P. H., C. E. Greene, R. L. Regnery, M. Dryden, A. Marks, J. Brown, J. Cooper, B. Glaus, and R. Greene.** 1995. Seroprevalence of *Rochalimaea henselae* in pet cats throughout regions of North America. *J. Infect. Dis.* **172:**1145–1149.
31. **Kirkpatrick, C. E., F. M. Moore, A. K. Patnaik, and H. E. Whiteley.** 1989. Argyrophilic, intracellular bacteria in some cats with idiopathic peripheral lymphadenopathy. *J. Comp. Pathol.* **101:**341–349.

32. **Koehler, J. E., C. A. Glaser, and J. W. Tappero.** 1994. *Rochalimaea henselae* infection: a new zoonosis with the domestic cat as reservoir. *J. Am. Med. Assoc.* **271:**531–535.

33. **Kordick, D. L., and E. B. Breitschwerdt.** 1995. Intraerythrocytic presence of *Bartonella henselae. J. Clin. Microbiol.* **33:**1655–1656.

34. **Kordick, D. L., and E. B. Breitschwerdt.** 1997. Relapsing bacteremia after blood transmission of *Bartonella henselae* to cats. *Am. J. Vet. Res.* **58:**492–497.

35. **Kordick, D. L., and E. B. Breitschwerdt.** 1998. Persistent infection of pets within a household with three *Bartonella* species. *Emerg. Infect. Dis.* **4:**325–328.

36. **Kordick, S. K., E. B. Breitschwerdt, B. C. Hegarty, K. L. Southwick, C. M. Colitz, S. I. Hancock, J. M. Bradley, R. Rumbough, J. T. McPherson, and J. N. MacCormack.** 1999. Coinfection with multiple tickborne pathogens in a Walker Hound kennel in North Carolina. *J. Clin. Microbiol.* **37:**2631–2638.

37. **Kordick, D. L., T. T. Brown, K. Shin, and E. B. Breitschwerdt.** 1999. Clinical and pathologic evaluation of chronic *Bartonella henselae* or *Bartonella clarridgeiae* infection in cats. *J. Clin. Microbiol.* **37:**1536–1547.

38. **Kordick, D. L., E. J. Hilyard, T. L. Hadfield, K. H. Wilson, A. G. Steigerwalt, D. J. Brenner, and E. B. Breitschwerdt.** 1997. *Bartonella clarridgeiae,* a newly recognized zoonotic pathogen causing inoculation papule, fever, and lymphadenopathy (cat scratch disease). *J. Clin. Microbiol.* **35:**1813–1818.

39. **Kordick, D. L., M. G. Papich, and E. B. Breitschwerdt.** 1997. Efficacy of enrofloxacin or doxycycline for treatment of *Bartonella henselae* or *Bartonella clarridgeiae* infection in cats. *Antimicrob. Agents Chemother.* **41:**2448–2455.

40. **Kordick, D. L., B. Swaminathan, C. E. Greene, K. H. Wilson, A. M. Whitney, S. O'Connor, D. G. Hollis, G. M. Matar, A. G. Steigerwalt, G. B. Malcolm, P. S. Hayes, T. L. Hadffeld, E. B. Breitschwerdt, and D. J. Brenner.** 1996. *Bartonella vinsonii* subsp. *berkhoffii* subsp. nov., isolated from dogs: *Bartonella vinsonii* subsp. *vinsonii;* an emended description of *Bartonella vinsonii. Int. J. Syst. Bacteriol.* **46:**704–709.

41. **Kordick, D. L., K. H. Wilson, D. J. Sexton, T. L. Hadfield, H. A. Berkhoff, and E. B. Breitschwerdt.** 1995. Prolonged *Bartonella* bacteremia in cats associated with cat-scratch disease patients. *J. Clin. Microbiol.* **33:**3245–3251.

42. **Kosoy, M. Y., R. L. Regnery, T. Tzianabos, E. L. Marston, D. C. Jones, D. Green, G. O. Maupin, J. G. Olson, and J. E. Childs.** 1997. Distribution, diversity, and host specificity of *Bartonella* in rodents from the southeastern United States. *Am. J. Trop. Med. Hyg.* **57:**578–588.

43. **Lappin, M. R., and J. C. Black.** 1999. *Bartonella* spp. infection as a possible cause of uveitis in a cat. *J. Am. Vet. Med. Assoc.* **214:**1205–1207.

44. **Maurin, M. R., R. J. Birtles, and D. Raoult.** 1997. Current knowledge of *Bartonella* species. *Eur. J. Clin. Microbiol. Infect. Dis.* **16:**487–506.

45. **Mehock, J. R., C. E. Greene, F. C. Gherardini, T. W. Hahn, and D. C. Krause.** 1998. *Bartonella henselae* invasion of feline erythrocytes in vitro. *Infect. Immun.* **66:**3462–3466.

46. **Mgeno, N., H. Oda, K. Yoshiie, M. R. Wahid, T. Fujimura, and S. Matayoshi.** 1999. Live *Bartonella henselae* enhances endothelial cell proliferation without direct contact. *Microb. Pathog.* **27:**419–427.

47. **Minnick, M. F.** 1996. Cell entry and the pathogenesis of *Bartonella* infections. *Trends Microbiol.* **9:**343–347.

48. **Mitchell, S. J., and M. F. Minnick.** 1995. Characterization of a two-gene locus from *Bartonella bacilliformis* associated with the ability to invade human erythrocytes. *Infect. Immun.* **63:**1552–1562.

49. **O'Reilly, K. L., R. W. Bauer, R. L. Freeland, L. D. Foil, K. J. Huges, K. R. Rohde, A. F. Roy, R. W. Stout, and P. Triche.** 1999. Acute clinical disease in cats following infection with a pathogenic strain of *Bartonella henselae* (LSU 16). *Infect. Immun.* **67:**3066–3072.

50. **Pappalardo, B. L., M. T. Correa, C. C. York, C. Y. Peat, and E. B. Breitschwerdt.** 1997. Epidemiologic evaluation of the risk factors associated with exposure and seroreactivity to *Bartonella vinsonii* in dogs. *Am. J. Vet. Res.* **58:**467–471.
51. **Regnery, R. L., B. E. Anderson, J. E. Clarridge, M. C. Rodriguez-Barradas, D. C. Jones, and J. H. Carr.** 1992. Characterization of a novel *Rochalimaea* species, *R. henselae* sp. nov., isolated from blood of a febrile, human immunodeficiency virus-positive patient. *J. Clin. Microbiol.* **30:**265–274.
52. **Regnery, R. L., J. A. Rooney, A. M. Johnson, S. L. Nesby, P. Manzewitsch, K. Beaver, and J. G. Olson.** 1996. Experimentally induced *Bartonella henselae* infections followed by challenge exposure and antimicrobial therapy in cats. *Am. J. Vet. Res.* **57:**171–179.
53. **Sander, A., C. Buhler, K. Pelz, E. von Cramm, and W. Bredt.** 1997. Detection and identification of two *Bartonella henselae* variants in domestic cats in Germany. *J. Clin. Microbiol.* **35:**584–587.
54. **Sander, A., M. Ruess, S. Bereswill, M. Schuppler, and B. Steinbrueckner.** 1998. Comparison of different DNA fingerprinting techniques for molecular typing of *Bartonella henselae* isolates. *J. Clin. Microbiol.* **36:**2973–2981.
55. **Sobraques, M., M. Maurin, R. J. Birtles, and D. Raoult.** 1999. In vitro susceptibilities of four *Bartonella bacilliformis* strains to 30 antibiotic compounds. *Antimicrob. Agents Chemother.* **43:**2090–2092.
56. **Ueno, H., T. Hohdatsu, Y. Muramatsu, H. Koyama, and C. Morita.** 1996. Does coinfection of *B. henselae* and FIV induce clinical disorders in cats? *Microbiol. Immunol.* **40:**617–620.
57. **Welch, D. F., K. C. Carroll, E. K. Hofmeister, D. H. Persing, D. A. Robison, A. G. Steigerwalt, and D. J. Brenner.** 1999. Isolation of a new subspecies, *Bartonella vinsonii* subsp. *arupensis,* from a cattle rancher: identity with isolates found in conjunction with *Borrelia burgdorferi* and *Babesia microti* among naturally infected mice. *J. Clin. Microbiol.* **37:**2598–2601.
58. **Yamamoto, K., B. B. Chomel, R. W. Kasten, C.C. Chang, T. Tseggai, P. R. Decker, M. Mackowiak, K. A. Floyd-Hawkins, and N. C. Pedersen.** 1998. Homologous protection but lack of heterologous protection by various species and types of *Bartonella* in specific pathogen free cats. *Vet. Immunol. Immunopathol.* **65:**191–204.
59. **Zangwill, K. M., D. H. Hamilton, B. A. Perkins, R. L. Regnery, B. D. Plikaytis, J. L. Hadler, M. L. Carter, and J. D. Wenger.** 1993. Cat scratch disease in Connecticut: epidemiology, risk factors, and evaluation of a new diagnostic test. *N. Engl. J. Med.* **329:**8–13.

Emerging Diseases of Animals
Edited by C. Brown and C. Bolin

Chapter 13

Plague

James R. Swearengen and Patricia L. Worsham

HISTORICAL SIGNIFICANCE

During these times there was a pestilence by which the whole human race came near to being annihilated.

Procopius, A.D. 542 (48)

In the year of our Lord 1348 there occurred in the city and contado of Florence a great pestilence and such was its fury and violence that in whatever household it took hold, whosoever took care of the sick, all the carers died of the same illness, and almost nobody survived beyond the fourth day, neither doctors nor medicine proving to any avail. . . . those symptoms were as follows: either between the thigh and the body, in the groin region, or under the armpit, there appeared a lump, and a sudden fever, and when the victim spat, he spat blood mixed with saliva, and none of those who spat blood survived. Such was the terror this caused that seeing it take hold in a household, as soon as it started, nobody remained: everybody abandoned the dwelling in fear, and fled to another; some fled into the city and others into the countryside. . . . sons abandoned fathers, husbands wives, wives husbands, one brother the other, one sister the other. The city was reduced to bearing the dead to burial.

Marchionne di Coppo di Stefano Buonaiuti, 1327–1385 (22a)

James R. Swearengen • Veterinary Medicine Division, United States Army Medical Research Institute of Infectious Diseases, Fort Detrick, MD 21702-5011. **Patricia L. Worsham** • Bacteriology Division, United States Army Medical Research Institute of Infectious Diseases, Fort Detrick, MD 21702-5011.

More than any other infectious disease, the specter of plague has evoked terror throughout recorded history. Until modern times, the sources and transmission of this rapidly spreading, fulminating disease were unknown and often attributed to the supernatural. Only a little more than 100 years ago was the causative agent identified and the link between rodents, their fleas, and human disease described.

The first documented plague pandemic, the Justinian plague, began in the busy port of Pelusium in Egypt during the sixth century, ultimately spreading to Mediterranean Europe and Asia Minor. The number of lives lost during this pandemic cannot be accurately determined, but there have been estimates of 15 to 40% mortality for a particular epidemic or location (45). The historian Procopius wrote that, at the height of the epidemic in Byzantium, more than 10,000 deaths occurred each day (35, 48).

The second pandemic, originating in central Asia in the early 14th century, spread along trade routes from China, eventually encompassing the Mediterranean basin, the Middle East, and most of Europe. Genoese trade ships infested with plague-infected rats carried the disease from Kaffa, a port city on the Black Sea, to Italy and Constantinople (41). The first European epidemic, which began in Messina in 1347, is thought to have killed approximately 30 to 40% of the European population and became known as the Great Dying, the Great Pestilence, and, more recently, the Black Death (41, 45). For hundreds of years, this second pandemic ravaged Europe; epidemics continued late into the 17th century, having far-reaching effects on European culture and economy. The loss of a large portion of the labor pool and the disruption in trade generated economic instability and discord between social classes. Some felt that the epidemic was a punishment from God and scourged themselves in public places as penance (flagellants) or sought protection from "plague saints" such as St. Roch and St. Sebastian. Jews were often thought to be responsible for spreading the plague and were widely persecuted and slaughtered (43). Long-lasting effects on European religion, politics, art, and language resulted from the continuing waves of plague during the second pandemic. The death of so many ecclesiastics and the failure of the religious authorities to explain the origins of the disease led some to a loss of faith in the clergy; this is thought to have contributed to Luther's success. It has also been suggested that the tremendous population loss suffered by Spain during the 16th and 17th century plague outbreaks contributed to its decline as a world power. Even the rise of written languages other than Latin was likely due in part to the demise of individuals who maintained the knowledge of the ancient language (43). One of the most notable effects of the plague was its influence on the development of modern medicine. Controversies over the origin of plague eventually led physicians away from the prevalent theories of the role of miasma or "corrupted air" in disease toward the modern concept of contagion. Quarantine of ships and goods became common during the second pandemic, as did the practices of isolating the sick and disinfecting their possessions (35, 43).

The current (modern or third) pandemic most likely began in China, reaching Hong Kong and other Asian ports in 1894. In just a few years, plague had been disseminated via rat-infested steamships to ports worldwide, leading to an estimated 26 million plague cases and 12 million deaths during the first 35 years of the pandemic (21). India was particularly hard hit during this period. It was during the early years of the modern pandemic that *Yersinia pestis* was introduced to new locations in North and South America, southern Africa, Australia, the Philippines, and Japan (21).

ORGANISM

Taxonomy

Y. pestis, the causative agent of plague, is a gram-negative coccobacillus belonging to the family *Enterobacteriaceae.* The genus was named in honor of Alexandre Yersin, the scientist who originally isolated *Y. pestis* during a plague outbreak in Hong Kong in 1894; the species name *pestis* is derived from the Latin for plague or pestilence. Previous designations for this species have included *Bacterium pestis, Bacillus pestis, Pasteurella pestis,* and *Pesticella pestis* (6). This species is closely related to two other pathogens of the genus *Yersinia, Y. pseudotuberculosis* and *Y. enterocolitica.* Recent studies suggest that *Y. pestis* evolved from *Y. pseudotuberculosis* 1,500 to 20,000 years ago (2). The extensive genetic similarity (>90%) between *Y. pseudotuberculosis* and *Y. pestis* led to a recommendation that *Y. pestis* be reclassified as a subspecies of *Y. pseudotuberculosis* (7). This proposal was not well received, primarily due to fear that this change in nomenclature would cause confusion and increase the potential for laboratory-acquired infections (61). Other members of the genus include a pathogen of rainbow trout, *Yersinia ruckeri,* and seven other Yersinia species: *Y. intermedia, Y. frederiksenii, Y. kristensenii, Y. aldovae, Y. rohdei, Y. mollaretii,* and *Y. bercovieri.* Other than *Y. ruckeri,* these species are primarily environmental isolates, although some have been isolated from sick and healthy animals, including humans (3).

Morphology

Use of Wayson's or Giemsa stain for *Y. pestis* typically reveals a short bacillus (0.5 to 0.8 μm by 1.0 to 3.0 μm) with a characteristic "safety pin" bipolar staining pattern. Depending on growth conditions, this species can exhibit marked pleomorphism, with rods, ovoid cells, and short chains present. A gelatinous envelope, known as the F1 capsular antigen, is produced by the vast majority of strains at a growth temperature of 37°C. Unlike the other mammalian pathogens of the genus, which produce peritrichous flagella at growth temperatures of <30°C, *Y. pestis* is nonmotile (3, 6).

Growth Characteristics

Y. pestis is capable of growth at a broad range of temperatures in the laboratory (4 to 40°C), with an optimal growth temperature of 28°C. Although it grows well on standard laboratory media such as sheep blood agar, MacConkey agar, and heart infusion agar, growth is slower than that of *Y. pseudotuberculosis* or *Y. enterocolitica;* more than 24 h of incubation is required to visualize even pinpoint colonies. The appearance of colonies can be hastened by growth in an environment containing 5% CO_2. The round, moist, translucent, or opaque colonies are nonhemolytic on sheep blood agar and exhibit an irregular edge. A fried-egg appearance is common in older colonies and is more pronounced in certain strains (3, 6, 45).

Biochemistry

Y. pestis is a facultative anaerobe, fermenting glucose with the production of acid. An obligate pathogen, it is incapable of a long-term saprophytic existence, due in part to complex nutritional requirements, including a number of amino acids and vitamins. It also lacks certain enzymes of intermediary metabolism that are functional in the closely related, more rapidly growing species such as *Y. enterocolitica* and *Y. pseudotuberculosis. Y. pestis* strains have traditionally been separated into three biovars, based on the ability to ferment glycerol and reduce nitrate (45). Molecular methods of typing, such as ribotyping and restriction fragment length polymorphisms of insertion sequence locations, support this division of strains (2, 29). Biovar orientalis (Gly^- Nit^+) is distributed worldwide and is responsible for the third (modern) plague pandemic. It is the only biovar present in North and South America. Biovar antiqua (Gly^+ Nit^+) is found in Central Asia and Africa and may represent the most ancient of the biovars. Biovar mediaevalis (Gly^+ Nit^-) is geographically limited to the region surrounding the Caspian Sea. There are no apparent differences in pathogenicity between the biovars (45).

Genetics

Most strains of *Y. pestis* carry three plasmids: pMT (or pFra), which encodes capsule and murine toxin; pCD, which is responsible for the synthesis of a number of antihost factors, and pPCP, which carries the gene for the virulence factor plasminogen activator (45). Two of these plasmids, pMT and pPCP, are unique to *Y. pestis;* the pCD plasmid is shared by *Y. pseudotuberculosis* and *Y. enterocolitica.* Often the term pYV, for "*Yersinia* virulence," is also used for the pCD plasmid. Long-term laboratory passage of *Y. pestis* or short-term growth under less than optimal conditions is associated with irreversible genetic changes leading to attenuation. These changes include the deletion of a large chromosomal pathogenicity island that encodes factors necessary for growth in both the flea and the mammalian host and the loss of one or more virulence plasmids (33, 45).

Much work has been done in mapping the genetic code of this organism; the genome of *Y. pestis* strain CO92 biovar orientalis has been sequenced by the Sanger Centre and is available on the World Wide Web (http://www.tigr.org/).

EPIDEMIOLOGY

The life cycle of plague is maintained in complex enzootic and epizootic episodes that involve numerous rodent reservoirs and flea vectors throughout the world. While more than 200 mammalian species have been reported to be naturally infected with *Y. pestis,* rodents are the primary reservoir hosts that ensure the ultimate survival of the organism. Humans are accidental hosts, and although they are not important for the continued existence of *Y. pestis* in nature, the long history of plague pandemics is evidence of the impact of this disease on human populations. During the course of human history, pandemics of plague have occurred under two geographic settings, classified as either urban or sylvatic. Urban plague is transmitted from rodents typically found in population centers, primarily rats, and sylvatic plague is transmitted from a variety of wild rodents in rural environments (45). There is growing concern regarding the role of tree squirrels in urban plague due to identification of *Y. pestis* in these rodents in cities in Colorado and Wyoming and most recently in 1993 in Dallas, Tex. (11). Sylvatic plague is currently the predominant form of plague found in the United States, with the last known urban outbreak occurring in 1924 in Los Angeles, Calif. (39).

Plague occurs worldwide, and human cases are reported in North America, South America, Africa, Asia, and the very southeastern portion of Europe. There were a total of 18,739 cases of human plague reported to the World Health Organization (WHO) from 1980 through 1994 (average, 1,087 cases per year) from 20 countries (13). Plague foci are present in much of Africa, especially in the southern portion of the continent and in the eastern portion south of Uganda. Recent plague outbreaks have occurred in the southern African countries of Madagascar, Namibia, Mozambique, Botswana, and Malawi, as well as in the east-central African countries of Uganda, the Democratic Republic of the Congo (Zaire), Kenya, and Tanzania. Currently, plague in Africa accounts for the majority of human cases worldwide. Foci also extend from the Caucasus Mountains in the southeastern portion of Europe through much of the Middle East, China, Southeast Asia, and Indonesia; sizeable human outbreaks have occurred in Vietnam, India, China, and Myanmar (Burma). Active plague foci in South America include Brazil and the Andes Mountain regions of Bolivia, Peru, and Ecuador. Plague is not present in Australia or western Europe (21). Since arriving on the North American continent in San Francisco at the turn of the 20th century, *Y. pestis* has spread via rodent communities through the western third of the United States, with human cases occurring as far east as Texas and Oklahoma. Surveillance of rodent populations indicates that *Y. pestis* is also present in southwestern Canada and northwestern Mexico. In the United States, an average of 10 to 15 human cases are reported to the U.S. Centers for Disease Control and Prevention (CDC) each year (21). Of the 350 cases reported in the United States from 1970 to 1997, approximately 80% were reported from the southwestern states of New Mexico, Arizona, and Colorado, 9% were reported from California, and nine other western states reported limited numbers of cases (14).

In the United States, plague is maintained in nature through enzootic and epizootic transmission cycles that involve wild mice (*Persomyscus* spp. and *Microtus* spp.) as the primary reservoir populations and other species of wild rodents (ground squirrels, prairie dogs, woodrats, chipmunks, marmots, brown rats, and tree squirrels) as amplifying hosts of the organism (39, 42). The maintenance of the organism in nature depends on both disease-resistant and disease-susceptible rodent populations. Resistant populations of rodents allow the enzootic cycle to exist, which provides a source of infection for fleas that transmit the disease to susceptible populations of rodents in the area, resulting in an epizootic. During an epizootic, susceptible rodents typically develop bacteremia and suffer a high mortality rate, resulting in a large population of infected fleas in search of other hosts. Most human cases of plague in the United States occur near a rodent epizootic during the summer months, when the risk of exposure to fleas is the greatest. These rodent epizootics occur in 4- to 5-year cycles which directly correlate to the number of human cases reported to the CDC each year. In 341 cases of human plague reported between 1970 and 1995 in the United States, flea bites accounted for 78% of the 284 cases for which the mode of transmission had been determined, and direct contact with infected animals and inhalation of infectious organisms accounted for 20 and 2% of the cases, respectively (13). Other animals may also have limited involvement in the transmission of *Y. pestis,* but with much less impact on human disease. Seropositive rabbits and wild boars have been identified as potential reservoir hosts in western Texas and southern California, respectively (31). Although carnivores such as coyotes, black bears, raccoons, badgers, skunks, mountain lions, bobcats, and dogs are accidental hosts capable of becoming infected, they are relatively resistant to disease and do not appear be a source of infection for feeding fleas (39). The domestic cat has become an epidemiological concern in the United States since the first identified case of human plague transmitted from a cat was described in 1977 (44). Cats were identified as the source of 19 human cases of plague from 1977 through 1997, with 5 of the cases resulting in death (21). Unlike most other carnivores, cats are susceptible to clinical disease, and infected animals are highly capable of directly transmitting the disease to humans (16). While the primary mode of transmission from rodents to humans is from the flea bite, approximately 75% of human plague cases transmitted from cats result from scratches, bites, or direct contact with infectious body fluids, resulting in the bubonic and less frequently the septicemic form of the disease. Probably more alarming is the fact that approximately 25% of these cases resulted from the inhalation of infectious respiratory droplets or other airborne materials leading to pneumonic plague, which is the most dangerous and fatal form of the disease (39).

The increasing popularity of the domestic ferret (*Mustela putorius furo*) as a pet might also raise concern about the rise of another potentially susceptible reservoir that has a close association with human households. This concern is tempered by the results of one study that involved eight domestic ferrets experimentally infected with *Y. pestis.* This study showed that domestic ferrets developed high serum antibody titers and showed no clinical signs of disease; *Y. pestis* could not be detected in tissues 21 days postinfection (58). The results indicate that domestic ferrets are resistant to plague and are probably not a significant concern for direct transmission of the disease to humans, although there has been a single report of plague in a black-footed ferret since the report of the study in domestic ferrets (59). Interestingly, the study was not performed to evaluate the potential for domestic ferrets to transmit plague to humans, but rather to assess the potential susceptibility of the nearly extinct black-footed ferret to *Y. pestis* with the domestic ferret and the Siberian polecat (*Mustela eversmanni*) as surrogate animal models.

Even though more than 150 species of fleas have been found to be naturally infected with *Y. pestis*, only a relatively few species play a significant role in the transmission of plague. The oriental rat flea (*Xenopsylla cheopis*) has been the primary vector involved in the long history of plague pandemics and is currently the most important vector of plague in all worldwide geographic foci of infection other than the United States (20). In the United States, a complex of rodent-flea associations are responsible for transmission of plague to humans during or shortly after rodent epizootics. The most treacherous vector is the rock squirrel flea (*Oropsylla montana*) due to its habits of promiscuous feeding on various ground squirrels, such as the rock squirrel (*Spermophilus variegatus*) and the California ground squirrel (*Spermophilus beechyi*), and its tendency to readily bite humans (20). Other identified rodent-flea associations include the prairie dog (*Cynomys* spp.) and *Opisochrostis hirsutus*, and Richardson's and the golden-mantled ground squirrels (*Spermophilus richardsoni* and *Spermophilus lateralis*, respectively) and *Opisochrostis labis*, *Opisochrostis idahoensis*, or *Thrassus bacchi* (42). While it has been proven that the oriental rat flea is an efficient vector of plague, this is not the case for most other species of fleas. The ability of other fleas to transmit *Y. pestis* through their bite appears to vary considerably due to differing abilities among species to become infected and transmit the disease before dying. One critical factor that affects the ability of a flea to transmit plague is the blocking phenomenon. When fleas feed on a bacteremic animal, the organism is taken with the blood meal into the midgut of the flea, where it multiplies, eventually forming a mass of aggregated bacteria that blocks the proventriculus, a valvelike structure leading to the midgut. This blockage starves the flea, which then makes repeated desperate attempts to feed. Because of the blockage, blood carrying *Y. pestis* is regurgitated into the bite wounds, thus spreading the disease to new hosts. The blocked flea, also a victim of the disease, eventually starves to death (33). The size and structure of the proventriculus vary between species of fleas, and this can have a significant impact on their ability to transmit plague. For example, it has been shown that the inability of the cat flea (*Ctenocephalides felis*) foregut to become blocked is the reason why it is a very poor vector of *Y. pestis* (45).

PATHOGENESIS

The persistence of plague in certain areas requires cyclic transmission between rodents and fleas; thus, *Y. pestis* has evolved to survive and replicate in two very different hosts. To maintain the transmission cycle, *Y. pestis* must multiply within the flea sufficiently to cause blockage and promote the infection of a new mammalian host. Equally critical is the ability to establish an infection and induce sufficient bacteremia in the mammal to infect fleas during the blood meal.

The milieu of the mammalian host is radically different from that of the midgut of the flea, yet clearly the organism successfully adapts to each host to complete its life cycle. The adaptation takes place through environmental regulation of virulence factors. For example, gene products necessary for growth in the flea are expressed most efficiently at the ambient body temperature of this host; presumably additional factors also cue the organism to recognize this environment and respond appropriately. Likewise, genes required for replication in the mammalian host are expressed at their highest levels at 37°C, and the synthesis of some proteins thought to be induced in the phagolysosome is also regulated by pH. Recent genetic analyses of *Y. pestis* and the other pathogenic yersiniae have begun to unravel the unique qualities that make *Y. pestis* a successful pathogen in both the flea and the mammalian host. For example, vector-borne transmission is now known to depend on a temperature-regulated, plasmid-borne phospholipase D activity that is essential for colonization of the flea midgut (34). Another protein, also synthesized most efficiently at 28°C, promotes bacterial aggregation, allowing the organism to persist in the proventriculus and establish the blockage essential for transmission of the disease to a mammalian host (33).

After the introduction of *Y. pestis* into the mammalian host, many of the organisms are killed by host phagocytic cells, but some organisms survive and reproduce within tissue macrophages. *Y. pestis* contains a variety of virulence determinants that are activated by certain signals once inside a vertebrate host, including increased temperature and contact with eukaryotic cells, synthesis of which is required for the organism to be fully virulent (45). Once these virulence determinants are expressed, the organism is capable of evading phagocytosis and can readily multiply extracellularly. Like a number of other gram-negative pathogens, *Y. pestis* possesses a potent antihost weapon, known as a type III secretion system, that enables an organism in close contact with host cells to deliver toxic proteins directly into the eukaryotic cell cytosol. Toxic activities against host phagocytic cells include disruption of the cytoskeleton, interference with phagocytic activity, prevention of proinflammatory cytokine synthsis, inhibition of the oxidative burst, induction of programmed cell death (apoptosis), and inhibition of neutrophil chemotaxis (17, 38, 57). A gelatinous envelope, known as the F1 capsule, is thought to protect the organism from host phagocytic cells by impeding complement-mediated opsonization (61). The F1 capsule is synthesized in large quantities in the mammalian host and when cultured in the laboratory at 37°C. Although the vast majority of natural isolates produce the capsular antigen, F1-negative strains have been isolated from rodent hosts in nature, from laboratory rodents, and reportedly from one human case (4, 55, 63). Some strains lacking F1 are virulent in animal models (18, 64); however, the fact that F1-negative strains are relatively rare in natural isolates suggests that the capsular antigen may play an important role in the maintenance of the disease in certain animal reservoirs. Historically, F1 has been of immense importance as a diagnostic reagent, as it is specific to *Y. pestis.* It is the major antigen recognized in convalescent-phase sera of humans and rodents (5, 8). Another virulence factor, plasminogen activator (Pla), promotes dissemination of the organism from peripheral sites of infection (52). The protease activity generated by plasminogen may enhance dissemination by degrading host tissue barriers of extracellular matrix or cleavage of fibrin deposits. There is also evidence that Pla may enhance bacterial attachment to host extracellular matrices (37). Like F1, Pla is *Y. pestis* specific and has been studied as a target for PCR-based identification of *Y. pestis.* However, a few strains that are Pla negative appear to be fully virulent for mice and have been identified as natural isolates or generated in the laboratory (36, 56). Presumably, these isolates synthesize other proteins that substitute for Pla function.

In humans, plague is generally classified as bubonic, septicemic, or pneumonic. The incubation period for the bubonic form of plague is typically 2 to 6 days. During the incubation period, the bacilli spread to lymph nodes proximal to the site of cutaneous inoculation, resulting in a suppurative lymphadenitis. From 6 to 8 h after the onset of sudden fever, chills, headache, nausea, and vomiting, the affected lymph nodes become extremely painful and are referred to as buboes. Within 24 h the buboes become visible as the lymph nodes, skin, and other surrounding tissues become swollen and erythemic. At times, bubonic plague leads to a secondary septicemia with hematogenous spread to other organs, including the liver, spleen, lungs, and, less commonly, the meninges. Primary septicemia can occur and is not usually associated with a clinically apparent lymphadenitis; therefore, it is not typically detected until late in the course of the disease, when general clinical signs such as nausea, vomiting, diarrhea, and abdominal pain are present. This makes primary septicemic plague a very dangerous form of the disease for the patient, as it is frequently misdiagnosed due to the nonspecific clinical signs that lead to misdiagnosis, late treatment, and a relatively high fatality rate (28%) (20). Pneumonic plague can result from inhaling infectious organisms or can occur as a secondary complication of septicemic plague. Individuals with pneumonic plague are capable of transmitting the disease directly to other humans. The incubation period for primary pneumonic plague in humans is 3 to 5 days, and clinical signs seen at the onset include sudden chills, high fever, severe cough, and dyspnea (53). Sputum is frequently watery and frothy and occasionally bloody and has a high bacterial load. Without antibiotic treatment within 24 h of the onset of clinical signs, pneumonic plague is routinely fatal, with death the result of acute respiratory insufficiency and/or septicemic shock (40). The rapid progression of this form of plague is due to the presence of phagocytosis-resistant bacilli that are already expressing virulence determinants as a result of their previous infection of a vertebrate host immediately before inhalation (42). Pharyngeal plague, acquired by ingestion or inhalation of the organism, has also been described (46).

Infection of rodents with *Y. pestis* occurs primarily through flea bites, and although direct contact, ingestion, or inhalation might seem like logical routes of infection due to sharing of burrows by most rodents, there is no solid evidence that these are significant sources of infection in wild rodent populations. Experimental infection of guinea pigs with *Y. pestis* via flea bites exhibited a pathogenesis similar to that in humans, with enlargement of the draining lymph nodes, septicemia, and death at approximately 2 weeks postinfection (54). Similar findings have been seen in naturally infected rats and ground squirrels, and although most animals develop a bubonic form of the disease, it varies in degree of internal organ involvement and bacteremia (45). Plague infection in rodents may be acute, chronic, or inapparent, with lesions varying based on the course of the disease. Lesions found in disease-susceptible rodents that have died from acute infection include hemorrhagic buboes and splenomegaly, without gross pathologic changes to other internal organs. Buboes found in rodents that die from subacute infection are caseous instead of hemorrhagic, and punctiform necrotic foci are found in the liver, lungs, and spleen (1).

The pathogenesis of plague in the dog and cat varies considerably, and the differences have resulted in the cat's becoming an emerging, and dangerous, link between the usual rodent hosts in the United States and the human population. Naturally acquired infection of carnivores most likely results from the ingestion of infected rodents or from the bite of rodent flea vectors. Experimentally infected dogs developed a mild fever of short duration with no other clinical signs, but they did become bacteremic.

Investigators were able to culture *Y. pestis* from the oropharynx for 10 days (47, 49). The pathogenesis in naturally infected cats is an acute disease depicted by fever, lethargy, buboes, formation of abscesses, and occasionally pneumonia (16). A retrospective study of 119 cases of naturally acquired plague in cats in New Mexico from 1977 to 1988 found that 53% of the cases were the bubonic form only, 10% were pneumonic (one-third being secondary to the bubonic form), and 8% were confirmed as septicemic. Of the cats in this study that were diagnosed with bubonic plague, 75% had enlargement of the submandibular lymph nodes. Other lymph nodes that were less frequently involved included the cervical, external iliac, popliteal, prescapular, and axillary lymph nodes. Of the 11 cats that were necropsied, 3 were diagnosed as having bubonic plague, 4 with pneumonic plague, and 4 presumed to be septicemic. Gross pathologic findings for those cats with bubonic plague included emaciation and caseous lymphadenopathy with peripheral edema of the surrounding skin and tissues. Microscopically the abscesses showed some fibrosis around the periphery, but they did not appear well encapsulated. The animals with pneumonic plague exhibited focal lung lesions from one to several centimeters in diameter, and histologic examination of the lung lesions revealed indistinct borders and a dense infiltrate of neutrophils and small coccobacilli in the alveoli and small bronchioles. No gross or microscopic lesions that could be specifically attributed to plague were found in the cats with the septicemic form of the disease (24). Cats with experimental infection showed a mortality rate of 38% and displayed the same clinical course of disease as cats infected naturally (50).

REEMERGENCE

In addition to the long natural persistence of this global disease, several other factors indicate that we should be concerned about the future of plague and its impact on both animal and human populations. The ability of infectious organisms to adapt and the everchanging face of human demographics have allowed many diseases to emerge and reemerge despite scientific and medical advances that were unimaginable only a few decades ago. Evidence of these concerns is seen in the recent reappearance of plague epidemics in Madagascar. Human plague existed as an endemic disease in Madagascar for 30 years with only 20 to 30 cases annually, but since 1990, annual epidemics occur with more than 200 confirmed or presumptive cases reported each year (15). In addition to the elevation of plague to an epidemic status, a multidrug-resistant strain of *Y. pestis* has recently been identified in Madagascar and exhibits resistance to drugs commonly prescribed for plague, including streptomycin, tetracycline, and chloramphenicol. The plasmid containing the genetic material responsible for the antibiotic resistance is highly transferable in vitro to other strains of *Y. pestis* and to *Escherichia coli*, which causes obvious concern about spread through wild-type strains of *Y. pestis* (27). The concern regarding the adaptation of *Y. pestis* is furthered by the report of three new ribotypes of *Y. pestis* that tend to migrate geographically being identified in one of the most active plague foci in Madagascar. Although the recent epidemics in Madagascar are possibly due to hampering of the national plague control program by economic difficulties, there is still much concern about the adaptations seen in *Y. pestis* (15).

To this day, a report of plague evokes panic. In 1994, India reported its first cases of plague in 28 years. Within a month of the initial case, an epidemic of pneumonic plague had been reported in Surat with more than 300 unconfirmed cases of pneumonic plague and 36 deaths. One hundred thousand people fled to Delhi and other cities. Tetracycline supplies were exhausted by individuals stockpiling the antibiotic and/or taking it prophylactically. Hundreds of tons of insecticides, including DDT, were used, even in areas with no evidence of plague. India was quarantined internationally; as a result, airline flights, tourism, and trade were disrupted (22). In the United States, the CDC developed and implemented an enhanced surveillance system to supplement existing measures designed to prevent the importation of human plague in direct response to the epidemic in India (25). Ironically, a recent report suggests that, although earlier cases seen in August in the village of Mamla were bubonic plague, the well-publicized Surat illnesses in September that elicited so much terror were misdiagnosed and were, in reality, due to other diseases (22, 28).

Since 1944, there has been a gradual increase in the number of states in the United States reporting human plague cases, as well as in the number of states reporting plague in animals. Three states reported human cases of plague between 1944 and 1953 (21). As of 1997, 13 states have reported human cases of plague, four more states have reported plague in animal populations only, and Kansas reported the identification of *Y. pestis* in prairie dog fleas for the first time since 1950 (14). Most organisms that have animal hosts, environmental limitations, arthropod vectors, or complicated life cycles are typically less prone to migrate to other geographical areas or populations (62). Plague meets all of these criteria but still manages to continually diversify and spread to new geographical areas. The ability of plague to manifest in the pneumonic form allows the disease to bypass many of these limitations and creates additional concerns from an epidemiological perspective. The eastward movement of plague and the ubiquitous presence of susceptible rodent populations coupled with the rapid urban expansion occurring throughout the United States bring together all the components necessary for epidemics to occur.

DIAGNOSIS

A laboratory experienced in the identification of *Y. pestis* should perform diagnostic tests for plague. Suspect animals should be treated with an insecticide before specimen collection. The organism may be cultured from the blood by direct inoculation into tryptic soy broth; lymph node or abscess aspirates may be transferred to a sterile swab and transported in Cary-Blair medium. In humans, recovery from the blood is enhanced by the collection of a series of blood samples over 45 min. If attempts at aspiration of lymph nodes are unsuccessful, sterile saline can be injected into the central portion of the suspect lymph node and aspirated. Samples for culture at necropsy should include the spleen, liver, enlarged lymph nodes, abscesses, and lung. Fleas may also be submitted for culture. For bacteriological staining of samples, air-dried and heat-fixed slides should be prepared. Care should be taken to avoid the generation of aerosols when handling suspect material (3, 46). Bacterial filamentation of *Y. pestis* has been shown to occur in infected mice treated with β-lactam antibiotics such as ampicillin, ceftazimide, and aztreonam. This phenomenon could result in histological misdiagnosis of filamentous actinomycosis from a lymph node aspirate or sputum sample containing filamentous *Y. pestis* obtained from a person or animal treated with β-lactam antibiotics (19).

Identification of the organism by biochemical methods is time-consuming; furthermore, many automated identification systems fail to identify *Y. pestis* correctly (23). A rapid and accurate presumptive diagnosis of plague can made by use of a fluorescent antibody test to detect the plague-specific F1 capsular antigen. Because the F1 antigen is produced only at temperatures of >33°C, this method requires a relatively fresh sample from the animal or from a laboratory culture incubated at the appropriate temperature. Flea samples will be negative, as will be samples refrigerated for more than 30 h (3). The test is performed at some public health and veterinary diagnostic laboratories and by the CDC; air-dried slides should be submitted for fluorescent antibody testing. Lysis by a species-specific bacteriophage confirms the identification of *Y. pestis* (3, 46). Acute- and convalescent-phase (3 to 4 weeks after onset) sera should also be collected. The standard method of serodiagnosis is the passive hemagglutination assay or enzyme-linked immunosorbent assay with F1 antigen. A recent review of methods for the isolation and identification of *Y. pestis* from clinical samples and animals is available (3).

In the future, it may become more common to identify *Y. pestis* by genetic methods, such as PCR. Such assays have been promising in identifying experimentally infected fleas and animals and can be used to detect *Y. pestis* when cultures and serum are not available (32). The use of more than one plague-specific primer set in diagnostic PCRs will allow the identification of rare genetic variants such as F1-negative and Pla-negative strains as well as wild-type strains. A rabbit polyclonal antiserum has also been developed that identifies both F1-positive and F1-negative strains of *Y. pestis,* although there is some cross-reactivity with *Y. pseudotuberculosis* (J. W. Ezzell, personal communication).

Suspect samples may be submitted to the CDC, Fort Collins, CO 80522. For consultation and assistance in diagnostic testing, the CDC should be contacted by telephone (970-221-6400) or fax (970-221-6476).

Investigations of human disease outbreaks in areas where plague is endemic can be enhanced by performing serologic tests of domestic dogs and cats located in the vicinity of plague victims, as they develop antibody titers to plague that are detectable for several months postinfection. Routine serologic tests of domestic dogs on military bases and veterinary clinics in plague-endemic areas has not proven an effective means of monitoring, but testing free-roaming dogs on Native American reservations was considered a good indication of the exposure risk for humans (12, 16). This difference is most likely due to the tendency of free-roaming dogs to hunt for food and ingest rodents. Serologic tests of wild carnivores, including coyotes and black bears, and of wild boars are used as a monitoring system for plague surveillance. Coyotes are the preferable wild carnivore for surveillance because they cover a large area and are present even in semiurban areas (51).

PREVENTION AND CONTROL

The avoidance of plague in both humans and domestic pets requires that several preventive measures be carried out simultaneously. In addition to the education of health care officials and the inhabitants of areas where plague is endemic, all pet owners should maintain an effective flea control program for their pets and prevent them from roaming freely (10). Although the common dog and cat fleas (*C. canis* and *C. felis,* respectively) are not good vectors for *Y. pestis,* carnivores are capable of transporting rodent fleas from one area to another (26). Areas around residences that attract plague-susceptible rodents, such as debris piles and other potential harborage, should be cleared, and garbage should be well contained. If burrows of susceptible rodents are located near population centers, they should be treated with insecticides, ensuring that all applicable environmental laws and regulations regarding their use are followed. Rodenticide use should be authorized and monitored by public health officials, and rodenticides should only be used in conjunction with insecticides. The use of rodenticides alone in an area where plague is enzootic could result in massive migration of infected fleas from dead rodents to other more highly susceptible rodents and to pet or human populations (20). Veterinarians located in areas where plague is enzootic should practice preventive measures with all feline patients that present with clinical signs of plague. Cats with suspected or confirmed infection should be isolated immediately, contaminated surfaces should be promptly disinfected, and all persons maintaining contact with the animal should wear appropriate protective clothing and both eye and respiratory protection (49).

Suspicion or confirmation of the pneumonic form of the disease necessitates extreme adherence to respiratory protective measures. The standard surgical masks used in most veterinary hospitals do not afford adequate respiratory protection. In addition to half- and full-face respirators with high-efficiency particulate air filters, disposable masks approved by the National Institute of Occupational Safety and Health for use with tuberculosis patients would provide adequate protection from aerosol transmission of *Y. pestis* (R. J. Hawley, personal communication). Treatment of cats has been shown to be effective for all three forms of plague, with the percentage of cats that survive with and without treatment being 91 and 24%, respectively (24). Tetracycline, streptomycin, gentamicin, trimethoprim-sulfamethoxazole, and doxycycline have all been used in treating human cases of plague. Tetracycline at a dose of 25 mg/kg of body weight every 8 h is recommended for use in infected cats not showing clinical signs (24). Human dosages for other antibiotics recommended for various forms of plague are available for consideration for use in animals, and it is recommended that treatment of plague in cats be continued for at least 5 days after the animal's body temperature has returned to normal (47). Recent studies have identified several antibiotics that are effective in treating experimentally induced pneumonic plague in mice. Treatment with gentamicin, netilmicin, ciprofloxacin, and ofloxacin, both early and late in the course of disease, was as effective as treatment with streptomycin, the historical gold standard for treating plague in humans. The same study showed that β-lactam antibiotics (ampicillin, ceftriaxone, cefotetan, cefazolin, ceftazidime, and aztreonam) were not effective in treating pneumonic plague when given late in the course of disease with this model, and most actually resulted in an earlier death endpoint than the saline control (9). This apparent acceleration of the disease process in mice calls into question the use of β-lactam antibiotics for the treatment of pneumonic plague.

A formalin-inactivated plague vaccine previously approved for use in the United States was no longer available at the time this chapter was written. This was a whole-cell vaccine with the F1 capsule being the major protective antigen (30). When available, the vaccine was only approved for use in high-risk individuals, and there was no evidence that the vaccine protected against the respiratory route of exposure to aerosolized *Y. pestis* or infectious droplets. The vaccine was not recommended as a routine precaution for persons living in areas where plague is enzootic (13). Vaccines for plague are currently available in England, Australia, and the former Soviet Union.

SUMMARY

Plague is a worldwide disease that is not expected to be eradicated in the near future. The tremendous rodent reservoir and vector potential indigenous to all parts of the world make control of this disease a true challenge to public health personnel. Recent events in the adaptive traits of *Y. pestis* call for additional concern and caution. Education will be a key component of any successful endeavor to prevent serious epidemics from occurring, especially in light of the ease with which the world's population can move from one continent to another. *Y. pestis* has also been propagated for use as a biological threat agent that could be used for biological warfare or bioterrorism (42). This unfortunate possibility brings with it a great need for professional education. In the event of a bioterrorism incident involving *Y. pestis,* it will be critical that human health care first-responders, veterinarians, and public health officials be familiar with the normal geographic distribution, clinical presentation, and diagnosis of the disease in both humans and animals. As with many infectious diseases, early diagnosis and treatment are critical in saving lives. Continuing research will be an essential part of developing more effective protection from both natural outbreaks and potential man-made biological events.

REFERENCES

1. **Acha, P. N., and B. Szyfres.** 1989. Plague, p. 131–140. *In Zoonoses and Communicable Diseases Common to Man and Animals,* 2nd ed. Pan American Health Organization, Washington, D.C.

2. **Achtman, M., K. Zurth, G. Morelli, G. Torrea, A. Guiyoule, and E. Carniel.** 1999. *Yersinia pestis,* the cause of plague, is a recently emerged clone of *Yersinia pseudotuberculosis. Proc. Natl. Acad. Sci. USA* **96:**14043–14048.
3. **Aleksic, S., and J. Bockemuhl.** 1999. *Yersinia* and other *Enterobacteriaceae,* p. 483–496. *In* P. R. Murray, E. J. Baron, M. A. Pfaller, F. C. Tenover, and R. H. Yolken (ed.), *Manual of Clinical Microbiology,* 7th ed. American Society for Microbiology, Washington, D.C.
4. **Anderson, G. W., P. L. Worsham, C. R. Bolt, G. P. Andrews, S. L. Welkos, A. M. Friedlander, and J. P. Burans.** 1997. Protection of mice from fatal bubonic and pneumonic plague by passive immunization with monoclonal antibodies against the F1 protein of *Yersinia pestis. Am. J. Trop. Med. Hyg.* **56:**471–473.
5. **Benner, G. E., G. P. Andrews, W. R. Byrne, S. D. Strachan, A. K. Sample, D. G. Heath, and A. M. Friedlander.** 1999. Immune response to *Yersinia* outer proteins and other *Yersinia pestis* antigens after experimental plague infection in mice. *Infect. Immun.* **67:**1922–1928.
6. **Bercovier, H., and H. M. Mollaret.** 1984. Yersinia, p. 498–503. *In* N. R. Krieg and J. G. Holt (ed.), *Bergey's Manual of Systematic Bacteriology,* vol. 1. Williams and Wilkins, Baltimore, Md.
7. **Bercovier, H., H. H. Mollaret, J. M. Alonsa, J. Brault, G. R. Fanning, A. G. Steigerwalt, and D. J. Brenner.** 1980. Intra- and interspecies relatedness of *Yersinia pestis* by DNA hybridization and its relationship to *Yersinia pseudotuberculosis. Curr. Microbiol.* **4:**225–229.
8. **Butler, T., and B. W. Hudson.** 1977. The serological response to *Yersinia pestis* infection. *Bull. W. H. O.* **55:**39–42.
9. **Byrne, W. R., S. L. Welkos, M. L. Pitt, K. J. Davis, R. P. Brueckner, J. W. Ezzell, G. O. Nelson, J. R. Vaccaro, L. C. Battersby, and A. M. Friedlander.** 1998. Antibiotic treatment of experimental pneumonic plague in mice. *Antimicrob. Agents. Chemother.* **42:**675–681.
10. **Centers for Disease Control and Prevention.** 1997. Fatal human plague—Arizona and Colorado, 1996. *Morb. Mortal. Wkly. Rep.* **46:**617–620.
11. **Centers for Disease Control and Prevention.** 1994. Human plague—United States, 1993–1994, surveillance summary. *Morb. Mortal. Wkly. Rep.* **43:**242–246.
12. **Centers for Disease Control and Prevention.** 1988. Plague in American Indians, 1956–1987. *Morb. Mortal. Wkly. Rep.* **37**(SS-3)**:**11–16.
13. **Centers for Disease Control and Prevention.** 1996. Prevention of plague: recommendations of the Advisory Committee on Immunization Practices (ACIP). *Morb. Mortal. Wkly. Rep.* **45**(RR-14)**:**1–13.
14. **Centers for Disease Control and Prevention.** 1998. Summary of notifiable diseases, United States 1997. *Morb. Mortal. Wkly. Rep.* **46:**1–88.
15. **Chanteau, S., L. Ratsifasoamanana, B. Rasoamanana, L. Rahalison, J. Randriambelosoa, J. Roux, and D. Rabeson.** 1998. Plague, a reemerging disease in Madagascar. *Emerg. Infect. Dis.* **4:**101–104.
16. **Chomel, B. B., M. T. Jay, C. R. Smith, P. H. Kass, C. P. Ryan, and L. R. Barrett.** 1994. Serological surveillance of plague in dogs and cats, California, 1979–1991. *Comp. Immunol. Microbiol. Infect. Dis.* **17:**11–123.
17. **Cornelis, G. R.** 1998. The *Yersinia* deadly kiss. *J. Bacteriol.* **180:**5495–5504.
18. **Davis, K. J., D. L. Fritz, M. L. Pitt, S. L. Welkos, P. L. Worsham, and A. M. Friedlander.** 1996. Pathology of experimental pneumonic plague produced by fraction 1-positive and fraction 1-negative *Yersinia pestis* in African green monkeys (*Cercopithecus aethiops*). *Arch. Pathol. Lab. Med.* **120:**156–163.
19. **Davis, K. J., P. Vogel, D. L. Fritz, K. E. Steele, M. L. Pitt, S. L. Welkos, A. M. Friedlander, and W. R. Byrne.** 1997. Bacterial filamentation of *Yersinia pestis* by beta-lactam antibiotics in experimentally infected mice. *Arch. Pathol. Lab. Med.* **121:**865–868.

20. **Dennis, D. T., and F. A. Meier.** 1997. Plague, p. 21–47. *In* C. R. Horsburgh, Jr., and A. M. Nelson (ed.), *Pathology of Emerging Infections*. American Society for Microbiology, Washington, D.C.
21. **Dennis, D. T.** 1998. Plague as an emerging disease, p. 169–183. *In* W. M. Scheld, W. A. Craig, and J. M. Hughes (ed.), *Emerging Infections 2*. American Society for Microbiology, Washington, D.C.
22. **Deohar, N. S., V. L. Yemul, and K. Banerjee.** 1998. Plague that never was: a review of the alleged plague outbreaks in India in 1994. *J. Public Health Policy* **19:**184–199.

22a. **di Coppo de Stefano Buonaiuti, M.** 1935. Cronaca fiorentina, p. 647–652. In R. Palmarocchi (ed.), *Cronisti del Trecento*. Translated by J. Usher. Rizzoli, Milan, Italy.

23. **Doll, J. M., P. S. Zeitz, P. Ettested, A. L. Bucholtz, T. Davis, and K. Gage.** 1994. Cat transmitted fatal pneumonic plague in a person who traveled from Colorado to Arizona. *Am. J. Trop. Med. Hyg.* **51:**109–114.
24. **Eidson, M., J. P. Thilsted, and O. J. Rollag.** 1991. Clinical, clinicopathologic, and pathologic features of plague in cats: 119 cases (1977–1988). *J. Am. Vet. Med. Assoc.* **199:**1191–1197.
25. **Fritz, C. L., D. T. Dennis, M. A. Tipple, G. L. Campbell, C. R. McCance, and D. J. Gubler.** 1996. Surveillance for pneumonic plague in the United States during an international emergency: a model for control of imported emerging diseases. *Emerg. Infect. Dis.* **2:**30–36.
26. **Gage, K. L., J. A. Montenieri, and R. E. Thomas.** 1994. The role of predators in the ecology, epidemiology, and surveillance of plague in the United States, p. 200–206. *In Proceedings of the 16th Vertebrate Pest Conference.* University of California, Davis.
27. **Galimand, M., A. Guiyoule, G. Gerbaud, B. Rasoamanana, S. Chanteau, E. Carniel, and P. Courvalin.** 1997. Multiple antibiotic resistance in *Yersinia pestis* mediated by a self-transferable plasmid. *N. Engl. J. Med.* **337:**677–680.
28. **Gaval, S. R., S. N. Shrinkhande, S. K. Makhija, N. S. Tankhiwale, A. A. Pathak, and A. M. Saoji.** 1996. Study of suspected plague cases for isolation and identification of *Yersinia pestis. Ind. J. Med. Sci.* **50:**335–338.
29. **Guiyoule, A., F. Grimont, I. Iteman, P. A. D. Grimont, M. Lefevre, and E. Carniel.** 1994. Plague pandemics investigated by ribotyping of *Yersinia pestis* strains. *J. Clin. Microbiol.* **32:**634–641.
30. **Heath, D. G., G. W. Anderson, Jr., J. M. Mauro, S. L. Welkos, G. P. Andrews, J. Adamovicz, and A. M. Friedlander.** 1998. Protection against experimental bubonic and pneumonic plague by a recombinant capsular F1-V antigen fusion protein vaccine. *Vaccine* **16:**1131–1137.
31. **Henke, S. E., D. B. Pence, S. Demarals, and J. R. Johnson.** 1990. Serologic survey of selected zoonotic disease agents in black-tailed jack rabbits from western Texas. *J. Wildl. Dis.* **26:**107–111.
32. **Higgins, J., J. Ezzell, J. B. Hinnebush, M. Shipley, E. Henchal, and M. S. Ibraham.** 1998. 5′ nuclease PCR assay to detect *Yersinia pestis. J. Clin. Microbiol.* **36:**2284–2288.
33. **Hinnebush, B. J.** 1997. Bubonic plague: a molecular genetic case history of the emergence of an infectious disease. *J. Mol. Med.* **75:**645–652.
34. **Hinnebush, J., P. Cherepanov, Y. Du, A. Rudolph, J. Dixon, T. Schwan, and A. Forsberg.** 1999. Murine toxin of *Yersinia pestis* shows phospholipase D activity and is essential for bacterial colonization of the flea midgut. *In Dangerous Pathogens 99.* Winchester, United Kingdom.
35. **Hirst, L. F.** 1953. *The Conquest of Plague.* Cambridge University Press, Oxford, United Kingdom.
36. **Kutyrev, V. V., A. A. Filippov, N. Y. Shavina, and O. A. Protsenko.** 1989. Genetic analysis and modeling of the virulence of *Yersinia pestis. Mol. Genet.* **8:**42–47.

37. **Lähteenmäki, K., R. Virkola, A. Sarén, L. Emödy, and T. K. Korhonen.** 1998. Expression of plasminogen activator Pla of *Yersinia pestis* enhances bacterial attachment to the mammalian extracellular matrix. *Infect. Immun.* **66:**5755–5762.
38. **Lee. V. T., and O. Schneewind.** 1999. Type III secretion machines and the pathogenesis of enteric infections caused by *Yersinia* and *Salmonella* spp. *Immunol. Rev.* **168:**241–255.
39. **Madon, M. B., J. C. Hitchcock, R. M. Davis, C. M. Myers, C. R. Smith, C. L. Fritz, K. W. Emery, and W. O'Rullian.** 1997. An overview of plague in the United States and a report of investigations of two human cases in Kern County, California, 1995. *J. Vector Ecol.* **22:**77–82.
40. **McCrumb, F. R., S. Mercier, J. Robic, M. Bouillat, J. E. Smadel, T. E. Woodward, and K. Goodner.** 1953. Chloramphenicol and terramycin in the treatment of pneumonic plague. *Am. J. Med.* **14:**284–293.
41. **McEvedy, C.** 1988. The bubonic plague. *Sci. Am.* **258:**118–123.
42. **McGovern, T. W., and A. M. Friedlander.** 1997. Plague, p. 479–502. *In* F. R. Sidell, E. T. Takafuji, and D. R. Franz (ed.), *Medical Aspects of Chemical and Biological Warfare, Part 1: Warfare, Weaponry, and the Casualty.* Office of the Surgeon General, Washington, D.C.
43. **McNeill, W. A.** 1976. *Plagues and Peoples.* Anchor Press/Doubleday, New York, N.Y.
44. **Patronek, G. J.** 1998. Free-roaming and feral cats—their impact on wildlife and human beings. *J. Am. Vet. Med. Assoc.* **212:**218–226.
45. **Perry, R. D., and J. D. Fetherston.** 1997. *Yersinia pestis*—etiologic agent of plague. *Clin. Microbiol. Rev.* **10:**35–66.
46. **Poland, J. D., and D. T. Dennis.** 1998. Plague, p. 545–558. *In* A. S. Evans and P. S. Brachman (ed.), *Infectious Diseases of Humans: Epidemiology and Control.* Plenum Medical Book Co., New York, N.Y.
47. **Poland, J. D., T. J. Quan, and A. M. Barnes.** 1994. Plague, p. 93–112. *In* G. W. Beran and J. H. Steele (ed.), *Handbook of Zoonoses,* 2nd ed. CRC Press, Inc., Boca Raton, Fla.
48. **Procopius.** 1914. *History of the Wars,* vol. 7, p. 451–473. Translated by H. B. Dewing. Modernization of text by J. S. Arkenberg. Harvard University Press, Cambridge, Mass.
49. **Rosser, W. W.** 1995. Bubonic plague, p. 34–37. *In* N. W. Leveque (ed.), *Zoonosis Updates,* 2nd ed. American Veterinary Medical Association, Schaumburg, Ill.
50. **Rust, J. H., D. C. Cavanaugh, R. O'Shita, and J. D. Marshall.** 1971. The role of domestic animals in the epidemiology of plague. I. Experimental infection of dogs and cats. *J. Infect. Dis.* **124:**522–526.
51. **Ryan, C. P.** 1992. Plague surveillance in Los Angeles County. *Am. J. Public Health* **82:**897.
52. **Sodeinde, O. A., Y. B. V. K. Subrahmanyam, K. Stark, T. Quan, Y. Bao, and J. D. Goguen.** 1992. A surface protease and the invasive character of plague. *Science* **258:**1004–1007.
53. **Tieh, T. H., E. Landauer, F. Miyagawa, G. Kobayashi, and G. Okayasu.** 1948. Primary pneumonic plague in Mukden, 1946, and report of 39 cases with 3 recoveries. *J. Infect. Dis.* **82:**52–58.
54. **Wayson, N. E., C. McMahon, and F. M. Prince.** 1946. An evaluation for three plague vaccines against infection in guinea pigs induced by natural and artificial methods. *Public Health Rep.* **61:**1511–1518.
55. **Welkos, S. L., K. M. Davis, L. M. Pitt, P. L. Worsham, and A. M. Friedlander.** 1995. Studies on the contribution of the F1-capsule-associated plasmid pFra to the virulence of *Yersinia pestis. Contrib. Microbiol. Immunol.* **13:**299–305.
56. **Welkos, S. L., A. M. Friedlander, and K. J. Davis.** 1997. Studies on the role of plasminogen activator in systemic infection by virulent *Yersinia pestis* strain CO92. *Microb. Pathog.* **23:**211–223.

57. **Welkos, S. L., A. M. Friedlander, D. McDowell, J. Weeks, and S. Tobery.** 1998. V antigen of *Yersinia pestis* inhibits neutrophil chemotaxis. *Microb. Pathog.* **24:**185–196.
58. **Williams, E. S., E. T. Thorne, T. J. Quan, and S. L. Anderson.** 1991. Experimental infection of domestic ferrets (*Mustela putorius furo*) and Siberian polecats (*Mustela eversmanni*) with *Yersinia pestis. J. Wildl. Dis.* **27:**441–445.
59. **Williams, E. S., K. Mills, D. R. Kwiatkowski, E. T. Thorne, and A. Boerger-Fields.** 1994. Plague in a black-footed ferret (*Mustela nigripes*). *J. Wildl. Dis.* **30:**581–585.
60. **Williams, J.** 1983. Warning on a new potential for laboratory-acquired infections as a result of the new nomenclature for the plague bacillus. *Bull. W. H. O.* **61:**545–546.
61. **Williams, R. C., H. Gewurz, and P. G. Quie.** 1972. Effects of fraction I from *Yersinia pestis* on phagocytosis in vitro. *J. Infect. Dis.* **126:**235–241.
62. **Wilson, M. E.** 1995. Travel and the emergence of infectious diseases. *Emerg. Infect. Dis.* **1:**39–46.
63. **Winter, C., W. Cherry, and M. Moody.** 1960. An unusual strain of *Pasteurella pestis* isolated from a fatal human case of plague. *Bull. W. H. O.* **23:**408–409.
64. **Worsham, P. L., M. P. Stein, and S. L. Welkos.** 1995. Construction of defined F1 negative mutants of virulent *Yersinia pestis. Contrib. Microbiol. Immunol.* **13:**325–328.

Emerging Diseases of Animals
Edited by C. Brown and C. Bolin

Chapter 14

Reemergence of Tuberculosis in Animals in the United States

Diana L. Whipple and Mitchell V. Palmer

Mycobacterium tuberculosis and *Mycobacterium bovis* are members of a closely related group of organisms referred to as the *M. tuberculosis* complex, which also includes *Mycobacterium africanum* and *Mycobacterium microti*. *M. tuberculosis* is the primary cause of tuberculosis in humans and, world wide, is the leading cause of death from a single infectious agent (56). Nearly one third of the Earth's population is infected with *M. tuberculosis*. Tuberculosis caused by *M. tuberculosis* also has been diagnosed in various species of animals, including nonhuman primates, elephants, cats, birds, dogs, and horses (60). *M. bovis* has a broader host range than *M. tuberculosis* and can cause disease in a variety of domestic and wild animal species, including cattle, deer, elk, bison, llamas, pigs, badgers, ferrets, possums, bear, raccoons, coyotes, dogs, and cats (6, 43; C. S. Bruning-Fann, S. M. Schmitt, S. D. Fitzgerald, J. S. Fierke, P. D. Friedrich, J. B. Kaneene, K. A. Clarke, K. L. Butler, J. B. Payeur, D. L. Whipple, T. M. Cooley, and J. M. Miller, submitted for publication). Human beings are also susceptible to *M. bovis*. Prior to pasteurization of milk and implementation of control programs, it was common for people to become infected with *M. bovis* by ingestion of milk from cattle with bovine tuberculosis (21, 43).

Because of the economic losses associated with *M. bovis* infections in cattle (bovine tuberculosis) and for the public health, the United States initiated a program for eradication of tuberculosis from cattle in 1917, when the prevalence of disease was approximately 5% (15, 19). Initially, the pro-

Diana L. Whipple and Mitchell V. Palmer • Bovine Tuberculosis Research Project, Bacterial Diseases of Livestock Research Unit, National Animal Disease Center, Agricultural Research Service, U.S. Department of Agriculture, Ames, IA 50010.

gram was based on area testing, which involved tuberculin skin testing of 15% of the cattle herds in each state each year. Cattle that reacted to the skin test were slaughtered. Because of the success of the program, the prevalence of bovine tuberculosis declined to less than 0.5% in all states by 1940 (15). In 1965, the program was changed from area testing to one based on slaughter surveillance and traceback of confirmed cases to the herd of origin. Meat inspectors examine carcasses and viscera as cattle are slaughtered. Tissues with macroscopic lesions that are consistent with tuberculosis are submitted to a laboratory for histopathologic examination and bacteriologic culturing. An epidemiological investigation is initiated if tuberculosis is confirmed by isolation of *M. bovis* or by the presence of microscopic lesions characteristic of tuberculosis (49) with acid-fast organisms that are confirmed to be in the *M. tuberculosis* complex (37). In most cases, once it is determined that a herd has cattle infected with *M. bovis,* the herd is depopulated. During the investigation, epidemiologists attempt to identify the source of *M. bovis* that infected the cattle in the herd. They also trace the movement of cattle from the infected herd to determine if other animals and herds were exposed to *M. bovis* by contact with infected animals. Animals with potential exposure to *M. bovis* through contact with infected cattle are skin tested and, in some cases, slaughtered to prevent the spread of tuberculosis to new herds. Using these methods, tuberculosis has almost been eliminated from domestic livestock in the United States, where the prevalence of disease is less than 0.0002%.

REEMERGENCE OF TUBERCULOSIS IN ANIMALS

Although bovine tuberculosis has nearly been eradicated from the United States, the disease has persisted and the prevalence has increased in recent years. Factors that contributed to the resurgence of tuberculosis in animals in the 1990s are importation of tuberculous cattle from Mexico, the persistence of low levels of *M. bovis* infection in large dairy herds in Texas and New Mexico, the presence of tuberculosis in captive deer and elk that are raised for agricultural purposes, the presence of tuberculosis in zoo animals, game parks, and other exotic animal collections, and the presence of tuberculosis in wildlife (17).

Tuberculosis in Mexican Cattle

In fiscal year 1992, there were 613 confirmed cases of bovine tuberculosis in the United States, which was the highest number in more than

20 years. Of these cases, 523 were identified as feedlot animals and 436 were cattle of Mexican origin (17). Livestock producers, along with state and federal animal health officials from Mexico and the United States, have been working together closely to develop programs to reduce the number of cattle with tuberculosis that are imported into the United States. In 1993, the Mexican government initiated a national bovine tuberculosis eradication program similar to the one used in the United States. In addition, the United States-Mexico Binational Tuberculosis Committee was formed in 1993 to assist both countries in eradication of bovine tuberculosis. As a result of these initiatives, there has been extensive testing of cattle for tuberculosis in Mexico, especially in the northern states that border the United States. The number of tuberculous cattle imported from Mexico into the United States has decreased dramatically since 1992 (17).

Tuberculosis in Large Dairy Herds of Cattle

The presence of tuberculosis in large dairy herds in Texas and New Mexico continues to be a problem (55). Depopulation of large herds with bovine tuberculosis has not been possible because of the lack of funds to compensate the herd owners for the loss of their animals. Cattle in these herds are tested on a regular basis, and animals with positive test results are removed (55). However, current diagnostic tests lack sensitivity and may not be detecting all of the cattle that are infected with *M. bovis*. Therefore, these animals remain in the herd and are potential sources of infection for other animals. It also is possible that cattle in these herds become infected from outside sources, such as cattle that are purchased from other herds (55). Although designated tuberculosis epidemiologists work closely with herd owners to control and eliminate tuberculosis from these herds, improved diagnostic tests are needed to detect all animals infected with *M. bovis*.

Tuberculosis in Captive Cervidae

In the last two decades, there has been a rapid expansion of the captive cervid industry in the United States, where deer and elk are raised for agricultural purposes. Tuberculosis caused by *M. bovis* in captive elk (*Cervus elaphus*) was first detected in the United States in late 1990, when an outbreak of tuberculosis in captive elk in Canada was traced to elk imported from the United States (15). Since then, tuberculosis has been diagnosed in 36 captive cervid herds in 17 states. In 1994, the Uniform Methods and Rules with guidelines for eradication of tuberculosis from captive members of the family Cervidae were implemented (1). The number of captive cervid herds with tuberculosis has decreased since implementation of the program guidelines.

Tuberculosis in Zoos, Game Parks, and Exotic Animal Collections

Tuberculosis caused by *M. bovis* and *M. tuberculosis* has been reported in zoos, game parks, and other exotic animal collections in the United States and throughout the world (3, 28, 38, 58, 61, 63, 65, 66; S. K. Mikota, R. S. Larsen, and R. J. Montali, submitted for publication). In the United States, exotic animal collections have been identified as the major source of *M. bovis* for 28 of the captive cervid herds with tuberculosis that have been discovered since 1991 (17). Epidemiological investigations indicate that *M. bovis* spread from some of the captive cervid herds and animal collections to herds of domestic cattle, bison, and other Cervidae (17). In addition, tuberculosis in these populations is considered a threat to public health. Animal caretakers, trainers, veterinarians, and other people that have direct contact with infected animals and animal waste have become infected with *M. bovis* and *M. tuberculosis* (18, 36, 58, 65).

Tuberculosis caused by *M. tuberculosis* in elephants in the United States has been reported and recently reviewed (Mikota et al., submitted). In 1996, tuberculosis was diagnosed in two Asian elephants (*Elaphus maximus*) that died after a period of weight loss and general poor health. Both elephants originated from the same private facility and had been in contact with other elephants both at the facility and during exhibition in a different state. The remaining elephants in this group were tested for tuberculosis to determine the extent of the problem. Tuberculosis was diagnosed by isolation of *M. tuberculosis* from trunk washings of one live elephant, and an epidemiological investigation revealed that two other elephants from this facility had died with tuberculosis in 1983 and 1994 (36, 51). The elephants from this facility had been at several exhibits throughout the United States prior to the diagnosis of tuberculosis in 1996.

Twenty-two animal handlers that worked at this facility were tuberculin skin tested, and 11 had responses that were classified as positive. *M. tuberculosis* was isolated from a sputum sample from one of the handlers, and the DNA fingerprint of the organisms matched the fingerprint of *M. tuberculosis* that was isolated from the elephants (36). The epidemiological data suggest transmission of *M. tuberculosis* from the elephants to the animal handler rather than from the animal handler to the elephants.

As a result of this outbreak and the problem of tuberculosis in zoos and other wildlife, the National Tuberculosis Working Group for Zoo and

Wildlife Species was formed in 1997 (17). This group, in conjunction with the U.S. Department of Agriculture (USDA), developed the Guidelines for the Control of Tuberculosis in Elephants (http://www.aphis.usda.gov/ac/awainfo.html). One of the recommendations in the guidelines is annual testing of all elephants in the United States. The definitive diagnosis of tuberculosis in live elephants is isolation of *M. tuberculosis* complex organisms from samples of trunk washes collected on three separate dates. As of January 2000, samples from approximately 540 elephants have been processed, and *M. tuberculosis* has been isolated from 15 of them (Mikota et al., submitted; J. B. Payeur, personal communication). The infected elephants are located in different areas throughout the United States and are part of groups that are in zoos, circuses, exotic animal collections, and animal sanctuaries. Many of the infected elephants have been successfully treated using antituberculosis drugs and are in good health (Mikota et al., submitted). Although DNA fingerprinting of the *M. tuberculosis* isolates from the elephants reveals animal-to-animal transmission in some cases, several of the cases do not appear to be epidemiologically related (36, 68). The original source(s) of *M. tuberculosis* associated with this outbreak of tuberculosis in elephants is not known. However, it is possible that some of the elephants became infected through contact with tuberculous human beings who were excreting *M. tuberculosis.*

Diagnosis of tuberculosis in zoo and other captive wildlife species is very difficult because the tests traditionally used for domestic animals have not been validated for other species and are not suitable for use (3, 24). For example, the tuberculin skin test is difficult to conduct in some species because the animals must be handled twice, once to administer the tuberculin and once to read test results. There is a risk of injury or death of the animal each time it is handled. Chemical restraint is required for many of these species, and complications from anesthesia may occur. In addition, once tuberculosis has been diagnosed in these species, destruction of the animal is not recommended in some cases because of the genetic value. This is especially true of animals that are endangered species. Additional work on diagnosis and control of tuberculosis in nondomestic and exotic species is needed to prevent the spread of disease within these animal species and to other animals and human beings.

Tuberculosis in Free-Ranging Wildlife

The first wildlife reservoir of *M. bovis* in the United States was recognized on the island of Molokai in Hawaii. Bovine tuberculosis in domestic cattle on the island was reported in the 1940s and in other small domestic herds and feral cattle in the 1960s (16). In 1978, bovine tuberculosis was confirmed in a large beef cattle herd that was located at the eastern end of Molokai. Subsequently, *M. bovis* was isolated from free-ranging axis deer (*Cervus axis axis*) in 1970, 1971, 1972, and 1981 that were harvested from the same area as the domestic cattle herds with bovine tuberculosis (16, 52). In 1980, *M. bovis* was isolated from 15% (9 of 60) of feral swine that were removed from eastern Molokai during a wildlife survey. It was concluded that *M. bovis* had become established in the feral swine population in eastern Molokai (16).

Because of the continued problem of bovine tuberculosis, depopulation of domestic cattle from Molokai was initiated in September 1985, and the island was declared free of domestic cattle on 15 January 1987 (25). Molokai remained free of domestic cattle for 1 year, followed by introduction of sentinel cattle over 2 years. Once the sentinel cattle were determined to be free of bovine tuberculosis, breeding cattle from herds accredited free of tuberculosis were moved to Molokai (64). In 1997, *M. bovis* was isolated from a cow that was on the same premise as one of the herds with tuberculosis that was depopulated in the 1980s (64). In addition, *M. bovis* was isolated from a feral swine that was removed from eastern Molokai in 1999 (R. M. Meyer, personal communication). Because wildlife are considered a possible source of infection for the most recent case of *M. bovis* infection in cattle on Molokai, surveys are being conducted to determine the prevalence of tuberculosis in the wildlife.

The greatest threat to the effort to eradicate tuberculosis from domestic livestock in the United States is the presence of *M. bovis* in a population of free-ranging white-tailed deer (*Odocoileus virginianus*) in northern Michigan (17). Although there have been other reports of *M. bovis* in wildlife, which were recently reviewed by Schmitt (54), this appears to be the first time that the infection has been maintained in a wildlife population and a reservoir of *M. bovis* has been established on the mainland of the United States (12).

In 1975, *M. bovis* had been isolated from a white-tailed deer that had lesions characteristic of tuberculosis. The deer was harvested from Alcona County, which is in the northeastern region of the lower peninsula of Michigan (Fig. 1) (54). This was considered an isolated case, and no additional work was done to determine if tuberculosis was present in other deer or livestock (54). During the fall hunting season of 1994, a hunter harvested a white-tailed deer with lesions suggestive of tuberculosis from Alpena County, which is adjacent to Alcona County. *M. bovis* was isolated from tissue samples collected from the deer (54). A survey of hunter-harvested white-tailed deer was conducted during the fall and winter hunting season of 1995, and *M. bovis* was isolated from 27 of 814 deer that were sampled (http://www.dnr.state.mi.us/wildlife/division/roselake/bovineTB/

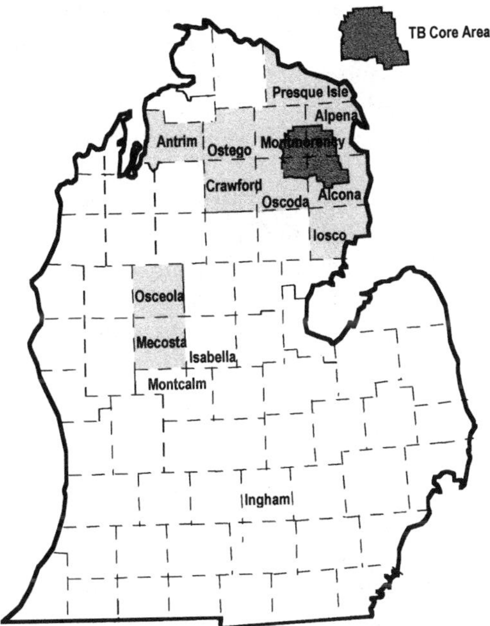

Figure 1. Map of the lower peninsula of Michigan. *M. bovis* has been isolated from free-ranging white-tailed deer in the counties that are shaded. The TB (tuberculosis) core area is the region with the highest prevalence of *M. bovis* in the white-tailed deer population.

wildlife.htm). The infected deer originated from a 650-km^2 region that included portions of Alpena, Alcona, Montmorency, and Oscoda Counties in the northeast region of the lower peninsula of Michigan. This region is referred to as the tuberculosis core area (Fig. 1). Although it was thought that the problem was confined to these four counties, subsequent surveys have revealed a more widespread problem. *M. bovis* has been isolated from free-ranging white-tailed deer harvested from Presque Isle, Otsego, Iosco, Crawford, Antrim, Mecosta, and Osceola Counties in addition to the four counties that were originally identified (C. Bruning-Fann, personal communication; website cited above). A risk assessment conducted by the USDA in 1996 concluded that *M. bovis* has been present in this population of free-ranging white-tailed deer for up to 40 years (12). *M. bovis* also has been isolated from six coyotes (*Canis latrans*), two raccoons (*Procyon lotor*), a black bear (*Ursus americanus*), and a red fox (*Vulpes vulpes*). Microscopic lesions with acid-fast organisms consistent with tuberculosis were also observed in tissue samples collected from another coyote and a bobcat

(*Felis rufus*) (Bruning-Fann et al., submitted; website cited above). Organisms belonging to the *M. tuberculosis* complex were detected by a PCR-based assay (37) in the tissue sample collected from the bobcat. Tissue samples from the coyote with lesions were not processed for PCR analysis. These animals were all harvested from the four counties that are included in the tuberculosis core area.

In 1997, *M. bovis* was isolated from a captive white-tailed deer harvested from a private hunting ranch in Presque Isle County (45). The herd was established in 1992 by enclosing a 1,500-acre parcel of land. Approximately 108 white-tailed deer were inside the enclosure at that time and were purchased by the land owner. It is likely that some of the deer that were captured when the herd was established were infected with *M. bovis*. The white-tailed deer herd was depopulated between January 1998 and March 1999.

Prior to 1993, the last known case of bovine tuberculosis in Michigan was in 1974 in a herd of cattle located in Ingham County (Bruning-Fann, personal communication [PC]). The state was declared free of bovine tuberculosis by the USDA in 1979 (12). In 1993, *M. bovis* was isolated from a cow located in a herd of dairy cattle in Isabella County (Fig. 1). The dairy herd had been established by purchasing cattle from various sources. Although the traceback was not definitive, there was evidence to suggest that the cow with tuberculosis originated from a herd in Alpena County near Turtle Lake, which is inside the tuberculosis core area (Fig. 1) (Bruning-Fann, PC).

In 1995, *M. bovis* was isolated from a heifer from Wisconsin that had lesions consistent with tuberculosis. Through the epidemiological investigation, it was discovered that the heifer had been in a herd near Turtle Lake in Michigan for several months. The DNA fingerprint of the *M. bovis* isolates from the 1993 Michigan cow and the 1995 Wisconsin heifer matched the fingerprint of most of the isolates from the white-tailed deer (68). These results indicate that the cattle were infected with the same strain of *M. bovis* as was isolated from the white-tailed deer and suggest an epidemiological link between the cases. In 1994, *M. bovis* was also isolated from a captive elk that originated from a herd in Montcalm County, Michigan. However, the DNA fingerprint of the *M. bovis* isolate from the elk did not match the fingerprint for the deer and cattle isolates (68, 70). Therefore, the captive elk was infected from a source other than the Michigan white-tailed deer.

The Michigan Department of Agriculture and the USDA have been skin testing the cattle herds in the areas of Michigan where tuberculosis has been found in the wildlife. Since 1998, tuberculosis has been detected in six herds of beef cattle and one herd of dairy cattle (12; Bruning-Fann,

PC). Five of the beef cattle herds have been depopulated, while the disposition of the other two herds is pending. Results of DNA fingerprinting indicate that the cattle and deer are infected with a common strain of *M. bovis*.

The discovery of tuberculosis in cattle herds in northeast Michigan has led to the suspension of the "accredited free of bovine tuberculosis" status for the state of Michigan (67). The official status of the state relative to bovine tuberculosis is currently pending while regulatory officials continue to gather data to determine the extent of the problem. The estimated economic impact on agricultural producers in the state of Michigan as a result of losing the accredited free status is approximately $16 million annually (30). These losses are associated with extra testing requirements for the sale of animals to other states and the loss of markets when other states and countries refuse to accept animals from Michigan. The full economic impact of tuberculosis in wildlife and domestic livestock in Michigan is not known at this time but will most likely be considerably higher than original estimates. The presence of a wildlife reservoir of tuberculosis in Michigan may prevent the United States from achieving the goal of eradication of tuberculosis in animals. Other countries that have a wildlife reservoir of tuberculosis, such as New Zealand with the brushtail possum (*Trichorsurus vulpecula*) and the United Kingdom and Republic of Ireland with badgers (*Meles meles*), have not been able to eradicate the disease from domestic livestock (39, 43). Once a wildlife reservoir of tuberculosis has been established, it is very difficult to eliminate. Results of a recent risk assessment indicate that *M. bovis* will continue to exist in the free-ranging white-tailed deer population of northern Michigan in 2010 even with aggressive efforts to eliminate the disease (11).

PATHOGENESIS AND TRANSMISSON OF TUBERCULOSIS IN ANIMALS

To comprehend the pathogenesis of a disease, it is important to understand the routes of infection and mode of transmission between hosts as well as the host response and characteristics intrinsic to the pathogen. Tuberculosis in animals is chiefly acquired by inhalation or ingestion. Inhaled mycobacteria can reach the pulmonary alveoli in microdroplets less than 5 μm in size, containing one to three bacilli (50). Cilia, which are essential in removing inhaled foreign material from the respiratory tract, are not present beyond the terminal bronchioles (14). Moreover, inhaled mycobacteria can establish pulmonary infection by being resistant to killing by nonactivated alveolar macrophages. Mycobacteria are phagocytosed by macrophages that exit the airway into the pulmonary parenchyma or into lymphatic vessels that carry infected cells to draining lymph nodes. Mycobacteria reside in macrophage phagosomes that fail to acidify (2, 13). Failure of acidification is partly due to prevention of phagosome-lysosome fusion. Lack of phagosome-lysosome fusion has been associated with sulfatides released by mycobacteria (20). Lack of phagosome acidification in spite of fusion with lysosomes also occurs and has been associated with exclusion of the lysosomal vesicular proton ATPase (59), urease production by the mycobacteria (8), and phagocytosis of bacteria via complement- or mannose-binding receptors rather than Fc receptors (53).

Nonactivated macrophages are unable to effectively kill mycobacteria. Mycobacteria multiply, eventually killing the macrophage, and are released to infect other macrophages or become disseminated through lymph or blood to other sites in the body. Days to weeks later, activated $CD4^+$ T lymphocytes, responding to antigens presented by previously infected macrophages, secrete gamma interferon and other cytokines that enable macrophages to effectively kill intracellular mycobacteria. Mycobactericidal reactive nitrogen intermediates such as NO, NO_2, and HNO_3 are also produced by macrophages (41). Activated macrophages, most of which are derived from the circulation as blood monocytes, are attracted to the foci of infection by T-cell cytokines. In infected tissues, macrophages proliferate and assume a distinctive morphology termed epithelioid macrophages because of their morphologic similarity to epithelial cells. Epithelioid macrophages have large, round, vesicular nuclei, abundant eosinophilic cytoplasm, numerous cytoplasmic organelles, and a plasma membrane that interdigitates with neighboring cells. These macrophages often contain bacteria and are in an activated mycobactericidal mode. Among the epithelioid macrophages are variable numbers of Langhans-type multinucleated giant cells formed by the fusion of macrophages. This mixture of epithelioid macrophages and giant cells forms the early granuloma. At the periphery of the granuloma is a zone of lymphocytes, plasma cells, and monocytes. As the duration of infection increases, peripheral fibroplasia develops around the granuloma (44). The degree of peripheral fibroplasia varies among species. Species with more innate resistance to mycobacterial infection typically develop more intense fibroplasia. Delayed-type hypersensitivity results in central caseous necrosis of the granuloma as $CD8^+$ T lymphocytes kill mycobacterium-infected macrophages. Direct killing of the macrophages by the bacteria also contributes to necrosis. Mycobacteria are unable to grow within the necrotic core due to the low oxygen content and acidic nature of the caseum. The necrotic core often undergoes dystrophic mineralization that is evident microscopically and can be appreciated grossly by the gritty nature of many granulomas. Centrifugal growth of the original granuloma can proceed as satellite granulomas are formed by spread of bacilli from the original granuloma. Numerous granulomas may coalesce to form large expansive caseonecrotic granulomas that may obliterate normal tissue architecture. In a susceptible host, bacilli may spread rapidly, free or in macrophages, through lymphatics or blood vessels. Hematogenous distribution is often the result of invasion of a vessel wall by an expanding caseonecrotic granuloma. Proteinases and nucleases present in the caseum may cause liquefaction of the necrotic core. Liquefaction enhances transmission by spread of liquefied material to adjacent airways, blood vessels, or lymphatics and accelerates bacterial multiplication due to increased availability of oxygen (10). Spread can also occur from one bronchus to another by coughing, or spread to the intestines may occur by swallowing infected sputum.

Infection may spread within an animal, resulting in disseminated tuberculosis. Alternatively, disease may be halted at an early stage and the animal may remain clinically normal but infected. A proportion of these animals will later reactivate a dormant case of tuberculosis that then progresses to disseminated disease. The outcome of any individual case will depend on the immune status of the host and the virulence of the infecting strain. In disseminated tuberculosis, many organs may develop granulomas, including the kidneys, liver, spleen, peripheral lymph nodes, meninges, and bone marrow. These are in addition to the more commonly affected tissues, including the lungs and the pulmonary, cranial, and mesenteric lymph nodes.

Although inhalation of droplet nuclei and primary pulmonary involvement have been thought to be the major route of infection in humans and animals, in cattle and deer, tonsillar processing of *M. bovis* and infection of the medial retropharyngeal lymph node may play an equally important role. Inhaled or ingested mycobacteria may be processed by the tonsil. Recent reports suggest that infection of the bovine tonsil with *M. bovis* may be more common than was previously thought (7, 33). Although tonsils have no afferent lymphatics, the primary efferent lymphatics drain to the medial retropharyngeal lymph node. Infection of the tonsil and medial retropharyngeal lymph node may represent a primary complex, much like the combination of lesions in the lung and pulmonary lymph nodes. Processing of mycobacteria by the tonsil, whether introduced to the oropharynx or nasopharynx by ingestion or inhalation, would likely lead to infection of the primary efferent lymph node, i.e., the medial retropharyngeal lymph node. The medial retropharyngeal lymph nodes, thoracic lymph nodes, and lungs are the most commonly affected sites in tuberculous cattle (31, 62, 69). Recent reports of *M. bovis* infection in red deer (*Cervus elaphus*) (34, 35) and white-tailed deer (54) show the medial retropharyngeal lymph node to be the most commonly affected site in naturally occurring tuberculosis, suggesting frequent exposure of the nasopharynx or oropharynx to *M. bovis.*

Species differences exist in susceptibility to infection with *M. bovis.* Eight species of carnivores and omnivores have been sampled in northeast Michigan as part of an ongoing surveillance program to determine the extent of *M. bovis* infection among wildlife. Coyotes, raccoons, a red fox, a black bear, and a bobcat have been found to be infected with *M. bovis.* However, lesions have rarely been identified in these hosts, and small numbers of acid-fast bacteria have been seen within lesions (6; Bruning-Fann et al., submitted). Without extensive lesion development and an avenue for bacterial excretion, it is thought that transmission of disease from these species to other susceptible hosts would be unlikely.

A distinction has been made between two categories of hosts. Spillover hosts require an external source of reinfection to maintain disease within the population, and reservoir hosts show disease persisting within the population without an external source of reinfection (23). Although the status of some scavenging species such as ferrets and feral pigs in New Zealand is still under debate (23), possums and wild red deer in New Zealand, badgers in the United Kingdom, and white-tailed deer in the United States are considered reservoir hosts. Presumably coyotes, raccoons, bobcats, bears, and red foxes are spillover hosts, and elimination of *M. bovis* infection from the white-tailed deer reservoir would result in the gradual elimination of infection in these spillover hosts. It is hypothesized that these scavenger species become infected by feeding on carcasses of *M. bovis*-infected white-tailed deer. It also is hypothesized that infection infrequently leads to lesion development due to a relative resistance to tuberculosis in these hosts compared to white-tailed deer (Brunning-Fann et al., submitted). Alternative explanations for the absence of lesions may be related to the dose of *M. bovis* ingested or the time between infection and examination. Restriction fragment length polymorphism analysis has confirmed that the isolates from these scavenger species in Michigan are identical to that found in the majority of white-tailed deer (70).

Spillover hosts were also identified in Montana, where *M. bovis* was isolated from a free-ranging mule deer (*Odocoileus hemionus*) and two coyotes that were harvested from an area near a herd of captive elk that had tuberculosis (47, 71). There was evidence that coyotes had entered the premise where the captive elk were located by digging under the fence (47). It was hypothesized that the coyotes had become infected by scavenging on carcasses of tuberculous elk. It was assumed that the mule deer became infected by fenceline contact with infected elk. Restriction fragment length polymorphism analysis showed that the free-ranging mule deer and coyotes were infected with the same strain of *M. bovis* that was isolated from the captive elk. Follow-up surveys of the wildlife in the area revealed no further evidence of tuberculosis in any of the wild animals that were examined (48). Likewise, surveys conducted in states other than Michigan have not revealed *M. bovis* in wildlife species (57, 64).

Differences in host response to *M. bovis* exist even among species of Cervidae. New Zealand red deer develop external draining sinuses from infected lymph nodes. These suppurative draining lesions are thought to be responsible for most deer-to-deer transmission among red deer (34). Similar draining lesions are common in the brushtail possum in New Zealand and also are thought to be important in disease transmission (27). Draining lesions have not been identified in hundreds of naturally and experimentally infected white-tailed deer examined, eliminating this method of disease transmission among white-tailed deer.

Shedding of *M. bovis* from infected white-tailed deer occurs through contaminated saliva, nasal secretions (46), urine, and feces (M. V. Palmer, unpublished data). These infectious materials may be present in aerosol form or may contaminate feed or water that may be used by other animals. Free-ranging roe deer (*Capreolus capreolus*) have been implicated in the transmission of *M. bovis* to cattle through contaminated feed (4). Experimentally, white-tailed deer efficiently transfer *M. bovis* to uninfected penmates. In such settings, feed samples shared by infected and uninfected deer, as well as urine and feces from infected deer, are found to contain *M. bovis* and are possible sources of infection (Palmer, unpublished). Cattle housed in pens where experimentally infected white-tailed deer have been housed and fed readily become infected with *M. bovis* and develop lesions typical of naturally occurring tuberculosis in cattle (D. L. Whipple and M. V. Palmer, unpublished data).

CONCLUSIONS AND SUMMARY

Emerging diseases result as newly recognized pathogens arise or as known pathogens cause disease in previously unrecognized hosts. As such, the establishment of a wildlife reservoir of *M. bovis* infection in the United States represents a serious emerging disease threat. Controversy exists over whether *M. bovis* in Michigan originated with cattle or with white-tailed deer. Although academically interesting, the question is pointless, as either scenario is unacceptable to U.S. agriculture.

Factors leading to the current situation in Michigan were likely set in place many years ago. Supplemental winter feeding of deer has been common and popular in northern Michigan. Most supplemental feeding is practiced by hunters to maintain large numbers of deer, but even nonhunters engage in supplemental feeding for the pleasure of viewing wildlife and satisfaction received from the perception that the wildlife have benefitted from the practice. Supplemental feeding has allowed deer densities in Michigan to be maintained above the natural carrying capacity of the environment for many years. Moreover, feeding sites attract large numbers of deer together in an unnatural fashion for a prolonged period of time. Such close contact between deer allows *M. bovis* to be spread by aerosol or consumption of contaminated feed. This combination of factors created the environment necessary to maintain *M. bovis* infection in this susceptible host. Emerging human diseases have been attributed to several factors, including increased population density, enhanced mobilization, and environmental disruption (5). The long-term winter feeding of deer clearly increased population density and may also have altered deer movement patterns and social structures.

Schemes that have been successful in controlling domestic animal tuberculosis, such as test and slaughter or market surveillance, are not appropriate for wildlife. In the United Kingdom, attempts to test badgers and remove only infected ones have been unsuccessful due to the lack of highly sensitive diagnostic tests for badgers. Alternatively, efforts to remove all badgers from cattle-farming areas have resulted in a decline in bovine tuberculosis (32). Similarly, no widespread eradication of a vertebrate host has ever been successful in New Zealand. Aerial distribution of possum baits containing poison have achieved 90% possum death rates in some areas. In these areas, tuberculin reaction rates in cattle herds have decreased temporarily, only to return to elevated levels as possum numbers recover through breeding and immigration from surrounding areas (42).

The best long-term approach to control of tuberculosis in the United Kingdom appears to be the development of a vaccine for cattle and improved diagnostic tests to discriminate infected from vaccinated cattle (29). However, vaccination of badgers to reduce excretion of mycobacteria and thereby decrease transmission to cattle has also received attention (26, 40). Many countries, including New Zealand and the United States, have adopted some form of test-and-slaughter control for tuberculosis in captive Cervidae (1, 9), and vaccination of captive Cervidae also has been investigated (22). Control of tuberculosis in free-ranging Cervidae, however, is more problematic. Issues such as method of delivery, dose control, unintentional exposure of nontarget species, and environmental impact must be addressed. Wild animals are perceived as belonging to everyone, and therefore many public groups are affected by wildlife management decisions. Whether control of tuberculosis in wildlife involves vaccination or severe culling, it is likely to be unpopular with some interest groups. Therefore, the solution to wildlife reservoirs of tuberculosis will need to address social, political, environmental, and economic as well as purely scientific issues.

REFERENCES

1. **Anonymous.** 1999. *Bovine Tuberculosis Eradication: Uniform Methods and Rules.* APHIS Publ. No. 91-45-011. U.S. Department of Agriculture, Animal and Plant Health Inspection Service, Riverdale, Md.
2. **Armstrong, J. A., and P. D. Hart.** 1971. Response of cultured macrophages to *Mycobacterium tuberculosis,* with observations on fusion of lysosomes with phagosomes. *J. Exp. Med.* **134:**713–740.
3. **Bengis, R. G.** 1999. Tuberculosis in free-ranging mammals, p. 101–114. *In* M. E. Fowler and R. E. Miller (ed.), *Zoo and Wild Animal Medicine: Current Therapy 4.* W. B. Saunders Co., Philadelphia, Pa.
4. **Bischofberger, V. A., and A. Nabholz.** 1964. Tuberkulöses wild als Ursache von Neuinfektionen in Rindviehbeständen. *Schweiz. Arch. Tierheilkd.* **106:**759–777.
5. **Brown, C.** 1997. Emerging diseases—what veterinarians need to know. *J. Vet. Diagn. Investig.* **9:**113–117.
6. **Bruning-Fann, C. S., S. M. Schmitt, S. D. Fitzgerald, J. B. Payeur, D. L. Whipple, T. M. Cooley, T. Carlson, and P. Friedrich.** 1998. *Mycobacterium bovis* in coyotes from Michigan. *J. Wildl. Dis.* **34:**632–636.
7. **Cassidy, J. P., D. G. Bryson, and S. D. Neill.** 1999. Tonsillar lesions in cattle naturally infected with *Mycobacterium bovis. Vet. Rec.* **144:**139–142.
8. **Clemens, D. L., B.-Y. Lee, and M. A. Horwitz.** 1995. Purification, characterization, and genetic analysis of *Mycobacterium tuberculosis* urease, a potentially critical determinant of host-pathogen interaction. *J. Bacteriol.* **177:**5644–5652.
9. **Clifton-Hadley, R. S., and J. W. Wilesmith.** 1991. Tuberculosis in deer: a review. *Vet. Rec.* **129:**5–12.
10. **Converse, P. J., A. M. Dannenberg, J. E. Estep, K. Sugisaki, Y. Abe, B. H. Schofield, and M. L. M. Pitt.** 1996. Cavitary tuberculosis produced in rabbits by aerosolized virulent tubercle bacilli. *Infect. Immun.* **64:**4776–4787.
11. **Corso, B. A.** 1999. Risks associated with *M. bovis* in Michigan free-ranging white-tailed deer: an update to the 1996 report. U.S. Department of Agriculture, Animal and Plant Health Inspection Service, Veterinary Services, Centers for Epidemiology and Animal Health, Fort Collins, Colo.
12. **Corso, B., B. Wagner, D. Norden, H. S. Hurd, J. Belfrage, A. Hayek, C. Bruning-Fann, L. G. Paisley, O. Williams, D. L. Whipple, M. W. Miller, C. W. McCarty, S. M. Schmitt, and H. M. Chaddock.** 1996. Assessing the risks associated with *M. bovis* in Michigan free-ranging white-tailed deer. CADIA Technical Report No. 01-96. U.S. Department of Agriculture, Animal and Plant Health Inspection Service, Veterinary Services, Centers for Epidemiology and Animal Health, Fort Collins, Colo.
13. **Crowle, A. J., R. Dahl, E. Ross, and M. H. May.** 1991. Evidence that vesicles containing living, virulent *Mycobacterium tuberculosis* or *Mycobacterium avium* in cultured human macrophages are not acidic. *Infect. Immun.* **59:**1823–1831.
14. **Dellman, H.-D.** 1981. Respiratory system, p. 187–202. *In* H.-D. Dellman and E. M. Brown (ed.), *Textbook of Veterinary Histology,* 2nd ed. Lea and Febiger, Philadelphia, Pa.

15. **Essey, M. A., and M. A. Koller.** 1994. Status of bovine tuberculosis in North America. *Vet. Microbiol.* **40:**15–22.
16. **Essey, M. A., R. L. Payne, and E. M. Himes.** 1981. Bovine tuberculosis surveys of axis deer and feral swine on the Hawaiian island of Molokai, p. 538–549. *In Proceedings of the Eighty-Fifth Annual Meeting of the United States Animal Health Association.* United States Animal Health Association, Richmond, Va.
17. **Essey, M. A., and J. P. Davis.** 1997. Status of the national cooperative state-federal bovine tuberculosis eradication program fiscal year 1997, p. 561–581. *In Proceedings of the One Hundred and First Annual Meeting of the United States Animal Health Association.* United States Animal Health Association, Richmond, Va.
18. **Fanning, A., and S. Edwards.** 1991. *Mycobacterium bovis* infection in human beings in contact with elk (*Cervus elaphus*) in Alberta, Canada. *Lancet* **338:**1253–1255.
19. **Frye, G. H.** 1995. Bovine tuberculosis eradication: the program in the United States, p. 119–129. *In* C. O. Thoen and J. H. Steele (ed.), *Mycobacterium bovis Infection in Animals and Humans.* Iowa State University Press, Ames, Iowa.
20. **Goren, M. B., P. D. Hart, M. R. Young, and J. A. Armstrong.** 1976. Prevention of phagosome-lysosome fusion in cultured macrophages by sulfatides of *Mycobacterium tuberculosis. Proc. Natl. Acad. Sci. USA* **73:**2510–2514.
21. **Grange, J. M.** 1995. Human aspects of *Mycobacterium bovis,* p. 29-46. *In* C. O. Thoen and J. H. Steele (ed.), *Mycobacterium bovis Infection in Animals and Humans.* Iowa State University Press, Ames, Iowa.
22. **Griffin, J. F. T., J. B. Hesketh, C. G. Mackintosh, Y.-E. Shi, and G. S. Buchan.** 1993. BCG vaccination in deer: distinctions between delayed type hypersensitivity and laboratory parameters of immunity. *Immunol. Cell Biol.* **71:**559–570.
23. **Hickling, G.** 1995. Wildlife reservoirs of bovine tuberculosis in New Zealand, p. 276–279. *In* F. Griffin and G. W. de Lisle (ed.), *Tuberculosis in Wildlife and Domestic Animals.* Otago Conference Series No. 3. University of Otago Press, Dunedin, New Zealand.
24. **Hietala, S. K., and I. A. Gardner.** 1999. Validity of using diagnostic tests that are approved for use in domestic animals for nondomestic species, p. 55–58. *In* M. E. Fowler and R. E. Miller (ed.), *Zoo and Wild Animal Medicine: Current Therapy 4.* W. B. Saunders Co., Philadelphia, Pa.
25. **Hosker, R. L.** 1997. Status of the state-federal tuberculosis eradication program fiscal year 1987, p. 507–522. *In Proceedings of the Ninety-First Annual Meeting of the United States Animal Health Association.* United States Animal Health Association, Richmond, Va.
26. **Hughes, M. S., S. D. Neill, and M. S. Rogers.** 1996. Vaccination of the badger (*Meles meles*) against *Mycobacterium bovis. Vet. Microbiol.* **51:**363–379.
27. **Jackson, R., M. M. Cooke, J. D. Coleman, R. S. Morris, G. W. de Lisle, and G. F. Yates.** 1995. Naturally occurring tuberculosis caused by *Mycobacterium bovis* in brushtail possums (*Trichosurus vulpecula*). III. Routes of infection and excretion. *N. Z. Vet. J.* **43:**322–327.
28. **Keet, D. F., N. P. J. Kriek, M. L. Penrith, A. Michel, and H. Huchzermeyer.** 1996. Tuberculosis in buffaloes (*Syncerus caffer*) in the Kruger National Park: spread of the disease to other species. *Onderstepoort J. Vet. Res.* **63:**239–244.
29. **Krebs, J. R., R. M. Anderson, T. Clutton-Brock, C. A. Donnelly, S. Frost, W. I. Morrison, R. Woodroffe, and D. Young.** 1998. Badgers and bovine TB: conflicts between conservation and health. *Science* **279:**817–818.
30. **Leefers, L., J. Ferris, and D. Propst.** 1997. Economic consequences associated with bovine tuberculosis in northeastern Michigan. Report to D. Wyant, K. L. Cool, and J. K. Haveman. Michigan State University, Lansing.
31. **Lepper, A. W. D., and C. W. Pearson.** 1973. The route of infection in tuberculosis of beef cattle. *Aust. Vet. J.* **49:**266–267.

32. **Little, T. W. A., C. Swan, H. V. Thompson, and J. W. Wilesmith.** 1982. Bovine tuberculosis in domestic and wild mammals in an area of Dorset. II. The badger population, its ecology and tuberculous state. *J. Hyg.* **89:**211–224.
33. **Lugton, I. W.** 1999. Mucosa-associated lymphoid tissues as sites for uptake, carriage and excretion of tubercle bacilli and other mycobacteria. *Immunol. Cell Biol.* **77:**364–372.
34. **Lugton, I. W., P. R. Wilson, R. S. Morris, and G. Nugent.** 1998. Epidemiology and pathogenesis of *Mycobacterium bovis* infection of red deer (*Cervus elaphus*) in New Zealand. *N.Z. Vet. J.* **46:**147–156.
35. **Lugton, I. W., P. R. Wilson, R. S. Morris, J. F. T. Griffin, and G. W. de Lisle.** 1997. Natural infection of red deer with bovine tuberculosis. *N.Z. Vet. J.* **45:**19–26.
36. **Michalak, K., C. Austin, S. Diesel, J. M. Bacon, P. Zimmerman, and J. N. Maslow.** 1998. *Mycobacterium tuberculosis* infection as a zoonotic disease: transmission between humans and elephants. *Emerg. Infect. Dis.* **4:**283–287.
37. **Miller, J., A. Jenny, J. Rhyan, D. Saari, and D. Suarez.** 1997. Detection of *Mycobacterium bovis* in formalin-fixed, paraffin-embedded tissues of cattle and elk by PCR amplification of an IS*6110* sequence specific for *Mycobacterium tuberculosis* complex organisms. *J. Vet. Diagn. Investig.* **9:**244–249.
38. **Morris, P. J., C. O. Thoen, and A. M. Legendre.** 1996. Pulmonary tuberculosis in an African lion (*Panthera leo*). *J. Zoo Wildl. Med.* **27:**392–396.
39. **Morris, R. S., D. U. Pfeiffer, and R. Jackson.** 1994. The epidemiology of *Mycobacterium bovis* infections. *Vet. Microbiol.* **40:**153–177.
40. **Newell, D. G., and R. G. Hewinson.** 1995. Control of bovine tuberculosis by vaccination. *Vet. Rec.* **136:**459–463
41. **Nozaki, Y., Y. Hasegawa, S. Ichiyama, I. Nakashima, and K. Shimokata.** 1997. Mechanism of nitric oxide-dependent killing of *Mycobacterium bovis* BCG in human alveolar macrophages. *Infect. Immun.* **65:**3644–3647.
42. **O'Neill, B. D., and H. J. Pharo.** 1995. The control of bovine tuberculosis in New Zealand. *N.Z. Vet. J.* **43:**249–255.
43. **O'Reilly, L. M., and C. J. Daborn.** 1995. The epidemiology of *Mycobacterium bovis* infections in animals and man: a review. *Tubercle Lung Dis.* **76** Suppl. 1:1–46.
44. **Palmer, M. V., D. L. Whipple, J. C. Rhyan, C. A. Bolin, and D. A. Saari.** 1999. Granuloma development in cattle after intratonsilar inoculation with *Mycobacterium bovis. Am. J. Vet. Res.* **60:**310–315.
45. **Palmer, M. V., D. L. Whipple, J. B. Payeur, D. P. Alt, K. J. Esch, C. S. Bruning-Fann, and J. B. Kaneene.** Naturally occuring tuberculosis in white-tailed deer. *J. Am. Vet. Med. Assoc.*, in press.
46. **Palmer, M. V., D. L. Whipple, and S. C. Olsen.** 1999. Development of a model of natural infection with *Mycobacterium bovis* in white-tailed deer. *J. Wildl. Dis.* **35:**450–457.
47. **Rhyan, J., K. Aune, B. Hood, R. Clarke, J. Payeur, J. Jarnagin, and L. Stackhouse.** 1995. Bovine tuberculosis in a free-ranging mule deer (*Odocoileus hemionus*) from Montana. *J. Wildl. Dis.* **31:**432–435.
48. **Rhyan, J. C., K. Aune, D. R. Ewalt, J. Marquardt, J. W. Mertins, J. B. Payeur, D. A. Saari, P. Schladweiler, E. J. Sheehan, and D. Worley.** 1997. Survey of free-ranging elk from Wyoming and Montana for selected pathogens. *J. Wildl. Dis.* **33:**290–298.
49. **Rhyan, J. C., and D. A. Saari.** 1995. A comparative study of the histologic features of bovine tuberculosis in cattle, fallow deer (*Dama dama*), sika deer (*Cervus nippon*), and red deer and elk (*Cervus elaphus*). *Vet. Pathol.* **32:**215–220.
50. **Riley, R. L., C. C. Mills, W. Nyka, N. Weinstock, P. B. Storey, L. U. Sultan, M. C. Riley, and W. F. Wells.** 1959. Aerial dissemination of pulmonary tuberculosis. *Am. J. Hyg.* **70:**185–196.

51. **Saunders, G.** 1983. Pulmonary *Mycobacterium tuberculosis* infection in a circus elephant. *J. Am. Vet. Med. Assoc.* **183:**1311–1312.
52. **Sawa, T. R., C. O. Thoen, and W. T. Nagao.** 1974. *Mycobacterium bovis* infection in wild axis deer in Hawaii. *J. Am Vet. Med. Assoc.* **165:**998–999.
53. **Schlesinger, L. S.** 1993. Macrophage phagocytosis of virulent but not attenuated strains of *Mycobacterium tuberculosis* is mediated by mannose receptors in addition to complement receptors. *J. Immunol.* **150:**2920–2930.
54. **Schmitt, S. M., S. D. Fitzgerald, T. M. Cooley, C. S. Bruning-Fann, L. Sullivan, D. Berry, T. Carlson, R. B. Minnis, J. B. Payeur, and J. Sikarskie.** 1997. Bovine tuberculosis in free-ranging white-tailed deer from Michigan. *J. Wildl. Dis.* **33:**749–758.
55. **Schoenbaum, M. A., and R. M. Meyer.** 1995. Tuberculosis in large, confined dairy herds: approaches to elimination, p. 131–144. *In* C. O. Thoen and J. H. Steele (ed.), *Mycobacterium bovis Infection in Animals and Humans.* Iowa State University Press, Ames.
56. **Snider, D. E., Jr., M. Raviglione, and A. Kochi.** 1994. Global burden of tuberculosis, p. 3–11. *In* B. R. Bloom (ed.), *Tuberculosis: Pathogenesis, Protection, and Control.* American Society for Microbiology, Washington, D.C.
57. **Steffen, D. J., D. W. Oates, M. C. Sterner, and V. L. Cooper.** 1999. Absence of tuberculosis in free-ranging deer in Nebraska. *J. Wildl. Dis.* **35:**105–107.
58. **Stetter, M. D., S. K. Mikota, A. F. Gutter, E. R. Monterroso, J. R. Dalovisio, C. Degraw, and T. Farley.** 1995. Epizootic of *Mycobacterium bovis* in a zoologic park. *J. Am. Vet. Med. Assoc.* **207:**1618–1621.
59. **Sturgill-Koszycki, S., P. H. Schlesinger, P. Chakraborty, P. L. Haddix, H. L. Collins, A. K. Fok, R. D. Allen, S. L. Gluck, J. Heuser, and D. G. Russell.** 1994. Lack of acidification in Mycobacterium phagosomes produced by exclusion of the vesicular proton-ATPase. *Science* **263:**678–681.
60. **Thoen, C. O.** 1994. Tuberculosis in wild and domestic mammals, p. 157–162. *In* B. R. Bloom (ed.), *Tuberculosis: Pathogenesis, Protection, and Control.* American Society for Microbiology, Washington, D.C.
61. **Thoen, C. O., W. D. Richards, and J. L. Jarnagin.** 1977. Mycobacteria isolated from exotic animals. *J. Am. Vet. Med. Assoc.* **170:**987–990.
62. **Thoen, C. O., and E. M. Himes.** 1986. Pathogenesis of *Mycobacterium bovis* infection. *Prog. Vet. Microbiol. Immunol.* **2:**189–214.
63. **Thoen, C. O., T. Schliesser, and B. Kormendy.** 1995. Tuberculosis in captive wild animals, p. 93–104. *In* C. O. Thoen and J. H. Steele (ed.), *Mycobacterium bovis Infection in Animals and Humans.* Iowa State University Press, Ames, Iowa.
64. **Thompson, D. L., and R. D. Willer.** 1997. Report of the Committee on Tuberculosis, p. 547–560. *In Proceedings of the One Hundred and First Annual Meeting of the United States Animal Health Association.* United States Animal Health Association, Richmond, Va.
65. **Thompson, P. J., D. V. Cousins, B. L. Gow, D. M. Collins, B. H. Williamson, and H. T. Dagnia.** 1993. Seals, seal trainers, and mycobacterial infection. *Am. Rev. Respir. Dis.* **147:**164–167.
66. **Thorel, M. F., C. Karoui, A.Varnerot, C. Fleury, and V. Vincent.** 1998. Isolation of *Mycobacterium bovis* from baboons, leopards, and a sea-lion. *Vet. Res.* **29:**207–212.
67. **VanTiem, J. S.** 1999. Status of the state and federal cooperative bovine tuberculosis eradication program fiscal year 1999, p. 594–596. *In Proceedings of the One Hundred and Third Annual Meeting of the United States Animal Health Association.* United States Animal Health Association, Richmond, Va.
68. **Whipple, D. L., R. M. Meyer, D. F. Berry, J. L. Jarnagin, and J. B. Payeur.** 1997. Molecular epidemiology of tuberculosis in wild white-tailed deer in Michigan and elephants,

p. 543–546. *In Proceedings of the One Hundred and First Annual Meeting of the United States Animal Health Association.* United States Animal Health Association, Richmond, Va.

69. **Whipple, D. L., C. A. Bolin, and J. M. Miller.** 1996. Distribution of lesions in cattle infected with *Mycobacterium bovis. J. Vet. Diagn. Investig.* **8:**351–354.
70. **Whipple, D. L., J. L. Jarnagin, and J. B. Payeur.** 1999. DNA fingerprinting of *Mycobacterium bovis* isolates from animals in northeast Michigan, p. 3. *Proceedings IX International Symposium World Association of Veterinary Laboratory Diagnosticians.* College Station, Tex.
71. **Whipple, D. L., P. R. Clarke, J. L. Jarnagin, and J. B. Payeur.** 1997. Restriction fragment length polymorphism analysis of Mycobacterium bovis isolates from captive and free-ranging animals. *J. Vet. Diagn. Investig.* **9:**381–386.

Index